The E

The Book of Wheat

An Economic History and Practical Manual of the Wheat Industry

Peter Tracy Dondlinger

Routledge
Taylor & Francis Group

First published in 1908 by Orange Judd Company

This edition first published in 2018 by Routledge
2 Park Square, Milton Park, Abingdon, Oxon, OX14 4RN
and by Routledge
711 Third Avenue, New York, NY 10017

Routledge is an imprint of the Taylor & Francis Group, an informa business

© 1908 by Taylor & Francis

Publisher's Note
The publisher has gone to great lengths to ensure the quality of this reprint but points out that some imperfections in the original copies may be apparent.

Disclaimer
The publisher has made every effort to trace copyright holders and welcomes correspondence from those they have been unable to contact.

A Library of Congress record exists under ISBN: 8023519

ISBN 13: 978-1-138-60490-2 (hbk)
ISBN 13: 978-1-138-60495-7 (pbk)
ISBN 13: 978-0-429-46276-4 (ebk)

THE BONANZA HARVESTER, A PRIME FACTOR IN EXTENSIVE WHEAT PRODUCTION

THE BOOK OF WHEAT

An Economic History and Practical Manual of the Wheat Industry

By

PETER TRACY DONDLINGER, Ph. D.

Formerly Professor of Mathematics in Fairmount College

ILLUSTRATED

NEW YORK
ORANGE JUDD COMPANY

LONDON
KEGAN PAUL, TRENCH, TRUBNER & CO., Limited
1910

To

WILLIAM GRAHAM SUMNER, LL.D.

*Pelatiah Perit Professor of Political and Social Science
in Yale University,*

THIS BOOK IS DEDICATED

for his guiding instruction and the encouragement received from his
friendship and sterling character link the volume with the honored name of one of the greatest lights of our generation

ACKNOWLEDGMENTS

In addition to individual acknowledgments made throughout the volume, I wish to express my gratitude to those who have aided in the work by many kindnesses in the way of advice and suggestion, and by the furnishing of various data. My acknowledgments are due primarily and chiefly to Professor W. G. Sumner and Dr. J. Pease Norton—to the former for that general aid and counsel that can be offered only after wide historical research, and for reading and criticising a large portion of the work; to the latter for his indefatigable kindness in giving continuous aid in obtaining material, and in giving help of a technical nature. Much assistance and encouragement was given by Mr. M. A. Carleton, Cerealist of the United States Department of Agriculture; by Mr. Wm. Saunders, Director of the Central Experimental Farm, Canada; and by Mr. W. M. Hays, Assistant Secretary of Agriculture of the United States.

One of the most important of the several institutions which are now expending considerable financial resources in economic and industrial research is the Carnegie Institution of Washington, from which financial aid has been received in some of the investigations necessitated by the preparation of this volume. Through the kind offices of Mr. P. B. Smith, President of the Minneapolis Chamber of Commerce, material encouragement has also been received from the St. Anthony and Dakota Elevator Company and from the Washburn-Crosby Company.

For carefully reading over the typewritten manuscript of the book and suggesting many improvements in diction and phraseology, I am indebted to my former pupil, Miss Elizabeth Hodgson. She is not to be held responsible, however, for any imperfections in language that may yet remain, inasmuch as I made all final corrections and changes.—[The Author.

PREFACE

The great industries which have been essential to the rise of state or nation have never received the attention which their importance should command, and the chronicling of their events greatly extends the meaning of economics and history proper. Industrial history has indeed received a certain amount of consideration, but in the main it has been somewhat desultory, and the field is so new that only a few of the great basic industries, such as those of cotton, corn, alfalfa and coal, have even been attempted. It is my purpose in this book to add another volume to the industrial-economic literature which deals with industries in their entirety. While many important works are available that cover certain phases of the wheat industry very adequately, and a few which cover a number of phases very admirably for the limited space that is devoted to them, there. is, however, no general work treating the entire subject as completely and extensively as is merited by the industry which furnishes the most staple food of the civilized world. Unquestionably the need of such a book on wheat is patent.

A work of this nature is of direct or indirect interest to all consumers of bread. The historical or evolutionary aspect is of universal significance. Those directly interested in the wheat industry, whether as growers, dealers, or millers, not only should be familiar with the technicalities of the phase of the industry in which they are engaged, but they should have accessible a general knowledge of the whole industry. No agricultural college or experiment station should be without a text-book on the subject. The agricultural or economic section of every library should certainly contain a general reference book on wheat. The method of treating the subject demanded by these needs was one that would appeal to the popular reader as well as to the student, instructor and experimenter. Treated from the American point of view, the subject demanded a less detailed consideration for foreign countries.

The book is the result of fifteen years of personal experience in the wheat fields of our Northwest, and of a careful study of the works listed in the appended bibliography. Not a little additional information was obtained from several hundred letters written on phases of the subject with which I was not sufficiently familiar, and concerning which little material that was recent or reliable could be found in the literature. Space limited the references in footnotes to the most important ones. If more detailed information is desired on certain subjects than the limits of the book have permitted, references quite ampl' for all purposes will be found in the topical index of authors included in the bibliography.   P. T. D.

New Haven, Conn., May 1, 1908.

CONTENTS

ILLUSTRATIONS

THE BOOK OF WHEAT

CHAPTER I.

WHEAT GRAIN AND PLANT

ORIGIN.

The Word Wheat can be traced back through the Middle English *whete* to Old English *hwaete*, which is allied to *hwit*, white. The German *Weizen* is related to *weisz*, which also means white. The French *blé* suggests *blêmir*, to grow pale. Perhaps wheat was called white, to distinguish it from rye and other dark colored grains. *Triticum*, the botanical and classical name, doubtless comes from *tritus*, which is a participle from the Latin *terere*, to grind. The Italian *frumento*, and the similar French *froment*, are descended from the Latin word for corn or grain, *frumentum*, which originated in *frux*, fruit. The Spanish *trigo* has evolved through French and Latin from the Greek *trigonon*, which has for its roots *tri*, three, and *gonia*, a corner or angle. Thus the most widely used names of the wheat plant were determined by the characteristics of the seed, as color, shape, the property of having to be ground for food, and the natural relation of the seed to the plant.

The Geographical Origin of wheat has never been certainly determined. Such evidence as exists seems to point to Mesopotamia, but this is largely a matter of opinion. While wheat has been found growing apparently wild, the doubt always seems to remain that it may have simply escaped from cultivation. However, the belief that wheat once grew wild in the Euphrates and Tigris valleys, and spread from these to the rest of the world, has wider acceptance than any other. De Candolle's conviction rests largely on the evidence of Berosus and Strabo, while Lippert, in addition to the former, also cites Olivier and André Michaux. Darwin appears to have favored the same theory. From this center wheat is supposed to have spread to Phœnicia and Egypt. The Chinese considered it a gift from heaven. Homer and Diodorus Siculus say that it grew wild in Sicily. Humboldt denies the claim of Hermandez that a wheat native to Chili was found. The Egyptian historian, Manetho, attributed its discovery to Isis.

The Historical Origin of wheat is unknown. The most ancient languages mention it, and under different names. Whether we assume that these names, with the languages in which they are found, became differentiated from a common parent, or whether we assume that wheat evolved and spread over the Old World so independently of man that its name did not accompany its progress, in either case a period of time long enough to antedate our oldest languages will be required. The fact that it has been found in the prehistoric habitations of man, notably in the earliest Swiss lake dwellings, is proof of its antiquity.

The Swiss of the neolithic period cultivated four distinct species of wheat. Wheat seems to have been cultivated in China 3,000 years B. C., and was a chief crop in ancient Egypt and Palestine. The Bible first mentions wheat in Genesis, Chap. 30, v. 14.

Biological Origin.—The botanist calls wheat a grass. The evolutionist has ascended the biological stream one stage farther, and calls it a degenerate and degraded lily, using these terms, of course, in an evolutionary sense. He assumes a great group of plants of a primitive type from which sprang first the brilliantly colored lilies, then the degraded rushes and sedges, and lastly the still more degenerate grasses. From these grasses man developed the cereals, and among them

CLASSIFICATION OF THE GRASS FAMILY.[1]

GRAMINEÆ	Spikelets One Flowered	Maydeæ: Corn-Teosinte-Tribes Andropogoneæ: Sugar Cane-Sorghum Zoysieæ Tristegineæ Paniceæ: Millet-Hungarian Grass Oryzeæ: Indian Rice-Rice
	Spikelets Many- Flowered	Phalardieæ: Canary and Sweet Vernal Grass Agrostideæ: Timothy-Red Top Aveneæ: Oats Festuceæ: Blue Grass-Bromus-Orchard-Grass Fescues Chlorideæ: Grama and Buffalo Grass Hordeæ: Wheat-Barley-Rye-English Rye-Grass Bambuseæ: Bamboo

[1] Minn. Bul. 62, p. 392.

wheat. This is the hypothesis that accounts for most of the facts involved. All of the grass family, Gramineæ, are easily distinguished by having only one seed leaf, and for this reason they are known as monocotyledons.

The wild animal grasses, Aegilops, found in such abundance in southern Europe, and resembling true wheat in every point except in size of grain, are considered as the nearest kin to wheat. Efforts have been made to develop wheat from *ovata*, the most typical species. Fabre of Agde, France, claimed that in 1838 he began to improve this plant by selection, and that by 1846 he had obtained a very fair sample of wheat. His results have not been supported by other conclusive experiments, and scientists generally have not accepted them. There was doubtless cross-fertilization.

The accompanying figure represents different stages in the evolution of wheat.[1]

DEVELOPMENT OF THE WHEAT PLANT

The above sketch from a photograph shows: (1) *Ægilops ovata*, a small dwarfed specimen, but one grain of wheat in each head, found in Southern Europe; (2) The same species better grown and developed; (3) *Triticum spelta*, the cultivated spelt of Europe; (4) *Triticum Polonicum*, Polish wheat or giant rye; (5) Head of Nebraska wheat. While this is an instructive comparison, it is very questionable whether No. 5 could be developed from No. 1 in a reasonable number of years.

[1] Minn. Bul. 62, p. 81.

The results of recent investigations have shown that improvement by selection is relatively a slow process

DISTRIBUTION.

Longitudinal.—The migration of wheat has necessarily been closely connected with the migration of peoples, and especially with those of Europe. Consequently its general direction of spreading has been westward, though it is claimed that it spread eastward to China at a very early date.

In the United States, the meridian bisecting the wheat acreage passed through eastern Ohio in 1850, and was about 81 degrees. In 1860 it was 85 degrees 24 minutes, in 1870 88 degrees, and in 1880 it had reached middle Illinois, 88 degrees 45 minutes. The center of wheat production at the time of the census in 1900 was near the east central border of Iowa, the meridian of 95 degrees. This shows that the westward march of wheat proceeded at a much more rapid pace from 1880 to 1900 than from 1860 to 1880. During the last half of the nineteenth century, the center of wheat production moved west about 680 miles and north about 99 miles.

Latitudinal.—As European peoples and their descendants are meeting the demands of increasing population by continually subjecting to cultivation land of colder and of warmer latitudes, the domain of wheat is being extended on both sides of the temperate zones. In 1887 Sering published a map of North America in which he gave as the northern boundary of wheat growing territory a line beginning south of Lake Ontario running fully half way around it, a little north of the northern boundary of the other Great Lakes, through Lake of the Woods, through the southeast end of Winnipeg lake, northwest to the Athabasca river, following this to the Rockies, and beginning again in northeastern Washington.

In 1894 the editor of the *Social Economist* denied that wheat could be raised in Canada or Siberia north of the 55th parallel. This widespread notion that wheat could not be raised in the far north was gradually dissipated as wheat crept closer and closer to the Arctic circle. Wheat has frequently been matured at Sitka, Alaska, 56 degrees north latitude. At the Sitka station, winter rye, spring wheat, barley, oats and buckwheat matured both in 1900 and 1901. In the

Peace river valley, extending 700 miles north of the Canada border, 58 degrees north latitude, enough wheat, barley and oats have been grown to bring about the erection of a 100-barrel roller mill at Vermilion, on the Peace river. Spring wheat of the Romanow variety matured at the Kenai station in Alaska in 1899, 60 degrees north. Experiments have shown that winter wheat will ripen here in ordinary seasons. On the Mackenzie river wheat has been grown farther north than 62 degrees. Spring wheat and winter rye have matured perfectly 65 degrees 30 minutes north latitude at Rampart, about 200 miles from the Arctic circle, and at Dawson, equally as far north, over 1000 miles north of the United States. While wheat can be grown this far north, the chances of failure are, of course, much greater than in a climate more temperate. Barley, oats and rye will grow farther north than wheat.

Towards the equator the limits of wheat generally vary between 20 and 25 degrees north and south latitude. It thrives in southern Brazil, in Cuba, and in southern Rhodesia in South Africa at these latitudes.

Altitudinal.—Another very important factor in determining where wheat can be raised is the altitude, which may be considered as the complement of latitude. On the mountain plains of Colombia and Ecuador it grows on the equator. Thus wheat is raised in America from the equator, 10,000 feet above sea level, to Dawson and the Klondike river, 2,000 feet above sea level, and at least 65 degrees 30 minutes north latitude. In the United States the census shows that in 1880, over 80 per cent of the grain was grown at an elevation between 500 and 1,500 feet above sea level. In 1890 the altitudes at which wheat was raised varied from 100 feet below sea level to over 10,000 feet above sea level, and about 70 per cent was raised between 500 and 1,500 feet elevation. It cannot be raised successfully at great elevations in England. The plains and mountain slopes of Sicily produce wheat, the upper limit of its growth having been given in 1863 as 2,500 feet in altitude.

A member of the Manitoba legislature, Mr. Burrows, has claimed that fifteen years of history show that altitudes have very much to do with summer frosts, and that 800 to 1,300 feet above sea level is the best altitude for No. 1 hard wheat in Manitoba. Perhaps the greatest elevation at which wheat has been raised is in Asia on the Himalaya mountains, 11,000 feet above sea level. The four counties of Kansas occupying the center of its famous wheat region have an average elevation of about 1600 feet. The Colorado station has developed a type of wheat adapted to the higher altitudes of the mountain regions, those of 6,000 to 9,000 feet elevation.

Historical and Geographical.—In the western half of Asia, in Europe, and in northern Africa, wheat has since time immemorial occupied the first rank of cereals. It was one of the main crops of the Israelites in Canaan. None was grown in the New World before the sixteenth century. Humboldt says that a negro slave of Cortez found three or four grains of wheat in the rice which served to maintain the Spanish army. This was apparently sown before 1530, about the date when the Spaniards introduced wheat culture into Mexico. In 1547 wheat bread was hardly known in Cuzco, Peru. The first wheat sown in the United States was by Gosnold in 1602 on the Elizabeth Islands off the southern coast of Massachusetts. It was first cultivated in Virginia in 1611, and in New Netherlands before 1622. By 1648 there were several hundred acres in the Virginia colony. Missionaries first introduced it into California in 1769. Cuba saw its cultivation at least as early as 1808. It must have been early introduced into Canada, at least by the close of the eighteenth century, for in 1827 Canada raised over twenty million bushels. The first wheat successfully grown and harvested in the Red river valley was in 1820. Victoria wheat, which had been acclimated by growing 200 years in the tropics, was successfully grown in experiments on Jamaica and the Bahama Islands, 1834 to 1836. There was a prejudice against it, however, and Indian corn was grown in preference. Minnesota's first settlements date back to about 1845. Wheat raising became a regular branch of farming in Argentina in 1882. Such were the historical beginnings of the wheat industry in the western hemisphere. It has now become a more or less important industry over practically all of America lying outside of frigid zone climates.

IMPORTANCE.

Quantitative.—Both in the quantity produced and in its value, wheat is the world's king of cereals. Recent statistics show, however, that 800,000,000 persons, or 54 per cent of the inhabitants of the globe, derive their sustenance mainly from rice. The most important cereal produced in the United States, measured in bushels or dollars, is corn, and wheat stands second. From the census we find that the United States produced in 1899, including farm animals and their products, an aggregate value of nearly five billion dollars. Of this, animals brought 900 millions, corn 828, and wheat 370, over 7.4 per cent. In 1906 the corresponding figures for corn and wheat were 1,100 and 450. For at least several decades, corn has formed over 50 per cent of the total acreage of cereals in the United States. Wheat formed 29.8 per cent in 1880, 23.9 per cent in 1890, 28.4 per cent in 1900, and 27 per cent in 1905. In value, corn formed 55.8 per cent in 1900, and wheat 24.9 per cent. Cereals form 51 per cent of the value of all crops, which gives the value of wheat as nearly 13 per cent of that of all crops. Out of a total of over 5.5 million farms in the United States, over two million raise wheat. The world's annual production and consumption of wheat is nearly 3.5 billion bushels.

Qualitative.—Taking the civilized world as a whole, wheat forms the principal food of man. It is much more widely distributed than either its commercial rival, corn, or its rival food cereal, rice. It is a prime necessity of civilized life. The quantity of wheat milled is larger than that of all other cereals combined. Sixty-two per cent of all cereal products milled in the United States during 1900 were from wheat. It is essentially a bread cereal. Bananas, rice, potatoes, and other soil products will sustain a greater population on a given unit of land than wheat will, but they are not so well adapted to a high standard of living. Herein lies the present and increasingly great importance of wheat, for it seems to be the tendency of the civilized world to raise its standard of living. As the standard of living rises, wheat becomes a relatively more important part of human food. Rye and oats furnished the bread of the great body of people in Europe during the middle ages. Wheat was high-priced and not extensively grown. England early became a wheat eating nation. France and

the other Latin countries followed later. Rye is still extensively used in Germany, but is gradually being superseded by
wheat. Even Russia is using more wheat flour than she did
twenty years ago.

The great intrinsic food value of wheat; its ease of cultivation and preparation for use; its wide adaptation to different
climates and soils; its quick and bountiful return; and the
fact of its being paniferous and yielding such a vast number
and variety of products are all factors that enhance the value
of the wheat grain. Its combined qualitative and quantitative
importance gives to wheat a great superiority over any other
cereal, and causes it to be dealt in more extensively upon the
speculative markets than any other agricultural product. As
an essential part of the food of civilized man it becomes of an
importance so vital as to be dominating.

CLASSIFICATION.

The Classification of wheat seems always to have been in a
more or less chaotic state. This is especially true of the
nomenclature of varieties. Nor is the fault to be laid particularly at the door of science. We have seen that wheat has
been continually migrating for many centuries. It is a plant
that is easily influenced by environment and therefore particularly unstable in type. Since it has always been migrating to
new environments, a complete change in type often resulted,
though it was still known by the old name. This is further
complicated by the fact that the modern art of breeding wheat
has originated many new varieties. Add to this the fact that
wheat has been shipped all over the world, not only for
commercial purposes, but also for seed experiments, and it is
not surprising that the nomenclature of varieties is somewhat
tangled, that several varieties are known by the same name,
or that one variety may have several names, and may pass for
several varieties. It is among the most common wheats that
the difficulty has been most perplexing.

Classes and Distribution.—There are several kinds of the
less common wheats, such as Polish wheat, spelt and durum
wheat, which have very marked characteristics, and which
have perhaps not migrated so widely. In spite of some confusion in names, it is generally possible to determine to which

class they belong. Some of the most common and widely used classifications are those based on time of sowing, as spring and winter wheat; on firmness of structure of the grain, as hard and soft; on the products for which they are used, as bread and macaroni wheats; and on the color of the seed, as red and white. As will later be shown, wheat adapts itself to new environments so that any one of these classes may be transformed into any other, and as wheat is raised so widely as to embrace practically every kind of environment, these classes grade into each other so imperceptibly that even an expert can hardly determine to which class a certain wheat may belong. An approximate division has, however, been made. Mr. M. A.

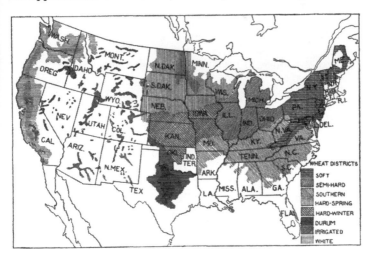

DISTRIBUTION OF WHEAT VARIETIES IN THE UNITED STATES

Carleton,[1] cerealist of the United States department of agriculture, has divided the wheat grown in the United States into eight classes, and has shown the distribution of these classes by districts in the accompanying map.

On the north Atlantic coast is the soft wheat district, south of the Great Lakes the semi-hard district, and south of these two districts is the southern district. The Red river valley is

[1] U. S. Dept. Agr., Div. Veg. Phys. & Path., Bul. 24.

the center of hard spring wheat, Kansas of hard winter wheat and north central Texas of durum wheats. White wheat is raised on the Pacific coast. The center of red wheat, not shown in this division, is from Kansas to the Red river valley. A still more general classification by the same author divides the United States crosswise into three divisions of approximately equal width, assigning the hard wheats to the northern states, the soft wheats to the states of the middle latitudes, and the durums to the southern states. About two-thirds of the wheat raised in the United States is winter wheat. Nearly 90 per cent of the wheat grown in Russia is spring wheat. In Canada, Manitoba raises spring wheat exclusively, but Ontario and Alberta raise some of the winter variety. In Germany, over 90 per cent of the wheat grown is of the winter variety, which is largely grown over southern Europe and on the British Isles. Spring wheat was once more generally called summer wheat, and winter wheat is often also called fall wheat.

Carleton, on a geographical basis, located groups of varieties having special qualities approximately as follows:

1. Starchy white wheats: Pacific coast and Rocky Mountain states, Chile, Turkestan, Australia and India

2. Amber or reddish grained wheats, also starchy: Eastern states, western and northern Europe, India, Japan and Australia.

3. Wheats with excellence of gluten content for making bread: Northern and central states of the plains, Canada, eastern and southern Russia, Hungary, Roumania and southern Argentina.

4. Wheats resistant to orange leaf rust: Southern Russia, Mediterranean and Black Sea regions, and Australia.

5. Wheats with excellence of gluten content for making macaroni: Southern Russia, Algeria, and the Mediterranean region in general.

6. Wheats with stiff straw, which prevents lodging: Pacific coast states, Japan, Turkestan, Mediterranean region and Australia.

7. Wheats with great yielding power (at least in proportion to size of head): Pacific coast states, Chile and Turkestan.

8. Non-shattering wheats: Pacific coast states, Chile, Turkestan, Germany (spelts), and East Russia (emmers).

9. Wheats of great constancy in fertility: Germany (spelts) and southern Europe.

10. Wheats of early maturity: Japan, Australia and India.

11. Wheats most resistant to drought and heat: East and South Russia, Kirghiz Steppes, Turkestan and southern Mediterranean region.

12. Wheats most resistant to drought and cold: East Russia.

Species.—There are eight principal types of cultivated wheat: Einkorn (*Triticum monococcum*); Polish wheat (*Tr. polonicum*); Emmer (*Tr. sativum dicoccum*); Spelt (*Tr. sat. spelta*); Club or Square-head wheat (*Tr. sat. compactum*); Poulard wheat (*Tr. sat. turgidum*); Durum wheat (*Tr. sat. durum*); and Common wheat (*Tr. sat. vulgare*).

Varieties.—In 1900, after five years of experimentation with about 1,000 varieties of wheat collected from the different wheat countries of the world, the United States department of agriculture decided that, tested by American conditions, there were 245 leading varieties. No one variety is best under all conditions, but climate, soil, and the purpose for which wheat is raised must in each case determine which variety is most profitable. If a variety can be secured that will yield more under the same conditions than other varieties do, then profits can be easily increased, for its production involves no additional expense, except possibly an extra outlay for seed. Prof. W. M. Hays estimates that Minnesota No. 169, a variety of wheat introduced by the Minnesota experiment station, has increased the yield of that state from 5 to 10 per cent.

The most widely and universally grown varieties of wheat in the United States are Fultz for soft winter, Turkey Red for hard winter, Fife and Blue Stem for hard spring, and Kubanka for durum wheat.

DESCRIPTION AND GROWTH.

Roots.—The first root appearing is called the radicle. This and the two other roots that soon appear form the whorl of three seminal or temporary roots. The crown of roots usually

ROOT SYSTEM OF A WHEAT PLANT AT HEADING-OUT TIME

grows about an inch beneath the soil, irrespective of the depth to which the grain was planted. From the crown are thrown out whorls of coronal or permanent roots. Any node of the wheat stalk under or near the soil may also throw out a whorl of permanent roots, somewhat similar to those of corn. There are four or five whorls with three to five roots each. The roots from the base of the crown strike directly downward, while those from the later whorls run at an angle for a few inches before taking a vertical direction. Most of the main roots penetrate to a depth of over 4 feet, perhaps 5 or 6 feet, provided the water-line is not closer to the surface than that distance, for below this the roots will not enter to any appreciable extent. The roots of wheat have been traced to a depth of 7 feet, and it has been found that if those of one plant were placed end to end they would reach 1,704 feet. The deep roots are all fine threads of practically uniform diameter throughout their entire length. They branch and re-branch freely to a depth of 18 or 20 inches, about eight branch roots occurring to an inch length of a main root. At a greater depth, branches are few or absent, and it is supposed that the deep roots are for securing moisture. The roots do not branch or feed much in the region just below that stirred by the plow, if that region is hard and gummy, as is often the case. The upper whorls give forth roots that are larger and coarser, and which resemble the brace roots in corn. It is said that the roots extend chiefly at their extremities, while the stem elongates equally, or nearly so, in all of its contiguous parts. The root development seems to be greatest in durum wheats. Early spring and summer rains cause shallow rooting. In the absence of these rains in the far west, a deeper root system, capable of resisting superficial droughts, is developed. Poor soil causes the roots to age rapidly.

Culms.—The culms of wheat are usually hollow, but in some varieties they are quite filled with pith. The length varies greatly in different varieties, soils and seasons, a fact which results in greater variation in size and yield of straw than of grain. Common wheat averages from three to five feet in height. The liability of lodging depends greatly on the culm, the length of which is also important in harvesting.

[1] Hunt. Cereals in Amer. (1904), p. 27.

Leaves.—When the internodes lengthen and the spike pushes upward, the wheat is said to shoot. Previous to this, the nodes are so close together that the plant seems to consist almost entirely of leaves. There are four principal parts to the leaf: The blade; the sheath, which clasps the stem and is split down the side opposite the blade; the ligule, also clasping the culm, and located where the blade and sheath join; and the leaf auricle, thin projections growing from the base of the blade. The first leaves of the wheat plant and the germ whorl of roots do not live through the winter in some varieties.

The Flower of Wheat is constituted collectively of the organs of reproduction, together with the two inclosing chaffy parts. The inner of these two parts is known as a palea, while the outer and lower one is the flowering glume. The latter often bears a long appendage, characteristic of bearded wheat. These awns or beards vary greatly in length even in the same spike, and in some varieties are deciduous upon ripening. Their color varies from light yellow to black.

Organs of reproduction in wheat; *A*, ovary, *o*, and stigma, *s*, just before flowering; *B*, the same at time of flowering; *C*, flower before opening, *a*, anthers, *f*, filament, *l*, lodicule; *D*, flower about to open.

The Spikelets.—Each consists of from two to five flowers encased within two hard oval chaffy coverings called outer glumes. In common wheat each spikelet generally matures two, and sometimes three, grains. The glumes vary greatly in form, color and size. The stem or rachis of the spike is of a zigzag form. On each of its joints or shoulders sits a single spikelet, attached by an exceedingly short rachilla. Arranged alternately on the stem, with flat sides toward the center, the spikelets usually give the head of wheat a square appearance when viewed endwise. Viewed from the side, the spike may be straight or curved; it may have uniform sides, or taper toward both ends, or only toward base or apex; or it may be clubbed at either end. The filling of the spikelets has much to do with the appearance of the spike, which varies much in different varieties. There is also great variation in compactness. Fifteen to twenty fertile spikelets, containing from 30 to 50 grains, are usually formed on a spike of wheat, the average length of which is between 3 and 4 inches. Humboldt said that in Mexico each spike of wheat averaged 90 grains, though some had as many as 160. Mummy wheat has been observed with ears containing nearly a dozen branches. There are 150 grains in one ear, and as many as 60 ears from one seed. Wheat has the advantage of extreme diminution of the number of seeds to each flower, giving richness in starch and gluten, combined with the advantage of numerous flowers on each plant, giving many seeds.

The Wheat Kernel is a dry, indehiscent, unilocular caryopsis. It is oval in shape, and has the appearance of being folded upon itself from two sides. A ventral crease marks the coming together of the two folds. At the base of the berry opposite the crease is found the embryo, germ, or chit. At the apex is a collection of minute hairs. The entire grain fills from 20 to 30 cubic millimeters of space, of which at least thirteen-fourteenths are occupied with the starchy endosperm. The latter almost surrounds the embryo, and its cells are very irregularly shaped. The embryo is composed of the absorbent organ (scutellum), and the miniature first leaves and roots. It forms about 6 per cent of the wheat kernel.

Coats of a wheat kernel; a, germ; b, starch cells; c, gluten cells; d, inner coat of bran; e, coloring matter of bran; f and g, outer coats of bran; h, epidermis of kernel.

The endosperm and embryo are completely enclosed by a single layer of aleurone or gluten cells. The weight of this layer is 8 per cent of that of the whole grain. The next covering is a single layer of collapsed cells, known as the tegmen. This is again surrounded by a third envelope, the testa, or episperm, which contains the greater part of the coloring matter of the grain. This coloring matter is of two kinds, one a palish yellow, and the other an orange yellow, and the degree in which one or the other predominates determines whether the wheat is known as white, yellow or red. The three layers just described constitute the envelope of the seed proper. They in turn are again inclosed by the pericarp, which is also composed of three layers, all colorless. The exterior of these three membranes, the cuticle, is easily removed by rubbing. Then come two layers of cellular tissue, the epicarp (from which spring the hairs above mentioned) and the endocarp. The tegmen and testa form about 2 per cent of the weight of the grain, and the pericarp forms fully 3 per cent. Thus the bran forms at least 13 per cent of the grain.

Germination.—The three conditions essential to the germination of wheat are moisture, warmth and oxygen. In the absence of any one of these the process will not begin, or if it has begun it will cease. Johnson defines the period of germination as lasting from the time when the rootlet becomes visible until the stores of the mother seed are exhausted and the young plant is wholly cast upon its own resources.

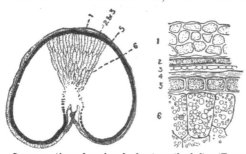

Cross section of grain of wheat on the left. (From micro-photograph by Tolman.) Transverse section, on the right, of an unripe grain enlarged about 100 times from drawing by Bessey. 1, ovary wall or pericarp; 2, outer integument; 3, inner integument; 4, remains of nucellus; 5, aleurone cells; 6, starch cells.

At 41° F., the time required for the rootlet to appear in wheat is about six days, which time corresponds to the more general idea of the period of germination. At 51° this time is shortened about one-half. The time required for the completion of germination is 40 to 45 days at 41 to 55° and 10 to 12 days at 95 to 100°. The lowest temperature at which wheat will germinate is 41°, the highest 104°, and that of most rapid germination, 84°. This is according to Johnson. Other authorities claim that wheat will germinate and grow on melting ice. It has also been said that it does not germinate successfully at a high temperature, and consequently should not be sown until cool weather in southern climates. Dissolved salts seem to aid germination under ordinary field conditions.

In germinating, wheat absorbs from five to six times its

weight of water. It loses 1.5 per cent of its own weight in 24 hours, 6.7 per cent in 90 hours, and 11.8 per cent in 144 hours. Besides the loss in weight, marked chemical changes take place which greatly decrease its value for bread baking purposes, and probably also as a food for stock. Great loss may thus be occasioned by the sprouting of wheat in field, shock, stack or bin. Experiments indicate that sprouted wheat will regerminate and form healthy sprouts until the stem (plumule) has reached a length of ¾-inch in the first germination, and an average of 80 per cent of all sprouted wheat with the length of the stem not exceeding ½-inch will again germinate.[1]

Stooling or Tillering.—Wheat, like other cereals, has the characteristic of throwing out side shoots after the plumule has appeared above the surface. These branches or culms may form at any node covered with soil. The number of such stalks from one seed varies much with conditions. There are usually at least six, but there may be from two to several dozen in extreme cases, 52 spikes having been observed. As a rule, the more favorable the conditions for plant growth, and the thinner the wheat is on the ground, the more it tillers. Cool weather during early development may result in a long period of subsequent growth which encourages tillering. Time of seeding also has great influence, for late sown wheat may not have time to stool. The habit varies quite materially in different varieties. While thinner sown wheat may tiller more, a greater amount of seed per acre often increases the yield, even though there are fewer stools. Pliny is said to have declared that it was not

A stool of wheat. The culms are from a single seed originally at *a.*

uncommon in northern Africa and in Italy to find from 200 to 400 stalks of wheat growing from a single kernel. Humboldt put on record that in Mexico each grain of wheat produced 40 to 70 stalks. It is probable that each of these men was seeing with the eyes of an enthusiast.

The Growth of a Wheat Plant is the aggregate result of the enlargement and multiplication of the cells which comprise it.

[1] Rept. N. D. Sta., 1901, p. 107.

Generally cells reach their full size in a brief time, and continuous growth depends mainly upon the constant and rapid formation of new cells. The essentials to growth are light, air, moisture, heat and food. In the absence of any one of these, the plant dies, and in their disproportionate combination, growth is sickly. In germination, food is furnished by the seed, and light is not essential. Over light man has no control. He can increase the amount of air that has access to the plant by loosening the soil around its roots. An adaptive control of heat is exercised by sowing during the warm season. By selecting soils, fertilizing and changing existing foods from unavailable to available forms, food can in a great measure be regulated, and water, acting as a solvent and vehicle, can be very largely regulated as to amount by drainage and irrigation. That the growth and multiplication of cells involves a migration of material within the plant has long been recognized. In wheat, as in many other plants, there is a comparatively large development of roots soon after the first leaves appear. Only some low-lying leaves are put forth while the great complex of roots is being formed. In a wheat plant only 23 days old, the roots had penetrated the soil over 1 foot in depth. When the system of roots has been formed, the stalk suddenly shoots up almost to mature stature. Perhaps the roots are completely developed by the time that the formation of grain has begun.

The leaves of the wheat plant, with their chlorophyl cells, have been considered as little laboratories elaborating vegetable matter. Under the influence of light they are able to extract carbonic acid from the atmosphere. This acid is one of the raw materials of these little factories. They decompose it, eliminate the oxygen, and from the residue they manufacture sugar, cellulose, straw-gum, vasculose, and all the ternary matters composed of carbon, oxygen and hydrogen. A perfect system of canals penetrates every part of the plant. These canals are filled with water, which enters at the roots, for leaves do not absorb water to any appreciable extent, and is in constant motion until it is exhaled from the leaves. During one hour of insolation a leaf of wheat exhales an amount of water equal to its own weight. Upon these highways of moving water are borne raw materials destined for the little cell factories, such as nitrates, phosphoric acid, potash and silica. They, too, are reduced. If there is an abundance of rain, the cells continue work long, elaborate much vegetable matter, and. the plant grows.

If the water supply is insufficient and the soil parsimonious, this prodigious consumption cannot be supplied, and dessication of organs takes place. This begins in the oldest leaves, and nearly always the little leaves at the base of the stem become soft, flabby, and withered. Analyses have been made which show that these leaves let escape some nitrogenized matter, phosphoric acid and potash, which they contained when living, green and turgescent. Thus the closing of one of these groups of little cell factories by the dessication of a leaf is a very important process to the plant, for less vegetable matter is elaborated than if it had continued its work. In dry years a shortening of the stems and a comparatively small amount of straw results.

The dying of leaves involves not only the closing of these workshops, but the transportation of much of the finished product stored in them. Metamorphosis of the nitrogenized matter which forms the protoplasm, the living part of the cell, takes place, and it assumes an itinerant property which enables it to pass through membranous walls and migrate over the liquid highways to new leaves. With it are carried phosphoric acid and potash. Some of the elaborated material is thus continually being transported from lower to upper leaves during the entire period of vegetation. Flowering takes place when enough material has been elaborated to nourish the appearing seeds. This migration of substance can take place only when there is plenty of water, and the crop fails when it is too dry. Too much water is also injurious, for it causes a tendency to keep up growth indefinitely. The Minnesota station found that the wheat plant produced nearly one-half its dry and three-fourths its mineral matter by the end of 50 days. This included 75 per cent of the potash, 80 per cent of the phosphoric acid, and 86 per cent of the nitrogen. At 65 days, 65 per cent of dry and 85 per cent of mineral matter had been produced, as well as most of the fiber, which suffered a loss after 81 days.[1] Compared with the processes observed in nitrogenized matter, phosphorus and potash, the

[1] Minn. Bul. 29. pp. 152-160.

formation of starch is yet quite a mystery. Its accumulation in the leaves cannot be detected in wheat as it can be in a large number of other species. Neither are reserves of saccharine matters to be found there. It is not until the last stage of vegetation that starch is formed. Consequently climatic influences at the close of the growing period have a marked effect on the amount of starch produced, and cause it to vary greatly from year to year.

The process of transporting elaborated material begins in the planted seed, and does not cease until the wheat is dead ripe. This is the explanation of wheat ripening after it is cut. It also explains the fact that wheat straw, as well as many other straws, is not as well liked by animals, and is not as nutritious, after it is ripe as when green, or when cut before ripe.

Fertilization.—The one-seeded ovulary is a little greenish swelling. It is surmounted by the stigmas, two erect and adjacent aigrettes of plumes. There are three stamens, and the anthers are compactly arranged about the ovulary. At flowering the filaments to which the anthers are attached elongate rapidly. As the anthers are pushed upward, they suddenly overturn, and the pollen falls upon the stigmas, which have

4-40 A.M. 4-43 A.M. 4-45 A.M. 4-47 A.M. 4-55 A.M. 5-08 A.M. 5-15 A.M. 5-18 A.M.

THE OPENING OF THE FLOWERS OF WHEAT. (AFTER HAYS)

now grow slightly divergent. These delicate operations all take place within the closed flower and generally wheat is thus essentially self-fertilized. The anthers are now pushed outside of the glumes, and the wheat is popularly said to be in flower. As soon as the pollen comes in contact with the stigmas, it germinates by sending out a long tube (called the pollinic branch) into the ovulary. This completes fertilization and the grain is formed. If fertilization in incomplete, the ovularies remain unfertilized, and the spikes bear sterile flowers in which no kernels are formed. It seems that the crop is thus injured when fertilization takes place in rainy weather. The water probably finds its way within the involucre, and the pollen grains are either imperfectly retained, or their germination is irregular. The process of fertilization generally occurs early in the morning, and may require less than an hour of time. After its completion the ovule (seed) grows very rapidly to maturity. The embryo develops first, and then the endosperm.

The Most Favorable Ripening of wheat requires a mild temperature and a slightly clouded sky. A high temperature the month before wheat is ripe diminishes the yield, and in particular prevents the formation of starch. There is a real, though small, loss in wheat from the period when it is "ripe" to the time when it is dead ripe, and it is claimed that this loss does not result from careless handling, or from drying of the grain.[1]

Deherain offers the explanation that "all the organs of a plant respire by the aid of the oxygen of the air consuming some of their principles. In the seed the combustion chiefly affects the starch, and a crop which remains standing long diminishes in weight both by the loss of seeds that fall and by the slow combustion which continues as long as desiccation is not produced." What is lost in quantity, however, is perhaps more than gained in quality, for the best flour can be obtained from dead ripe wheat only. Such flour has a better color, and will take more water in bread-making. If the grain is cut before ripe, the most serious feature is increased acidity in the flour. This interferes with fermentation in bread-making, and is liable to make the bread sour or dark.

The Rate of Multiplication of Wheat.—Paley gave 300 grains harvested from one grain sown as a moderate estimate; 400 as a possible one; and 10 to 12 as a practical one. Herodotus said that on the irrigated land of Assyria, wheat yielded from two to three hundred fold, and grew to giant size. Fifty grains of wheat, selected from one spike, were planted, and the 30 grains which grew produced 14¾ ounces of wheat. This was sown the next year, and produced 5 pecks of grain, which in turn produced 45 bushels the subsequent year. The 45 bushels produced 537 bushels in another year, enough seed

[1] Kedzie, Rept. Mich. Board Agr., 1881-2, p. 337; Mich. Bul. 191, p. 160; Neb. Bul. 32, p. 97.

from one spike in four years to sow about 500 acres.[1] In ten years, one grain of North Dakota wheat, now known as Minnesota 163, without any attempt to increase it rapidly the first few years, actually produced about 300,000 bushels of wheat. One thousand acres of land south of Walla Walla in eastern Washington yielded 51,000 bushels in 1881. "This yield was made the subject of a careful measurement and reported to the Agricultural Department, where it stands today as the largest yield for a thousand-acre field ever reported."[2] The greatest wheat crop ever recorded in the world's history as being produced from unfertilized land was that of western Canada in 1901, where 63,425,000 bushels were harvested from a little over 2,500,000 acres; an average yield of over 25 bushels per acre.

Physical Properties.—The number of grains in a pound of wheat varies from 7,500 to 24,000; from 377 determinations the average was 12,000 grains. The number in a bushel has been given as varying from 446,580 to 971,940. The Winchester bushel (2150.42 cubic inches) used in the United States, has a standard and legal weight of 60 pounds. The measured bushel generally varies in weight from 54 to 65 pounds, and greater extremes occur. The Imperial bushel (2218.192 cubic inches) used in England, has a corresponding weight of 61.89 pounds. This is the reason why English wheat appears heavier than American grain.

The specific gravity of American wheat has been found to vary from 1.146 to 1.518. Lyon found high specific gravity associated with low nitrogen content. As a rule, the harder the grain, the higher is the gluten and nitrogen content, and the deeper red the color.

Viability of Wheat.—Experiments have shown the optimum period for germination to be the second year after harvest. Seed one year old often gives better results than fresh seed, but after the first year the viability generally diminishes rapidly from year to year. Ordinarily it is not advisable to sow wheat over two, or at the most three, years of age, at least not without testing its germinating powers, which have been found to vary from 15 to 75 per cent after five years. The longest

[1] Neb. Bul. 32, p. 84.
[2] Rept. Bureau of Statistics, Washington, 1903, p. 69.

period for which conclusive modern scientific experiments have shown wheat to be viable is ten years. During six successive years Saunders found the average viability of three varieties to be respectively: 80, 82, 77, 37, 15 and 6 per cent.[1] Varro, speaking of the granaries of the first century B. C., remarks that the vitality of wheat can be preserved in them for 50 years. Daubeny questioned this in 1857, and stated that wheat does not retain its vitality over 40 years. Humboldt states that for causes not well known, Mexican grain is preserved with difficulty for more than two or three years. The reported germination of wheat taken from Egyptian mummies thousands of years old is a modern myth originating in the impositions of fraud and cunning upon credulity.

The highest temperature at which dry wheat seed can retain its vitality is also an unsettled question. Chambers's Cyclopedia makes the statement that some dry seeds survive 212 ° F., and —248° F., but does not state what kind. Klippart gives —58° F. as the point at which wheat loses its vitality, and says that the germinating power is completely destroyed if the grain is steeped 15 minutes in water having a temperature of 122° F. According to the same writer, it could perhaps stand 170° F. in a dry atmosphere without serious injury. He gives this as a probable reason why wheat does not grow in the tropics, where the soil often has a temperature of 190° F. Recent experience has shown that steeping wheat ten minutes in water of 132 to 133° F. to kill smut germs does not injure its viability. In northern Canada, —52° F. has no injurious effect upon the vitality of dry and unplanted wheat. Beyond these temperatures, no scientific experiments have been found recorded by the author.

Time Required for Ripening.—The mean temperature required for the successful cultivation and ripening of wheat has been given as 65° F. for 45 to 60 days, and 55° F. for three or four months of the growing season. Of the wheat in the United States, according to the census of 1880, 67.5 per cent was grown where the mean annual temperature was between 45 and 55° F., and 62.7 per cent of it where the annual rainfall was between 35 and 50 inches. It has been claimed that the total amount of sunshine and heat units required to mature a crop

[1] Rept. Can. Exp. Farms, 1903, p. 44.

of wheat is the same for all latitudes, and that if these vary, the period of growth will vary in inverse proportion. In support of this position Cooke* collected the statistics given below.

Locality	Period of Growth	Mean Temperature of growing Period	Heat Units
Near Poona (India)...................................	115 days	74.0 degrees	8,510
At Alsace (S. France)...............................	137 "	59.0 "	8,083
Near Paris (N. France).............................	160 "	56.0 "	8,950
Near Edinburg (Scotland)........................	182 "	47.5 "	8,645

Cooke found the number of heat units required to be approximately the same for different countries, i. e., about 8,500. Experiments conducted at Fargo, N. D., to verify this failed of their purpose, and gave approximately 6,500 heat units. The period of growth was about 100 days. Recent observations have shown that the number of heat units decreases when the growing period shortens. In general, the growing period is shortest in the coldest climate.

The Weight of Different Materials entering into an acre of the wheat crop is shown in the table given below. All weights are in pounds. The grain and straw are given as air dry material, which contains about 15 per cent of water.

WEIGHT OF MATERIALS IN AN ACRE OF WHEAT.

	Average U. S. Crop[1]	Mont[2]	Canada[3]	6,600 lb. Crop[4]	Grain of 6,600 lb. Crop	Straw of 6 600 lb. Crop
Total of potash phosphoric acid & nitro'n	....	52.61	77.28	122.67	50.25	72.42
Potash..........................	....	13.69	19.11	40.17	8.55	31.62
Phosphoric acid.........	....	9.49	17.64	26.85	14.10	12.75
Nitrogen.................	....	29.73	40.53	55.65	27.60	28.05
Water evaporated....	522,000	....	600,000	....	....	....
Weight of grain........	720	840	1,440	1,500	....	....
Weight of straw........	1,500	1,680	2,200	5,100	....	....

[1] Yearbook U. S. Dept. Agr., 1894, p. 174.
[2] Rept. Mont. Exp. Sta., 1902, p. 61.
[3] Evidence of Saunders, 1900, p. 23.
[4] Neb. Bul. 19, p. 15.
* N. D. Bul. 47, p. 704.

Chemistry.—The five outer layers of the wheat grain are composed chiefly of cellulose, a woody, fibrous substance. The endosperm, the food part of the grain, contains large quantities of starch, a nitrogenous substance known as gluten, a little sugar, and the cellulose of its cell walls. The gluten content is greatest at the hard exterior of the endosperm. The softer center makes better flour, however, for it remains freer from the bran in the grinding. The germ is composed of cellulose, nitrogenous substances, and about 10 per cent of fat.

The following table gives in per cents of the entire weight the comparison of different kinds and commercial grades of wheat, and of wheat straw and chaff.

COMPARISONS OF GRADES OF WHEAT, WHEAT STRAW AND CHAFF.

Kind of Wheat	Water	Ash	Protein	Crude Fiber	Nitrogen free extract	Fat
310 Amer wheats, min[1]	7.1	0.8	8.1	0.4	64.8	1.3
310 Amer wheats, max[1]	14.0	3.6	17.2	3.1	78.6	3.9
310 Amer wheats, aver[1]	10.5	1.8	11.9	1.8	71.9	2.1
Amer No. 1 hard[2]	—	1.8	17.2	2.4	76.3	2.4
Amer No 1 northern[2]	—	2.2	17.9	3.4	74.0	2.5
Amer No 2 northern[2]	—	2.2	18.3	3.2	73.9	2.3
Amer No 3 northern[2]		2.4	20.7	3.2	71.3	2.4
Minn No 163[3]	7.9	2.2	20.3	2.3	65.2	2.1
Rysting's Fife[3]	10.3	1.9	17.3	2.5	65.3	2.8
Bolton's Blue Stem[3]	9.3	2.0	16.6	2.5	67.2	2.3
Kubanka (durum wheat)[3]	16.5	2.2	18.9	2.5	57.3	2.7
Winter wheat, grain[4]	14.4	2.0	13.0	3.0	67.6	1.5
Canada common wheat[5]	10.9	1.4	12.8	2.0	70.4	2.5
Indian wheat, aver[6]	12.5	1.7	13.5	2.7	68.4	1.2
Emmer kernels alone[7]	10.5	1.7	13.2	2.6	69.4	2.8
Emmer kernels and chaff[7]	9.5	3.6	10.7	10.8	62.9	2.5
Emmer chaff alone	6.4	10.5	2.6	37.9	41.1	1.6
7 Amer wheat straws, min[1]	6.5	3.0	2.9	34.3	31.0	0.8
7 Amer wheat straws, max[1]	17.9	7.0	5.0	42.7	50.6	1.8
7 Amer wheat straws, aver[1]	9.6	4.2	3.4	38.1	43.4	1.3
Winter wheat, straw[4]	14.3	5.5	2.0	48.0	30.2	1.5
Winter wheat, chaff[4]	14.3	12.0	4.5	36.0	33.2	1.4
Durum wheat bran[8]	10.9	5.3	12.3	10.8	54.9	5.9
Durum wheat shorts[8]	10.4	4.1	14.4	6.1	59.2	5.9
Common wheat bran[8]	11.5	5.4	16.1	8.0	54.5	4.5
Common wheat shorts[8]	11.8	4.6	14.9	7.4	56.8	4.5

[1] U. S. Dept. Agr., Office Exp. Sta., Exp. Sta. Bul. 11, pp. 106-118.
[2] N. D. Bul. 8, p. 6. Average results from many analyses.
[3] Rept. N. D. Exp. Sta., 1903, p. 26. Data from N. D. wheat crop of 1901.
[4] Johnson, How Crops Grow, pp. 386-387.
[5] Rept. Canada Exp. Farms, 1900, p. 182. 1899 crop grown in the N. W. Terrs.
[6] Church, Food Grains of India, p. 95.
[7] An average from Dak. Stations, U. S. Dept. Agr., Farm Bul. 139, p. 11.

The substances of which wheat flour is composed may be divided into three classes: (1) Nitrogenous, which include mainly gluten, fibrin, albumen, casein, cerealin, and modifications of some of these; (2) non-nitrogenous, embracing sugar and dextrin, but chiefly starch, fat and cellulose; and (3) the minerals, for the largest part alkaline phosphates and silicates, especially phosphate and silicate of potash.

WATER.—Wheat ordinarily contains from 10 to 15 per cent of moisture. Changes in the moisture content of the air cause corresponding variations in wheat, and consequently in its weight. Usually such fluctuations in weight do not exceed 6 per cent, but they may be as much as 25 per cent, and an increase of 9 per cent in 24 hours has been observed. When wheat is shipped, especially if it is transported long distances, this may be a matter of great commercial importance. Wheat transported from the dry atmosphere of the inland of California to ordinary temperate regions will invariably gain from 5 to 15 per cent in weight. In a voyage from San Francisco to Liverpool, the increase in weight due to the moisture absorbed en route may be sufficient to pay all expense of transportation. Every portion of the wheat grain is so susceptible to influence from hydroscopic conditions that all of the products of wheat exhibit similar oscillations in weight. Two days equalized the moisture content in samples of flour varying from less than 8 to over 13 per cent.[1]

ASH.—Lawes and Gilbert observed the composition of the ash of wheat grown on unmanured ground during 20 years. The average results are given in the table below.[2]

	Grain	Straw
Ferric oxide	0.645	0.69
Lime	3.175	5.075
Magnesia	10.48	1.525
Potash	33.345	15.355
Soda	0.18	0.265
Phosphoric anhydride (P^2O^5)	50.065	3.10
Sulphuric anhydride (SO^3)	1.42	3.84
Chlorine	0.05	2.13
Silica	0.655	68.505
Total	100.015	100.485

[1] Hunt, Cereals in Amer. (1904), p. 38; Mich. Bul. 191, p. 159-164.
[2] Hunt, Cereals in Amer. (1904) p. 39

Total	100.015	100.485
Deduct O=Cl	.015	.485
Total	100.00	100.00

There is great uniformity in the ash constituents of the grain of wheat when it is not subject to irregularities in ripening, and there is but slight deviation under normal variations in soil composition.

PROTEIN.—Osborne and Voorhees[1] recognized and investigated five proteids. Approximately they form the following per cent of the grain: A globulin, 0.65; an albumin, 0.35; a proteose, 0.30; gliadin, 4.25; and glutenin, 4.25. Gluten is composed of several nitrogenous compounds, chiefly gliadin and glutenin. Wheat bread owes its excellence to the peculiar properties of gluten, which makes it lighter and more digestible than bread made from the other cereals. The amount and quality of gluten determine the baking qualities of a flour. It is now claimed that 55 to 65 per cent of the total gluten should be in the form of gliadin. Hard wheats have a higher gluten content than soft wheats, and consequently yield better flour. Gluten generally forms from 12 to 14 per cent of the wheat grain. Dough washed with water will retain only the crude gluten. A short growing period or a season unfavorable to full maturity of the grain increases the amount of protein. The nitrogenous compounds are the most desirable part of the nourishment found in wheat, but they tend to give a yellowish tint to the bread, "against which fashion rebels," for the "unnatural demand of the times" is for a starchy, snow white flour.

NITROGEN FREE EXTRACT.—This forms the larger portion of both grain and flour, and is composed very largely of starch, the amount of which is easily influenced by the irregularity of seasons.

Composition Influenced by Seasons and Fertilizers.—A favorable season seems to give a high weight per bushel, a large percentage of starch, and a low ash and nitrogen content. The following table gives the results of the observations of Lawes and Gilbert at Rothamsted.[2]

[1] Amer. Chem. Jour., 15:392-471.
[2] Hunt, Cereals in Amer. (1904), p. 43.

	Wt. per bu. lb.	Grain to straw Per cent	Grain per acre lb.	Straw per acre lb.	Nitrogen in dry matter Per cent	Ash (pure) in dry mat'r %
Average of eight favorable harvests:						
Plat 2—Farm yard manure	62.6	62.5	2342	6089	**1.73**	**1.98**
Plat 3—Unmanu'd	60.5	67.4	1156	2872	**1.84**	**1.96**
Plat 10A—Ammonium salts alone	60.4	66.2	1967	4774	2.09	**1.74**
Average of eight unfavorable harv's:						
Plat 2—Farm yard manure	57.4	54.5	1967	5574	1.96	2.06
Plat 3—Unmanu'd	54.3	51.1	823	2433	1.98	2.08
Plat 10A—Ammonium salts alone	53.7	46.7	1147	3601	2.25	1.91

Composition as Affected by Light.—Light is essential for the formation of proteids. The following table shows the effect of differently colored glasses upon the nitrogen and albumen content of wheat.[1]

Percentages of	Black Glass	Green Glass	No Glass
Nitrogen	2.54	2.74	2.08
Albumen	15.87	17.12	13.00

Climate, soil and culture are also all factors that affect the chemical composition of wheat. They are treated more fully in subsequent chapters.

The Composition of Different Commercial Grades of wheat shows that the amounts of protein and ash decrease as the grade of wheat becomes higher, while the nitrogen free extract increases. Differences in protein, gluten or gliadin content do not seem to be an adequate basis, however, for the commercial grading of wheat. The grading seems to be based rather on the relative yield of first quality flour. The greater the weight of the kernel and the weight per bushel, the higher is the grade of the wheat.

Historically there has been little change in the chemical composition of wheat. It seems likely that the wheat of ancient Egypt did not differ more in composition from modern wheat of the same variety than one sample of modern wheat frequently differs from another.

[1] Sci. Amer., 93 (1905): 508.

CHAPTER II.

IMPROVEMENT OF WHEAT.

INSTITUTIONAL EVOLUTION.

Early Significance.—The culture of wheat has perhaps never been exclusively the subject of individual effort, but has also always been the subject of institutional essay, however vague and remote. Since the latter phase of wheat growing became scientific in the nineteenth century, it has been fraught with a significance of the widest and deepest interest. From an institutional point of view, the growers of wheat are not sufficiently differentiated from the agricultural element of society to warrant a distinctive treatment as a class proper. Only by a statement of such characteristics of the agricultural class as are apropos for a consideration of the institutional development relevant to the culture of wheat can the subject be approached.

By proverbial repute, the tillers of the soil are, comparatively speaking, independent, unprogressive, non-co-operative, and without marked tendency toward organization. Historically, they have been the last great class to be brought under a progressive regime of societal institutions. There are two main causes for this, neither one of which is inherent in the class. The first and fundamental cause is that agriculture is an occupation in nature and conditions such as to require isolation of those engaged in it, with comparatively little division of labor among them. It is an industry as broad as the land upon which it takes place, and admits of no concentration. On the other hand, taking the number of people adequately supported on a given area as a test, the industry is universally developed by a decrease in the size of the holdings of each individual, and by the diversification of labor consequent to this decrease. The second cause, more remote and less important than the first, is that in agriculture the influence of competition is necessarily indirect, and under certain conditions entirely inoperative. In civilized life competition in one form or another has always given a great impetus to organization and co-operation. In modern agriculture, especially if the farmer owns his land, the only point at which the influence of competition can enter is in the sale of farm products.

Other things being equal, a progressive farmer may be able to offer his wheat for sale at a price below the cost of production for the unprogressive grower. While this is competition, its point of incidence is mainly below the line of subsistence for the farmer, and as most farmers are above this line, much of the force of competition is lost. When a government guarantees to an individual the ownership of a certain area of land, he has a monopoly of that area as long as he raises enough produce from it to pay the taxes, or their equivalent, for the governmental guarantee, and to keep himself supplied with the necessaries of life. If he is unprogressive and isolated in his farming, he is quite free to continue so his whole life, and his son and his grandson are just as free to follow in his footsteps.

In the early days the farmer looked to better informed powers than those of human origin for the solution of difficult problems. Wily and insinuating shamans and medicine men astutely took a benevolent interest in him by unfolding, interpreting, and at times even creating, the knowledge and instruction which numerous deities dispensed through these, their agents, for the benefit of agricultural mankind. When to plow, sow, harvest, and when to sell his crop, were thus made manifest to him by the deities whose special business it was to know these things. The gifts of rain and sunshine were in their hands. They alone were the instrumentalities of fructification and bounteous harvests. With the advance of civilization, however, the deities became less communicative, the shaman's magic power waned and became less occult, while his usual recompense grew more burdensome to those who paid it, and his functions became differentiated and were gradually assumed by the botanist, the chemist, the agriculturist, the physicist, the miller, the speculator, the instructor, and above all, the experimenter. As the paternal concern of the gods and medicine men for the farmer became relaxed, little interest was taken in him for centuries, and he has never since been the object of such profound solicitude from any source. In the middle ages and during the conquests of the Goths, Vandals and other barbarians, agriculture in Europe ebbed to the lowest degree of respectability. It was revived by the Saracens of Spain, and by their successors, the Moors, it was carried to a height perhaps not surpassed in Europe before the last quarter of the nineteenth century.

While Plato, Socrates and Pliny took an interest in agriculture, it is claimed that the oldest of writers on husbandry whose works have survived is Cato, the Roman Censor (234-149 B. C.). In 1757, Home stated that Virgil and Columella were still the best authors on this subject. From the downfall of the Roman democracy until the dawn of English history, little was written on agriculture. At times it was encouraged in a general way and highly honored, as it always has been in China, but usually the farmer was left to work out his own salvation. This he did, and successfully, though it required centuries of time. He no longer relies for information upon the elucidations of subtle shamans revealing the will of elusive, evasive, and ever vanishing gods, creations of the fancy. In nearly every civilized country of the world he is supported by scientifically grounded institutions. As these are practically the scientific foundation of modern wheat raising, especially of some of its most recent and interesting phases, they are considered of sufficient importance here to be taken up briefly.

The National Governments of all of the principal wheat growing countries of the world are factors in an official capacity in the culture of wheat, and at times millions of dollars are expended by a single government in endeavoring to solve some problem of unusual importance. In the United States, Washington in 1796 suggested the establishment of a national board of agriculture. The first appropriation made by Congress for agricultural purposes was in 1839, $1,000. Lincoln approved the act which established our National Department of Agriculture in 1862. Under Cleveland, in 1889, it was raised to an Executive Department.

The development of the department has been surprising, especially in recent years. The things most characteristic of it have been its rapidly increasing magnitude, the study of questions most diversified in interests and far-reaching in importance, and the thorough, effectual and scientific methods employed. As new interests arose, were investigated, and increased in importance, they were assigned to a new bureau or division especially created for their research. The distribution of seeds and plants was begun in 1839. Since that time, over 20 divisions and bureaus have been created.[1] The importance that may be attached to the activities of the department is well illustrated by its work with durum wheat. By securing its introduction and its use in manufacturing macaroni in the United States, the department practically established a new industry, in addition to extending materially the wheat producing area.

Experiment Stations.—Liebig in Germany, Boussingault in France, and Lawes and Gilbert in England, were the greatest of the pioneers who blazed the path subsequently followed by the experiment station. The organization of scientific experimentation with governmental aid dates from 1851. The American stations are an adaptation of those of Europe to the conditions and requirements of this country, but one of their characteristic features is extensive co-operation. Their establishment naturally followed that of the agricultural colleges. In 1875 the first station in the United States was established at Middletown, Conn., for which the credit is due to Orange Judd, then editor and proprietor of the *American Agriculturist*. Seventeen stations had been established by 1887, when Congress passed the Hatch act, the great boon for American stations. In 1894, 55 stations were in operation. At some of the stations, especially that of Minnesota, new varieties of wheat and other cereals have been originated which increase the yield several bushels per acre over old varieties under the same conditions, giving to the farmer a pure gain of millions of bushels.

Agricultural Institutions of Learning.—The American agricultural colleges were organized under the land grant act passed in 1862, supplemented by an act of 1890. Under the provisions of these acts 65 institutions are in operation in the several states and territories. The movement for farmers' institutes, originating in various farmers' societies, has now become national in scope, and during the year ended June 30, 1905, institutes were held in nearly all of the states and territories.

[1] For a complete account of the department, state experiment stations and agricultural colleges, see Bulletin 112, Office of Experiment Stations, U. S. Department of Agriculture.

Economic Position of Wheat Growers.—The story of the agriculture of the wheat area in the middle west of our country is the oft repeated one of agriculture in a new country, a fact which bespeaks an economic justification. There was but one way in which the western pioneer could draw a draft that would be honored for the cost of buildings, machinery and live stock, and that was to draw it at the expense of the natural fertility of the soil. One-crop wheat farming and neglect of crop rotation and domestic animals resulted. For over half a century, "Uncle Sam is rich enough to give us all a farm," was a household phrase. The farm having been obtained, it was used and abused in every way that was supposed to yield the largest amount of immediate profit, regardless of all other considerations. In no other section was this so true as in the wheat raising areas. In the meantime, millions of acres of fresh land produced more grain than domestic consumption could utilize, and for years the very existence of the farmer was threatened by 40-cent wheat and 20-cent corn. Lack of capital and the hard conditions of frontier life soon resulted in debt. Often there was not the wherewithal to pay the high interest and to procure the necessaries of life. With the twentieth century came a change, a change of such moment and speed as to be without parallel in the economic history of agriculture. The prosperity of the middle west transformed a million agricultural debtors into financially independent farmers. Free land, free immigration, and free private enterprise in railroad construction were the chief factors that ultimately led, not only to financial independence, but also to a new dignity and to a higher standard of living. With the telephone, the daily mail and newspaper, and means for traveling, a new horizon of comfort surmounts the skyline of the farmers' economic strength. This recent era of rural prosperity augurs well for the nation's future.

IMPROVEMENT.

Wheat Improvement Proper consists of artificially increasing the natural variations of the wheat plant and its environment. Historically, it is unknown whether the plant or the environment was first the subject of improvement. The subsequent

HARVESTING A 32-BUSHEL-PER-ACRE CROP OF MINNESOTA BLUE STEM WHEAT

portion of this chapter is exclusively devoted to the plant, the treatment of which naturally comes first.

Variation.—It has been recognized for at least a century that wheat is capable of variations. These may be peculiar to the plant itself, and may occur although the environment remains constant. Variation in this sense became established only with the theory of evolution, and refers to those changes which tend to become permanent through inheritance. Such variations are assumed to be the manifestations of a natural tendency inherent to all organic life.

The theory of common descent for all living beings found its first great advocator in Lamark at the beginning of the nineteenth century. Fifty years later Darwin assembled enough evidence in support of the theory to enable it to gain general acceptation. Darwin assumed that the great variation involved in the theory proceeded in the main by slow and gradual changes. He recognized, however, that species may also originate in nature by leaps and sports. The theory that all variation occurs by sudden mutations has been held by a minority of scientists. Cope and De Vries[1] are among those who have most recently increased the evidence in this direction. A defence of discontinuous evolution has also been made by various other scientists, such as the paleontologist Dollo, the zoölogist Bateson, and the botanist Korshinsky. In general, it may be said that if the followers of Darwin have been open to the criticism of under-emphasizing sudden change, the supporters of the theory of mutations have certainly erred more widely in the opposite extreme.

Variations may also be induced. In this process two different methods may be used, hybridization and change of environment. Only those variations which may occur or be induced independently of environment are considered in this chapter. Others are treated in subsequent chapters. Variations may include differences in habit of growth, chemical composition, periods of development, appearance, form, yield, prolificacy, vigor, hardiness and stability of type. Whatever his concep-

[1] An able criticism of the theory of mutations has been made by Prof W. F. R. Weldon. "Professor De Vries on the origin of species." Biometrika, 1:365, 1902. A study of this theory is interesting in conjunction with the more elaborate theory of homotyposis developed by Prof Karl Pearson in his work at University College, England, but space forbids a discussion of the matter here.

tion of variation may be, the scientific wheat grower utilizes the process in two different ways, by the simple process of selection, or by the compound process of selection, hybridization and selection.

Selection is an unfailing means for the modification of form and tendency in organic life. It augments the power of variation by successively selecting the most marked variations in any direction. While conscious selection is a modern process which has attained commercial importance at a comparatively recent date, there is no doubt of selection having been one of the most powerful influences from the very first in developing wheat, although men were not aware of its operation. Whatever protection or cultivation early man bestowed upon the cereal plants was naturally bestowed upon the grasses and wheats which produced the most food in return, and not upon those comparatively less important as food. The very essence of the importance attached to wheat has always been its food yielding quality. It is a perfectly sound inference that those varieties of wheat which had this quality in the highest degree had an advantage which aided them to survive other varieties. This, however, is only the operation of the prime factor of selection, or, as Darwin calls it, the "law of the preservation of the favorable individual differences and variations, and the destruction of those which are injurious."

Selection and cultivation, in the ordinary sense, were the processes of domestication. After domestication, varieties continue to be propagated in a similar manner. The results have been attained none the less advantageously and certainly on account of the fact that man was unconsciously the selecting agent. To this force of artificial selection was added that of natural selection in early development, which was a result of the coincidence that the quality of wheat as a human food and the reproductive functions of the plant were both united in its seed. The plant producing the greatest number of seeds was most apt to survive, not only because man was most likely to give it his fostering care, but also because of the increased chances of reproduction. In wheat artificially sown, care must be exercised lest this force of natural selection operate disadvantageously, for fewer seeds are no longer a disadvantage in reproduction. If for any reason, such as being brought to a new climate, wheat shows an unusual tendency to vary, it changes, and if the better yielding plants are crowded out the change results in a lower yield. Virgil[1] mentions selection of seed before the time of Christ, and noticed its advantages. A Scottish agriculturist, Shireff, made discoveries pertaining to the selection of wheat as early as 1819. The Belgian horticulturist, Van Mons, scientifically practiced selection before 1835. The works of Le Couteur, the English breeder, show that selection in wheat was early practiced, but never long continued or repeated. One of the early experiments in selection of wheat was that of Hallett[2] in England, begun in 1857. He selected the best heads and kernels. The following table gives his results.

EXPERIMENTS IN SELECTION OF WHEAT.

Year	Grains	Length	No. of grains	No. of ears on one root
1857	Original ear..................................	4 3-8 inches	47	...
1858	Finest ear raised...........................	6 1-4 inches	79	10
1859	Finest ear raised...........................	7 3-4 inches	91	22
1860	Heads imperfect...........................		....	39
1861	Finest ear....................................	8 3-4 inches	123	52

Thus by means of repeated selection alone, the length of the ear was doubled, the number of grains per head was nearly trebled, and the tillering power was increased over fivefold. It is only within recent years that wheat experiments of this nature have been carried on in America. The most extensive and successful of these were begun in 1892 at the Minnesota experiment station under the direction of Prof. W. M. Hays. From 1891 to 1896 experiments were made in Kansas with light, common and heavy seed, and seed from selected heads. The light seed uniformly gave a lower yield, but common seed gave the highest yield during three years.[3] At the Minnesota station from 1895 to 1898, No. 169, a wheat selected on principles similar to those of Hallett, gave an average yield of 28.3 bushels per acre, while during the same years the un-

[1] Georgics I., lines 286-288.
[2] Neb. Bul. 32, p. 91.
[3] Kan. Buls. 20, 33, 40 and 59.

selected parent sort yielded only 22.5 bushels, an increase during four years of 5.8 bushels per acre. In ten years nearly 25 per cent in yield was gained.[1]

Ninety-six tests of selected wheat seed during the years 1900 to 1902 at the Canada experiment farms gave an average gain of about 3.6 per cent in favor of selection.[2] Principles differing somewhat from those usually followed in selection were

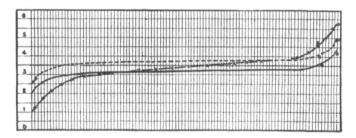

CROSSING AS A CAUSE OF VARIATION:

Yield in grain of 100 plants, showing greater variation in yield of hybrid than of parents. Yield of hybrid shown by x line. (After Hays.)

utilized by Lyon.[3] His selections were for quality rather than quantity. He experimented with the smallest and lightest kernels on account of their high nitrogen content. Heavy seed planted at the rate of 1.5 bushels per acre gave a greater yield of wheat the first year than light seed sowed at the same rate. Selecting heavy seed grown from the heavy wheat and light from the light wheat, the difference in yield in 3 or 4 years was small. After the first year of the separation, the light seed gave much the greater amount of proteids per acre. Lyon points out, however, that proteid nitrogen is no index to the amount of gluten, which is the better basis for improvement. It is not yet decided whether selection should be for plants with large heads or for plants with a large number of medium-sized heads. In general, the results of many experiments seem to favor the selection of large seed.[4]

[1] Hays, Plant Breeding, p. 10.
[2] Evidence of Wm. Saunders, 1903, p. 48.
[3] U. S Dept. Agr , Bu. of Plant Indus., Bul. 78 (1905)
[4] Hunt, Cereals in Amer. (1904), pp. 87-89.

While there may be a few slight or questionable exceptions to the general rule,[1] it can be said that enough results of scientifically conducted experiments are now at hand to prove conclusively that by means of selection alone the yield of wheat can be materially increased, even in a few years. The gain from the increased yield is much greater than the cost of making the selection, if the work is carried on systematically through a series of years.

A method frequently used is the selection of plump kernels from grain in the bulk. While there is doubtless some advantage in this method, it cannot give the best results, for many of the plump kernels may come from imperfectly filled heads, or from plants having few or weak suckers. Selection is a choosing of the individual, which, in the case of wheat, is a stool with several spikes and many seeds. When mutually antagonistic characters are desirable, such as earliness and productiveness, selection is very difficult and requires good judgment. By proper selection, not only may yield be increased, but all the other variations above mentioned may be influenced. Prolificacy of races may be fixed. Another important quality to be considered is vigor. Indeed, it has been held that the vigor and productiveness of the parent are far more important than its mere size.

The chance of improvement by selection increases as the number from which individuals may be chosen grows larger. The plant breeder has a great economic advantage over the animal breeder, for the expense of producing seeds for individual plants is so small that only a few of the best seeds are kept, while in animal breeding expense ordinarily forbids disregarding more than a small per cent as poor specimens. Prodigious variations may be induced by a long continuation of the selective process. Rigid selection systematically and scientifically practiced on a large scale by European seed growers in the last century has increased the sugar content of sugar beets more than 100 per cent.[2] Six years of selection at the Minnesota station increased the length of flax fiber over 20 per cent. Hays estimates that the farmers have increased the yield of corn 20 per cent by annually selecting the largest and best formed ears from among many thousands. The process has

[1] Thorpe, Harper's Mag., 15:302.
[2] Yearbook U. S. Dept. Agr., 1901, pp. 217-218.

also resulted in adapting corn to regions far north of its former habitat. Wheat perhaps has not been so generally improved by selection as corn has, but the wide practice of seed grading through the use of the fanning mill must have similar results. This is a slow process, however, and no great changes are effected at any one time.

Natural selection is also continually operative, especially in connection with such qualities as rust resistance and hardiness against heat, drought or cold. Thus wheat naturally tends to adapt itself to its environment. Such crude methods of seed selection as have been practiced in conjunction with natural selection have been the factors in evolving Turkey wheat so that it is more drought resistant than formerly, and has improved in hardiness so that it can be grown much farther north. Quality of the grain in any respect, yield, earliness in ripening, and non-shattering, in addition to the qualities just named above, are some of the most important characteristics that may be readily increased on any farm by selecting seed from those plants which exhibit these qualities in the highest degree. As these things cannot be properly determined after harvest, all selections for seed should be made in the field. Marked variations or sports possessing improved characters are occasionally met with in the fields. These are often carefully developed into valuable races by seed selection. Fultz, some of the Fife wheats, and many other well-known races have been originated in this way.

Hybridization consists in cross-fertilization. This may be simple, the fertilization of one race with another, resulting in a hybrid of two bloods, or it may be composite, the fertilization of a hybrid with another race or hybrid, resulting in a hybrid containing the blood of three or more races or species. Hybridization may be natural or artificial. Natural hybrids rarely occur. This is shown by growing different varieties of wheat side by side. Why varieties do not cross under these circumstances has not been fully explained. It is claimed that over half of the pollen from an anther is deposited into the air, and it would seem that it could readily find its way to adjacent flowers. Possibly the stigma is usually not receptive to foreign pollen.

In artificial cross-fertilization, self-fertilization must be prevented by removing the male organs, the anthers, from the flower before the plumes are open and the pollen shed. A good spike of wheat is prepared for hybridization by removing with sharp scissors all but one or two dozen strong flowers in the center of the spike. From these the anthers are removed while they are still green, or slightly tinged with yellow. To prevent accidental introduction of foreign pollen, the emasculated spike is wrapped about with tissue paper, tied above and below. Neighboring spikes of the same age show when the flowers are fully developed, usually in one or two days. Pollen brought from the variety chosen for the male parent is then inserted into the emasculated florets, and the cross-pollinated spike is again wrapped to exclude other pollen and to afford protection against pilfering birds and insects. The hybrid produced partakes of the characters of both parents. Saunders found that the crossbred kernel closely resembles that of the female plant, and that the modifications were not distinctly manifest until the second generation, when they appeared in a remarkable degree. Some races of wheat may differ so widely that they cannot be successfully crossed. If it is desired to combine the characteristics of the two, it can be done by first crossing each with an allied variety, when composite hybridization will succeed between the two hybrids produced.

The operation of cross-fertilization is by far the easiest part of the process for attaining results desired. We began with a "good spike." To secure this requires a ready knowledge and judgment of wheat. In hybridization, as in selection, any quality may serve as an ideal for the operator. To attain success, he must know for which qualities to seek and he must have the judgment which enables him to recognize these qualities and to select a foundation stock which possesses them in a high degree. This is truly a case of well begun, half done. Certain desirable or necessary qualities may be entirely lacking in a variety, which must then be improved by breeding into it the desired characteristics from some other variety possessing them to an unusual extent. In selecting a stock with which to begin, it is advantageous to draw from a variety already improved by selection, but the breeding of wheat should not be limited to the few very best wheats, for a fairly large number of varieties can be used profitably for special characteristics.

The great advantage of hybridization is shown in three effects, all of which aid in accomplishing more rapidly the results aimed at in selection. It makes it possible immediately and directly to combine the qualities of two different plants in one; it immensely increases that variation which alone makes selection possible; and it imparts greater vigor to the offspring. Hybridizing does not always give a progeny im-

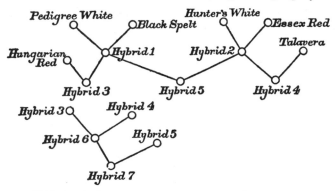

DIAGRAM SHOWING PEDIGREE OF GARTON'S HYBRID.

mediately averaging better than the parents. In many cases the first progeny will average much poorer than either parent. Its great value lies in throwing together qualities and multiplying variations, both of which may be developed by selection. This greatly increased variation has been explained on the ground that "the wheat plant being so closely self-fertile, there is within it, lying dormant, a wonderful power to vary (a power far greater than in plants cross-fertilized in nature), which is thrown into action when different varieties are artificially crossed."[1] As to these varieties Hays says: "The further they have departed from ancestral characteristics and formed diverse qualities, the more likely will their progeny exhibit new characteristics made up by combining those which have become so radically different in the two parents."[2] There

[1] Carleton, Basis for Improv. of Amer. Wheats, p. 73.
[2] Plant Breeding, p. 37.

also arise characteristics so new in kind and degree that they can hardly be considered as a mere combination of any characteristics found in the parents. All of the so-called "botanical" classes of wheat have been produced by hybridizing two varieties, a fact which "certainly indicates blood relationships between the classes of wheats."

Since the qualities originated by hybridization must be developed by selection, it may take years before the value of the hybrid can be determined. The advantage in large numbers lies in the fact that "only one individual in several thousand has marked power to produce a valuable strain." These individuals have been called the "Shakespeares of the species," and the labor of eliminating by selection all of the other individuals is 99 times that of producing the hybrids.[1] By the usual method, the seeds to be tested and selected are planted individually in rows 4 or 5 inches apart. If any promising plants develop, 100 seeds from each are again planted. These groups of plants from single parents are called centgeners. By means of selection, crossbred wheats can thus be reduced in four or five generations to a type so uniform that little or no variation will occur among plants in the field. Whether they will retain their acquired characteristics has been questioned. Hybrids originated by Hays and Saunders seem to do so. The question as to whether wheat will deteriorate under self-fertilization is still an open one.

In breeding a variety of wheat, the ideal to be held in mind constantly is "that it yield the largest possible amount of grain of the best quality for the purpose desired under given conditions."[2] In such a course botanical appearances seemingly will take care of themselves.

Historical.—The sexuality of plants was proved experimentally by Camerarius (1691). The first recorded hybrid was produced by Thomas Fairchild (1719), an English gardener, who crossed the carnation with the sweet william. The publications of Koelreuter (1761) paved the way for the work of Thomas Andrew Knight (1800), the eminent English plant physiologist, who has been called the father of plant breed-

[1] Hays, Yearbook U. S. Dept. Agr., 1901, p. 229.
[2] Scofield, Algerian Durum Wheats, p. 7.

ing. The first hybrid produced in the United States was probably a pear (1806). The importance of hybridization in relation to variation was demonstrated by Naudin and Nägeli (1865).

The pioneer producer of wheat hybrids in America was C. G. Pringle of Charlotte, Vt. He began his work in 1877, and several varieties have received his name, some of which have become standard. Pringle's Defiance has been a rust resistant variety of California since 1878. During his connection with the Colorado agricultural college A. E. Blount produced quite a large number of hybrids, some of which are now well known in the United States and are also among the most valuable varieties in Australia, both as field wheats and as parents of native hybrids. The most important are Amethyst, Improved Fife, Hornblende, Gypsum, Blount's No. 10, Felspar, Ruby and Granite.

The director of the experimental farm at Ottawa, Canada, Dr. William Saunders, began hybridizing wheats in 1888. His main object has been to procure early ripening varieties, and he has attained success by hybridizing American and Russian races. Preston and Stanley are two of his best productions. In the main these hybrids have been produced in the most simple way. A. N. Jones of Newark, N. Y., practicing composite crossing, though always with quite closely allied parents, has done the most important work in wheat hybridization in this country, and his varieties are now the most widely used of all recent American wheat hybrids. The two features characteristic of his work have been composite methods, and high gluten content as an ideal. The nature of the soil and climate of eastern United States is such as to produce soft and starchy wheats. His efforts have been to raise the standard of eastern varieties as to gluten, and he has largely succeeded. Winter Fife and Early Red Clawson were the two most popular of his first varieties. Early Genesee Giant, another well-known variety which he originated, is widely grown in New York and Pennsylvania. It has no ancestors outside of the common bread-wheat group. This seems to be a weak point in Jones' method of procedure, for the most advantageous composite crossing is supposed to be with varieties of entirely different wheat groups.

This gives by far the greatest variations in degree and number, and gives qualities not otherwise obtainable. For example, the highest degree of non-shattering must be obtained from spelt or emmer, while the quality of resistance to leaf rust is best acquired by crossing with the durums. Jones' Winter Fife could not be grown in the Palouse country on account of its shattering, though it yielded 60 to 65 bushels per acre.

The Garton Brothers of England and William Farrer of New South Wales have extensively practiced crossing the different wheat groups. Every variety and every intergradation results from such crossing. A local variety may acquire, not only rust resistance and tenacity of chaff by intercrossing with a spelt and a durum, but also greater fertility of the head drawn from the spelt, and increased vigor of the seed, which produce a higher yield. These, and increased hardiness and gluten content, are practical results attained by the Garton Brothers. William Farrer has done an immense amount of excellent work in improving Australian wheats, especially as to rust resistance. The most important work in breeding cereals on the continent has probably been done by W. Rimpau of Schlanstedt, Germany, though his work is not generally characterized by composite methods. The Vilmorins have also done work in this line. The Dattel, one of the most widely distributed varieties of wheat around Paris, was originated by them.

Breeding Experiments have been carried on in the Kansas wheat belt for some years, and extensive co-operative work in this line has been taken up with the experiment stations in different wheat growing states, particularly in Texas, Kansas, South Dakota, Minnesota and Maryland. Efforts are being made to secure a variety that will ripen a few days earlier, so that by sowing two varieties the harvest period can be lengthened, and the danger of green cutting and shattering be avoided. Wheat and rye have been successfully hybridized by a number of experimenters, but as yet with no valuable results.

Experience has taught that the most successful and practical way to fight disease is to aid natural selection in producing disease-resistant or immune plants, rather than to attempt to

cure the disease. "As a foundation for rational wheat im·
provement, a knowledge is required of the characteristics and
needs of different wheat districts, and the characteristic qualities of the natural groups of wheats." A century ago wheat
was wheat, but now thousands of varieties have been bred up
which thrive best under the local conditions for which they
were bred, and often they satisfy conditions, uses and tastes
not in existence a century ago. The entire wheat harvest of
the world is being improved. The value of this work in
proportion to its cost must appeal to everyone, and indicates
its permanency. Luther Burbank made the statement that if
a new wheat were bred that would yield only one grain more
to each head, Nature would produce annually, without effort or
cost for man, 15,000,000 extra bushels of wheat in the United
States alone.

The conclusions of scientists seem to be that varieties will
not wear out or materially change if the same conditions which
made them excellent are kept up. If special care was exercised
to produce an artificial variety, this care must be continued,
or it will deteriorate. The improvement of wheat by breeding
is no longer a theory, as in the time of Darwin, but an established fact.

CHAPTER III.

NATURAL ENVIRONMENT.

The individual plant is the complex resultant of two forces, heredity and environment. Those characteristics of wheat which are acquired from environmental influences and which are transmissible from generation to generation of plants may be considered as belonging to heredity, a subject fully treated in the preceding chapter. The natural environment, consisting of soil and climate, is a pronounced factor in the growth of wheat, independent of the artificial modifications known as cultivation. The latter subject is treated in a later chapter.

ENVIRONMENTAL INFLUENCES.

Soil.—There are mechanical and chemical differences in soil that exert a varying influence upon the quantity and quality of wheat. The effect upon yield is more pronounced than that upon quality. In North Dakota 39 different samples of Blue Stem and Scotch Fife wheats of known history were obtained from farms representing the varied soils of the state. Sown upon the same soil, all gave approximately the same results in yield and quality of grain and straw. They also matured at the same date, and had like periods of development. Another experiment was made in which seed raised from one soil was hand picked to uniformity, and then grown upon various types of North Dakota soil in différent portions of the state. The resulting grain and straw showed great variation.[1] Similar experiments were made in Indiana[2] and Maryland[3] with practically the same results.

The soil has been a great factor in determining the distribution of wheat. Much of the wheat of the United States is grown upon glacial drift soil. There are two general types of this soil: The uplands, which are usually of a light-colored,

[1] N. D. Bul. 17, pp. 89-95.
[2] Ind. Bul. 41.
[3] Md. Bul. 14.

tenacious clay; and the lowlands and prairies, which have a dark, loamy, organic, friable soil. Common bread wheats are usually grown on black soils. These soils are not well adapted to fall wheat, however, for it is apt to winterkill. Durum wheats thrive best in alkaline soils rich in nitrogenous matter. Sandy bottom land is best adapted to the production of soft wheat. Richardson attributed the low protein content of some American wheat to a deficiency in soil nitrogen. The ash of wheat stands next to the gluten in variability, and the factor most concerned in its variation is the soil.

DURUM WHEAT DISTRICTS OF THE UNITED STATES
(By Carleton)
The lined districts show where durum wheat will succeed best and the dotted districts where it may be grown with grain of less quality.

Climate.—Seasonal differences are included under this subject, because their effects are the same in kind as those of climatic differences. Certain climates produce certain corresponding characteristics in wheat, regardless of what the soil conditions are. The protein content of wheat, and correspondingly its moist and dry gluten, is extremely sensitive to environment of a meteorological nature. The starch content is also sensitive, but in an inverse ratio. Climate varies from year to year in any locality, and it is well known that this causes corresponding variations in wheat, even under similar soil conditions. In the gluten content is seen the first reflection of a change in environment. The claim has even been made that a number of varieties of wheat grown under uniform soil and meteorological conditions would yield relatively the same percentages of gluten, however much these might vary from the normal.[1]

Northern grown seed of spring wheats will mature plants earlier than southern-grown seed of the same variety, but the reverse is true of fall-sown grain, which ripens earlier from southern-grown seed.[2] Wheat raised on the sea coast develops special characteristics due, at least in part, to climate. In southern Russia Arnautka wheat attains its highest perfection only when grown within a limited area bordering the Azov sea. All wheat raised directly on the Pacific coast in western United States is soft, damp, dark and has a very thick skin. It shades off gradually to that grown inside of the coast range and protected from the fogs. This inland grain is bright, very hard and dry, and has a thin skin.

Regions having cold winters produce most of the world's wheat. Marked exceptions to this are California, Egypt and India. Small, hard, red grains having a high nitrogen content are usually found in a climate characterized by extremes of temperature and moisture. Climate and season both affect the length of the period of growth. This has an important influence on the chemical composition of wheat, for a short season of growth raises the percentage of protein and lowers that of starch. In Canada, a shorter period of time is required for maturity in northern latitudes. The growing season of Winnipeg is about one week longer than that prevailing 500 miles farther north.

It has been said that "other things being equal, varieties which have become acclimated are to be preferred." While this is true, it still leaves us the case where other things are not equal. Nearly every climate has its disadvantages for wheat growth, and, as we have seen, wheat always adapts itself to overcome these disadvantages. The greater they are, the

[1] O. Bul 129, p. 5.
[2] Yearbook U. S. Dept Agr., 1901, p. 235; S. C. Bul. 56, p. 12.

more highly developed must be the resisting qualities of wheat to overcome them. By the law of the survival of the fittest, a very detrimental condition in climate or soil develops in wheat a correspondingly great power of resistance. This is the scientific foundation of the importation of seed wheat. It has been recognized and taken advantage of to a certain extent, but not as fully as might have been, and consequently this point will merit subsequent mention. Advantageous importations are well illustrated by the introduction of hardy and drought and rust resistant varieties from the cold and the hot and dry parts of Russia into sections of the United States having a similar climate. A low altitude and an abundance of moisture seem to produce softer wheats.

Soil and Climate.—Many characteristics of wheat are due to the combined influences of soil and climate. Environments that differ widely are characterized by peculiar varieties of wheat varying in composition and physical appearance. Soft wheat repeatedly sown on heavy, black, upland soil tends to become hard, while hard wheat becomes soft after years of successive planting on bottom lands. Experiments have shown that the environment of Colorado affects the composition of wheat by increasing its gluten content at the expense of the starch content, while the environments of Oregon, California and North Carolina have the opposite effect. A study of the map[1] showing wheat districts will show the general effects of climate and soil in the United States. Broadly speaking, the hard, red wheats are found in the central, elevated plains, and the grain becomes softer and of lighter color as either ocean is approached. American, Russian and Algerian wheats have about 12 per cent of moisture, while those of Europe have about 14 per cent.[2] As early as 1884 it was determined by chemical analyses that the wheats of the Pacific coast in the United States have a smaller percentage of albuminoids than those of the rest of the country. In recent years there has also been a gradual deterioration in the gluten content of North Dakota wheats. The attention of the Department of Agriculture was called to these deteriorations, which are due to the combined effects of soil and climate, and extensive experiments were carried on to determine the exact causes and afford relief.

[1] See p. 9.
[2] Girard & Lindet, Le Froment et sa Mouture, pp. 86-93.

Wheats are easily changed as to the season in which they are sown, the winter to spring and the spring to winter varieties. The change is most readily effected in warm, arid climates, where irrigation is practically the sole source of moisture. It can also be accomplished by sowing the winter wheats later and the spring wheats earlier each season. Winter wheat may be sown in spring and spring wheat in the fall. Only a very few plants will ripen seed, but when this is continuously sown, in three years the spring variety will be changed to the winter, and vice versa. In 1857 Klippart wrote that red bearded wheat could be changed to white, smooth wheat, and vice versa. Kubanka, a yellowish-white spring wheat, found in its perfection east of the Volga on the Siberian border, developed into a red winter wheat in the Caucasus.[1] Red wheat is usually more hardy than white wheat, while bald wheat is usually not so well adapted to a hot, dry climate or alkali soil as bearded wheat. When seed from irrigated soft wheat has been planted without irrigation, it has been known to harden remarkably in a single year.

HEREDITARY INFLUENCES.

Seed Wheat—In each kernel of wheat are embodied the latent possibilities of its future development. Consequently, it is very important to select the seed which will bring the best results possible in the environment under which it must be grown. A knowledge of the importance of good seed wheat, and of the principles of its development, does not eliminate all of the practical difficulties involved in securing good seed. Frequently the grower is so situated that he must purchase his seed, and he should not follow the common practice of waiting to do this until the sowing season has arrived. It is then too late to ascertain the origin and history of the grain, or even to test its vitality. The speculative markets do not trade in seed wheat, and they are not a factor in determining its price. The great bulk of seed wheat does not move far, but is grown in the locality where it is to be used. Good seed of any of the different classes of wheat may generally be procured from the section in which that class is most commonly grown. For example, Turkey Red wheat should be bought in Kansas, Ne-

[1] Carleton, Macaroni Wheats, p. 11.

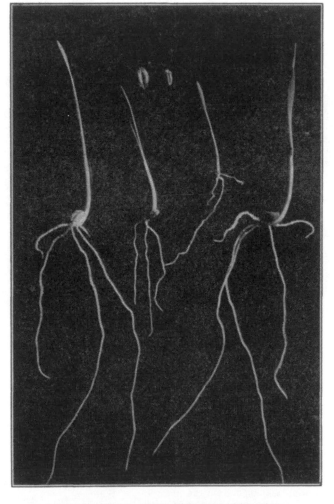

WHEAT PLANTS FROM GOOD AND POOR SEED

braska or Iowa; Sonora wheat in California; and hard spring wheats in the Dakotas or Minnesota.

Dealers in seed may be divided into three classes according to the methods which they pursue. One class buys seed in the open market and sells it as that variety for which it was bought. Such dealers should be required either to improve their methods or to seek a new occupation, for usually the most advantageous disposition that can be made of their seed is at the nearest grist mill. Another class buys by sample in the open market, using all care possible under the circumstances to secure correctly named seed of good quality. They make germination tests, reclean the seed if necessary, remove light and injured seeds, and offer for sale only those which are good and sound. There is a small class of dealers each of whom makes a specialty of some variety which he has grown under contract. Two points of great importance in regard to their seed is that it is usually of the highest quality and almost always true to name. The grower, in looking for a wheat with the proper origin and history, should have a knowledge of the main facts and laws set forth in this and the previous chapter, and should select seed according to the needs of his environment. He should buy by sample in the fall or winter before sowing. Its purity he can have determined, and a simple home germinating test will give its capacity in germination. Many farmers may not yet be able thus to procure intelligently the seed best suited to their environment, but they have a number of institutions at their command whose business it is to dispense information of exactly this nature. Some experimentation must always be engaged in.

It is only within a few years that the quality of commercial seeds has been subjected to any tests other than those made by the reliable seed firms. This shows a great lack of appreciation of one of the essential factors of agriculture. It is one of the most remarkable and unaccountable facts in connection with the development of the United States Department of Agriculture that it spent a neat fortune in distributing seeds for over a half century, and never once tested the quality of the seeds sent out. In 1895, "for the first time in the history of the department did its authorities know the real quality of the seeds they had distributed."[1]

[1] Yearbook U. S. Dept. Agr. 1897, p. 94.

There are no statistics to show how great is the loss resulting from the impositions of unreliable seed firms, but it must be
millions of dollars. The loss from sowing poor seed grown on
the farm is also great. Frequently, especially among the uned-
ucated classes, any wheat which is injured too badly for market
purposes, either by such diseases as smut, or by improper harvesting or storing, is used for seed purposes. Many experiments with immature seed wheat have been made. While its
germinating powers may be greater, the conclusion is that
smaller and less vigorous plants are produced, resulting in a
lower yield.[1]

It is very questionable whether wheat frozen in ripening, or
burned in the stack or bin, can be safely used for seed. It
certainly should not be sown if badly affected, and the only
way to determine its value is by a germination test. A low
germinating power often means a lack in quality as well as in
quantity, which makes the use of such seed very hazardous.

Whether seed will "run out," and whether it is profitable
occasionally to "change seed" or not, has long been a mooted
question. A change of seed, especially if the change is between very distant sections, is almost invariably accompanied
by some disadvantages. If it is merely a promiscuous exchange,
as is so often the case, it is very likely that the disadvantages
will greatly outweigh the advantages. The principles above
pointed out throw some light on this question and show that
there is one case in which a change of seed is advantageous,
namely, from an environment unfavorable in certain conditions
to one more favorable in these same conditions and having no
new disadvantages which counterbalance the good results. It
is obvious that if the transfer is from a favorable to an unfavorable environment, the wheat must, by a selective process,
adapt itself to the new conditions before it can yield as much
as that which is already adapted. This, of course, has reference only to the one variety which is under consideration.
There are also other considerations, however. In the first place,
the custom of changing seed is a costly one in actual expendi-
ture of cash. Farmers purchase annually many thousands of
bushels of seed wheat, paying fancy prices and freightage

[1] Rept. N. D. Agr. Col., 1902, p. 32; Yearbook U. S. Dept. Agr.,
1896, p. 306.

from distant points, and also paying duty on foreign varieties. Then, too, injudicious seed exchange is a source of weed and disease dissemination, as well as a powerful influence against proper methods of plant breeding. Selection cannot be successfully practiced in improving the quality of grain if the seed must be given up every few years for a strain grown upon other land.

Nothing has been found in the principles of wheat development which would indicate that wheat "runs out," or deteriorates, if continually grown on the same farm under rational methods of culture. Undoubtedly, any seed may deteriorate because of injuries arising from disease, improper cultivation or selection of seed, or from many other causes which militate against the production of a normal type of kernel. Grass wheat formerly grown in Kansas is a case in point. Experiments at different stations have shown that seed may be sown on the same land for many years, and yet give no appearance of running out.[1]

The idea that a change in seed gives good results has always been founded more upon opinion than upon well ascertained facts. It was doubtless first advanced by Columella shortly after the Christian era, and has been widely held ever since. The most striking argument in favor of the idea is put forth by Darwin, who reasons that since a change of residence is of undoubted benefit to convalescents, it may be that a change of soil is advantageous for wheat. This is only reasoning by analogy, however, and involves the comparison of abnormal animal life with normal plant life, certainly not a strong argument at best. Some good authorities still hold that wheat will run out if sown continually on the same land.[2] A great and preponderating amount of evidence has accumulated, however, to show that farmers should rely chiefly upon locally developed seed, and that they should give more attention to the production of their own seed. The importation of seed is profitable only when differences in the rigors of soil and climate exist. Such importations have greatly improved the standard of American wheats and have also extended the industry of raising them. Foreign wheats are one of the most important fac-

[1] N. D. Bul. 17, p. 98.
[2] Saunders, Evidence 1903, p. 46; Rept. Kans. State Bd. Agr., Vol. 21, No. 81, p. 7.

tors in hybridization, supplying precisely those qualities in
which American wheat is deficient.

So far the most valuable importations have been made from
Russia. The now world famous red winter wheat grown in
the section of which Kansas is the center was originally imported by the Department of Agriculture from the Crimea in
Russia. So successful has it proved that in 1901 Kansas growers individually imported over 15,000 bushels of this variety
for seed. Its superiority consists in higher yield, hardiness to
winter cold, better milling qualities and great rust resistance.
Four or five winter varieties obtained from eastern and southern Russia were tested by the department in 1901 and 1902.
They were much hardier than any varieties grown in this
country, and extended the winter wheat area farther north and
west. The better Russian varieties are late in maturing, while,
as a rule, Japanese sorts are early. Hybrids from the two ripen
early and possess the good qualities of the hardy Russian sorts.

Drought-Resisting Durum varieties adapted to alkali soils
have been introduced from Russia. They have proved themselves admirably suited to the region west of the 100th meridian,
from Texas to Dakota, where wheat growing was supposed to
be practically impossible. The area on which they may be
grown is shown on the accompanying map.[1]

Some Hungarian wheats have also been introduced, as well
as white wheats from Australia, Europe and the Orient, to obtain a higher grade of wheat with which to replace the de-
teriorating white wheat of California. The Weissenburg, a
very promising variety from Hungary, is the source of the flour
that sells on the Liverpool market for $1.00 more per barrel
than any other flour.

The introduction of spelt and emmer must also be mentioned
here, both for a crop and for hybridizing wheats. In all this
work of introducing and breeding wheat, disease resistance is
kept in mind, and some sorts remarkably free from rust have
been procured. There are two needs which are common to the
whole country. Greater yielding power is, of course, always
desirable, and for one reason or another, the same is also quite
generally true of earlier maturity, whether it is to escape
drought, rust, insects or frost. The new environment may

[1] See p. 48.

cause changes in the imported variety, and it may require several years to determine its merits.

Instead of scientific agriculture having "almost reached the limit of its development," it has just fairly begun to develop. This statement is especially applicable to the careful selection of seed. "In Germany, where the percentage of sugar in sugar beets is high, they deem it necessary to adopt the following plan to improve the standard. Ten thousand beets say, all perfect, are selected from a field where the choicest strain was sown and carefully tended. A small section is taken from each beet and tested to determine the percentage of sugar it contains. The hundred beets of the highest quality are selected and planted the next season for seed. The seed from these, is, of course, very valuable, representing hundreds of dollars worth of work, and it is used simply for growing seed beets. From the seed beets thus grown only one hundred of the best are again selected as stock to grow seed beets from, while the rest of the 10,000, though grown from the same strain of seed, are considered only good enough for growing seed for the man who raises sugar, and not sugar beet seed."[1] Seed which is good enough for growing beets for seed is considered much too valuable to use in growing beets for sugar. The beet grower has made more progress in this respect in one century than the wheat grower has in many centuries. It should, nevertheless, be said that the improvement in beets was partly the unforeseen result of European legislation. By a peculiar tax, whatever sugar above a certain per cent was extracted from beets paid no tax, or a smaller rate. To increase profits by increasing this excess proved such an additional stimulus to improve the sugar content of beets that the legislators apparently could not modify the laws fast enough to keep pace with the advance.[2]

While the past importance of introducing new varieties is conceded, it is said that "the time will soon arrive when there will be no further varieties to introduce better than we already have." Unquestionably, the breeding of wheat will have an increasing importance. As wheats may be developed, so they may deteriorate, on account of soil and climate, and in such cases there must perhaps be a periodical importation of seed.

[1] Proc. Tri-State Grain Grow. Ass'n, 1900, p. 170.
[2] Handwörterbuch der Staatswissenschaften. 7:997-1009.

CHAPTER IV.

CULTIVATION OF WHEAT

SOIL PREPARATION

The cultivation of wheat is coeval with agriculture itself. No land was alone the ground of Ceres where the human race learned to plow and sow, nor was it the task of any one race or nation to tutor mankind in the agricultural arts. How the nations of antiquity tilled, sowed and reaped would be of great interest, but the records which time has bequeathed to us are all but silent upon these homely topics.

Climatic Effects.—The soil in nature is more uniformly covered with vegetation throughout the year than it is under cultivation. Dark, exposed soil absorbs more heat than soil covered with vegetation. Thus cultivation is supposed to moderate climate and there is a widely prevalent opinion that it also lessens frosts, humidity and rainfall.

Plant and Soil Effects.—Soil alterations of the highest importance are made by means of tillage, fertilizing and irrigation. The resulting variations in wheat are quantitative rather than qualitative, except in the case of irrigation, which is practiced on a comparatively small part of the wheat area. This process and that of fertilizing are discussed in later chapters. Cultivation, as here used, refers only to the mechanical operations connected with raising wheat in the natural environment treated in the previous chapter. It renders this natural environment artificial to the degree in which it alters the mechanical arrangement of the soil in nature and eliminates from the competition of life other species of plants which would naturally compete with wheat in the struggle for obtaining the sustenance held by soil and atmosphere. This sets the wheat plant free from many natural conditions which tend to destroy unfit variations and to force wheat to assume one type. Thus cultivation, while it has no direct influence in increasing variation, by removing conditions which exert a selective influence, is indirectly the means by which a greater number of variations survive. In cultivation itself, as above defined, there is no selection.

The principal effect of cultivation on the growth of wheat is through its influence upon the physical condition of the soil, to which great importance is attached. By physical condition is meant friability or openness, capacity for absorbing and retaining water and heat, and permeability to roots. Air, which is necessary to the roots, is excluded by hard, water-soaked, baked or puddled soils, and such soils are also impermeable to roots. Stirring or cultivating the soil enables the air to circulate and the roots to penetrate through it. Tillage has been known to increase the yield of wheat over eight bushels per acre.[1] Richardson claimed that it increased the nitrogen content of wheat. Generally about 50 per cent of the volume of soils is empty space. That is, in one cubic foot of soil there is about a half a cubic foot of space into which air and water can enter.

The Motor Power first utilized was the muscular energy of man himself. Its application requires the least intelligence. The abundance of human labor, and hence its cheapness, coupled with a lack of intelligence to utilize other forces, are conditions still existing in vast regions of the earth where it is impossible for any other motor power to compete successfully with man himself. India, largely using human labor, often at a cost of but 4 to 8 cents per day, has been so successful in competing with more civilized nations using other forms of power as to assume fourth rank among the wheat raising nations, and to be able to undersell many of them in the world markets. Hand labor is used almost exclusively in raising wheat in China, Japan, Siam, Syria and Colombia, and very extensively in Egypt and parts of Greece, Spain, Mexico, and some of the South American republics.[2]

Animal Power.—The first one of the forces of nature which man subdued and utilized in relieving himself of some of the drudgeries incidental to agriculture was that of the domesticated beast. There are no marked periods of progress in this. Animal power is by far the most universally used in agricultural operations. As a rule, oxen are found in communities

[1] S. C. Bul. 56, p. 12.
[2] U. S. Daily Consular Repts., Oct. to Dec., 1903.

A COMBINED STEAM PLOW, DISC HARROW AND SEEDER USED IN PACIFIC COAST BONANZA WHEAT GROWING

less developed agriculturally and horses in the more developed ones. For example, oxen were preferred to horses in England from 1250 to 1650.

Steam.—After nearly two centuries of projection and invention, steam was successfully used for agricultural operations in England in 1832. The system adopted was that of dragging the implements by the aid of pulleys and a cable revolved by a stationary steam engine. This method in improved form is still found in Europe. The movable engine appeared before 1850. In the United States, activity in the invention of steam plows began in 1861, and it was perhaps entirely confined to the use of the traction engine. On the Pacific coast, steam is used quite extensively in the cultivation of wheat, especially on the larger farms. In Germany and Hungary there was about one steam plow to every 10 small plows in 1900. There have also been experiments with electricity as a motor power in agriculture.

Plowing.—The first plow was simply a "sharpened piece of wood or the crotched limb of a tree," and was evolved from the hoe. Some of the earliest plows were drawn by two men, while two others kept them in the ground. The form represented on Egyptian monuments (3,000 B. C.) is an improvement on the hooked stick. Chinese historians say that the first plows in China were made 2,737 B. C. The plow described by Homer was a composite piece drawn by oxen or mules. The early Romans had no cast steel or iron, and their implement was essentially like that of the Germans. The first use of metal on a plow is unknown, but before the time of Christ the Romans yoked the steer to one with a "shining share." The ancient Egyptians and Assyrians had plows pointed or edged with iron. These primitive implements turned no furrow, but simply stirred the ground. Those of the Greeks 2,000 years ago had wheels supporting the beams, and similar forms are found depicted in Saxon manuscripts.

The rude primitive plow seems to have been almost universally the first agricultural implement drawn by beasts of burden. The constancy of the type among different peoples is remarkable. Under uncivilized or frontier conditions it nearly always appears, and its persistence is very great. In the United States plows were worked in Virginia as early as 1617.

Twelve years after the landing of the Pilgrims, the farmers about Boston had no plows. The first ones used by French settlers in Illinois were of wood with a small point of iron tied on with straps of rawhide. The oxen were yoked to them by the horns. This method of hitching was rivaled only in Saxony and Ireland, where the horses were fastened to the plow with their tails. An attempt was made to abolish this practice in Ireland by act of Parliament in 1634. Arthur Young (1741-1820) mentioned it in his time, however, and Gibbons maintained that it was still to be found in remote parts of Ireland as late as 1896.

In England, from one to eight oxen were used in the eleventh century, while four horses or oxen were usual in the seventeenth century. The first plow in California (about 1835) was a crooked branch with an iron toe. On the whole, the American form before 1767 was practically the same as that used by the Romans before the Christian era, and this type was still found in Europe in 1867. It was the only agricultural implement of France in the eleventh century, and of Sicily in 1863. In southern Greece many plows similar to those of the age of Pericles (450 B. C.) are still being used. Many of those in Russia are equally primitive, while the Spanish, South French and Italian forms resemble the Roman type.

There was little improvement in the plow during the middle ages, perhaps largely on account of legislative restraint. Many popular prejudices also existed. In England, for example, after the farmers had experimented with iron plows of good construction, they concluded that the iron made the weeds grow; and in America iron plows were supposed to poison the soil and to prevent the growth of crops. It was not until the end of the seventeenth century that plows began to be improved. The moldboard was then made of iron and steel and given its proper form. While the plow was always essentially a wedge-shaped instrument forced through the soil to loosen it, these improvements perfected it so that the draft was reduced by one-third and the implement was also much more complete in its operation on the topsoil, which it gradually loosened, raised and completely turned over to one side. Coulters were known in England at least as early as the eleventh century. Fitzherbert writes of different kinds of plows for different soils in

1534. The date of the first English patent on this implement is 1720. The cast iron plow was first patented in the United States in 1797. A patent on an adjustable cast iron point in 1818 marks the introduction of the most useful economy in plow manufacture, the interchangeability of parts.

Modern Plows are practically the same in principle as those described above. The only improvements which have been made are in minor details. The draft and friction have been reduced to a minimum, and forms have been invented which are best suited for different types of soil and for the application of different kinds of motor power. The common hand plow is undoubtedly most widely used, and the small farm rarely ever has any other. It is drawn by two horses. Another widely used form is the sulky plow, having two wheels to carry the beams, and a seat for the driver. Two or three horses are required. The acreage covered depends on the condition of the soil, and varies from one to two acres per day. These are the common forms used by all the large wheat raising countries. Another common type used on large farms is the gang plow, drawn by horses or steam. This is merely a number of common plows combined in one frame. A usual plow in the Red river valley is a gang cutting 16 inches in two furrows, drawn by five horses and turning 250 acres in from four to six weeks. Steam is not used, as mud was found to cut out the plow bearings when it was wet, and the expense of keeping horses is necessitated by other farming operations. In some parts of California, plows are set in gangs of as many as 14. They are drawn by eight mules, and plow three inches deep at the rate of 10 or 15 acres per day. A traction engine with large gangs of plows or discs is often used on the larger farms, accomplishing an enormous amount of work in a little time.

Special forms of plows adapted to the use of a stationary engine have been evolved in Europe. The Fowler plow is perhaps the best known and most effective of these. It consists practically of eight turnover plows yoked together, and is capable of plowing 40 acres of land a day and accommodating itself to the most uneven ground. The electric plow of Austria is also worthy of mention.

Time of Plowing.—In general, it may be said that in the spring wheat area of the United States, fall plowing slightly

THE TYPICAL FARM WHEAT DRILL WITHOUT FERTILIZER ATTACHMENT

increases the yield, is most destructive to weeds and insects, and is the most economical in farm management. For winter wheat, the ground is plowed as soon after harvest as is practicable. This destroys the weeds before they ripen their seed and gives time for a compact seedbed. The pulverized surface soil more readily retains and absorbs moisture, upon which, in the absence of vegetation, no demands are made by growth. The depth of plowing should vary with the climate and with the nature of the soil and the subsoil. The limits of the variations usually found advantageous are between four and eight inches in depth.

Subsoiling.—As the common plow is in effect a wedge passing through the soil on a horizontal plane, the uppermost layer of the subsoil is compacted at each plowing. This renders the subsoil more impervious to water and roots. Subsoiling consists in breaking up the subsoil, and does not necessarily involve changing the relative positions of subsoil and topsoil. Judged by experiment station results, it does not seem to be an economical operation.

The Seed Bed.—Soil, on account of its fine texture or wet condition, may be lumpy after plowing. The spaces in it are then very irregular in size, and the soil is in a poor condition to draw up water from below, or to furnish uniform germinating conditions for the seed. In such cases it is customary and advisable to work the soil with a harrow, roll, or other implement until the larger lumps are broken and the surface becomes smooth and even. The seed bed is then ready for the sowing. Thorough preparation conserves the moisture, diminishes winterkilling, and increases the yield. In both the spring and winter wheat districts of the Mississippi valley, it is a general practice to sow without plowing on land that has produced corn the preceding year. In the case of winter wheat, the grain may simply be drilled between the rows of corn, with a five-hoe drill, or the seedbed may first be prepared with a disc or tooth harrow. Corn ground for spring wheat also is often prepared by using an implement of the disc harrow type.

SEEDING.

Sowing.—There are three methods of sowing wheat: Broadcasting, which scatters the seed evenly over the ground; drilling, which places it in rows; and dibbling, in which a certain number of grains are dropped in each hill by means of a dibbling iron. Diverse means have been employed in each method. Dibbling, once quite extensively practiced in England, is never found now, unless it is with the experimenter. Nature's method, broadcasting, was also the first method of artificial seeding. The seed was simply scattered by hand. Of the three ways, drilling is now recognized as the most advantageous. The conclusion from station experiments is that the increase in yield will amply pay for any extra cost involved in drilling.[1] Less seed is required, for the wheat is more uniformly distributed and covered. If it is sown at an even depth in moist soil, quick germination results. This places weeds at a disadvantage, especially in spring wheat. Drilling also decreases the danger from drought, winterkilling, and the blowing of soil by the winds. The snow lodged in the furrows left by the drill affords protection and moisture.

Seeders.—After hand sowing came the seeder, which accomplished the same results mechanically. Such machines are by no means modern, though in England and Germany they can be traced only to the beginning of the seventeenth century. The ancient Chinese, Persians, Hindoos and Romans used them, as well as the drill, which was doubtless the next seeding machine to be invented. Ardrey maintains that the first historical knowledge of a seeder pertains to an Assyrian drill used many centuries before Christ. The Egyptians of 3000 B. C. sowed by hand, the method still widely followed all over the world where the farms are very small, or where the standard of farming is not high, as, for example, among the lower classes of Russian peasantry. In early England the wheat was sown into the plow furrow, often by a mere child, who carried a bag or wooden hopper (known as a seedlip or seedcod) full of grain in front of the horses or oxen drawing the plow. The same practice prevails in east central India, a woman taking the place of the child. By another method in India, the seed is thrown through a tube attached to the plow handles.

Jethro Tull introduced the drill in England in 1730. His first machine sowed three rows of wheat at a time. In 1851

[1] Hunt. Cereals in Amer (1904). p. 84.

Pusey wrote: "The sower with his seedlip has almost vanished from southern England, driven out by a complicated machine, the drill, depositing the seed in rows, and drawn by several horses." In America, the first patent granted on a seeding machine was in 1799. A slide broadcast seeder, which was a riding implement, was patented in 1835; the rotary broadcast seeder came in 1856; the grain drill in 1874; and the riding grain drill in 1884.

The Wagon Seeder is the best machine that has ever been devised for rapidly broadcasting wheat. It is mounted on a special tail-board, which, when the machine is used, is substituted for the tail-board of the wagon. It consists of a seed-hopper; a driving shaft connected by a sprocket chain with a sprocket wheel fastened to one of the rear wagon wheels; a rotating seed plate in the bottom of the hopper; and a

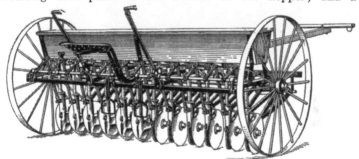

A MODERN PRESS DRILL WITH DISCS

distributing wheel shaped like a windmill. The grain falls upon the distributing wheel, from which it is effectively scattered by centrifugal force. In the ordinary force-feed seeder gravity does the distributing, but here the additional factor of the centrifugal force given by the distributing wheel is involved. Two men and one team can broadcast 100 acres a day with this machine.

The Press Drill is similar to the ordinary broadcast seeder in that it carries, parallel to the axis of its two wheels, a seed-box having a number of seed-cups in its bottom. From these cups the seed is brought by feed-wheels which are attached to a revolving shaft. This force feed was the first great improvement over utilizing gravity alone for the purpose of distributing the seed. The grain falls into a tube, which, instead of scattering it as in the seeder, carries it in a steady stream to the bottom of the shoe. The soil is pressed laterally by the shoe, and the seed finds a moist bed in which to germinate. It differs from the ordinary drill in that it presses a V groove instead of scratching a trench. The press or shoe drill has largely superseded the hoe drill, especially in the far west. Disc drills are also used, but they are not adapted to stony, hilly or wet land. Drills and broadcast seeders are made in standard widths of 8, 11 and 14 feet. The tendency in recent years is to drill in the wheat, except perhaps in California. In the Red river valley four-horse press drills covering 12 feet are used. About 30 acres a day are sown by one man, and no

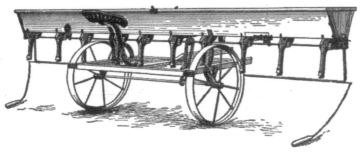

A TYPICAL FORCE FEED BROADCAST SEEDER

subsequent cultivation is necessary. By the old method of seeding by hand, one man could sow about 16 acres per day, and the wheat had to be cultivated into the ground after it was sown.

The Order in Which Seeders Have Evolved is somewhat as follows: (1) Sowing by hand; (2) the broadcast seeder, taking the place of the hand, the flow of the seed depending on gravity; (3) the broadcast seeder with force feed; (4) the ordinary drill with a force feed putting the grain in evenly in rows and deeper; (5) the press drill, which is now the best machine we have for seeding. In the absence of wind, the hand grass seeder can be used advantageously for broadcasting small areas.

Perhaps the only region in the world where nature still occasionally seeds the ground by her own methods so efficiently as to produce a crop is on the Pacific coast of the United States. Some wheat is nearly always ''shed'' or shelled out before or during harvest, and, if cultivated into the ground by harrowing or discing, produces what is known as a ''volunteer'' crop. If not enough has been shed, frequently a little more is scattered over the field, and instances are not uncommon where 25 to 30 bushels per acre have been yielded by such volunteer wheat lands.

The Amount of Seed required per acre varies with time and method of seeding, with soil and climate, with different varieties of wheat, and even with size and quality of seed of the same variety. One variety may have only half as many grains in a bushel as another. A bushel of shriveled wheat will have more grains than a bushel of plump wheat. The lower the germinating power, the more seed will have to be sown per acre. Less seed is required if the time is early, if the rainfall is light, if the soil is fertile, if the seedbed is well prepared, and if the grain is drilled. The yield, however, is not proportionate to the seed sown, for by tillering more or less, the wheat plant adjusts itself to its environment. The most usual amount sown per acre in the United States is about 5 pecks. It varies from 2 pecks in parts of California to 9 pecks in Ohio. The average amount sown per acre in the United States is $1\frac{3}{8}$ bushels in the winter wheat regions, $1\frac{1}{2}$ bushels in the spring wheat regions, 7 to 9 pecks in the Middle Atlantic states, 6 to 8 pecks in the Mississippi and Ohio valleys, and 3 to 8 pecks in California.[1]

The Time of Seeding varies so much with soil, climate and different varieties of wheat that, taking the world around, any time during the entire year is the best time for some particular locality. For the United States Carleton says: ''It is a pretty safe rule to follow the practice of sowing always at a date which is considered to be early in that locality. At the proper time the seeding should be done at once, without regard to weather conditions.''[2] Local conditions must always determine the time for any particular locality. For example, if an attack

[1] Hunt, Cereals in Amer. (1904), p. 86.
[2] Yearbook U. S. Dept. Agr., 1900, p. 541.

of Hessian fly is imminent in a certain region, the farmers should take concerted action for later sowing. Spring wheat should usually be sown as soon as the ground is in a condition for seeding. Winter wheat sown too late lacks the vitality needed to withstand the cold, and sown too early it produces a rank and succulent growth that is injured by freezing.

The Depth of Seeding varies with the nature of the soil, the amount of moisture, and the condition of the seedbed. In a dry, sandy or cloddy soil, it is necessary to sow deeper than in a wet, clay, or level soil. Ordinarily, the wheat should be covered with about one inch of moist soil.

Harrowing.—After the ground is once plowed, the implement most commonly used for further cultivation, either before or after sowing the wheat, is the harrow. There are three principal objects in harrowing: (1) To kill weeds and grass, which would otherwise absorb moisture and nourishment needed by the wheat; (2) to level the surface and to keep it covered with a loose, dry mulch, both of which also conserve moisture; and (3) to cover the seed. In drilled wheat the latter is performed in sowing. All three of these objects may be attained in one operation.

The most primitive method of harrowing was to drag over the ground the limb of a tree with extending branches. This implement, like the rude plow, is often found reappearing on the frontier of civilization. It is easily improved and widened by fastening together a number of branches so that it does better work and covers a wider area. In California, in 1835, the wheat was sown broadcast by hand and brushed in with the branch of a tree drawn twice over the ground. The writer can well remember when, as late as the middle eighties, he brushed wheat into the ground with a "drag" made from scraggy wild plum trees cut on the banks of the Dakota river. A similar implement was also used in other parts of the United States. Another very primitive method of covering the grain was that used in ancient Egypt, where it was trampled into the loose ground by the hoofs of animals.

The Romans used a kind of harrow before the Christian era. In 1534 harrows with iron teeth were used in England, as well as some with wooden teeth. In Northumberland, in 1650, "the harrow was constructed without joints and without iron, of branches of the mountain-birch, fixed together with wooden pegs, with tines of the tough broom.''[1] The oldest and simplest form of the harrow had a wooden frame with teeth of wood or iron. As it was drawn over the field, it combed or raked the surface quite level. Two improvements have since been made. It is constructed in two or more sections so that it can accommodate itself to uneven ground; and flexible steel bars are used in the frame so that by means of a lever the teeth can be set at any angle. Harrows 25 feet in width are now used on the large western farms of the United States. With such a harrow one man and four horses can cover 60 to 75 acres per day.

Various other forms of harrows have been devised. The principal ones are the spring tooth and the disc harrows. The latter consists of a main frame to which are pivoted two supplementary frames. Mounted within each one of these is a shaft carrying a series of concavo-convex discs, and the whole series is rolled over the ground. Adjusting levers swing the supplemental frames to any angle in relation to the line of draft. The soil is cut and thrown out in a degree proportional to the angle set. It was first used by the Japanese in ancient times. In the last decade the disc principle has been widely applied to harrows, plows and cultivators.

Cultivation by one Operation.—As early as 1618 a machine, worked by steam, was invented and patented in England which plowed and fertilized the land and sowed the seed, all at one operation.[2] There is no record of its having done any work. In the same country a ''double-hoppered drill-plough'' was advertised as a new machine in 1744. It drilled and covered wheat and fertilizer together. Perhaps the only instance where any practical and extensive results in this line have been obtained is in California and northwestern Canada. Gang plows are used, and a broadcast seeder attached to the rear of the plow sows the seed as fast as the ground is plowed. The seeder is usually followed by a harrow, also attached to the plow. A small outfit, operated by one man and drawn by a team of eight mules, will plow, sow and harrow-in the seed in one operation at the rate of from 10 to 15 acres per day. On the large farms

[1] Grey, Agr. in Northumb., p. 4.
[2] Perels, Bedeutung des Machinenwesens, etc., pp. 11-13.

this machinery is combined into great gangs drawn by a powerful traction engine, and such outfits may cover from 35 to 100 acres per day.

Cultivation Subsequent to Sowing.—As a rule, in most countries wheat receives no cultivation between sowing and harvesting. Occasionally, however, it is harrowed or rolled after the seed has germinated, or after it has made some growth and become firmly rooted. This is done to kill weeds or retard evaporation. Ordinarily, such cultivation has not been found of advantage in modern wheat growing. In Japan wheat is planted in rows and hoed, but vegetables are usually raised at the same time between the rows. In the time of Fitzherbert, a kind of wooden shears or hook was used in pulling the weeds out of wheat. In the eighteenth century when wheat was drilled in England, it was hoed with a mattock or hoe.

Pasturing winter wheat is practiced to a certain extent. This should never be continued late in the spring, or when the soil is not in suitable condition, for yield and quality of wheat will then be lowered. If judiciously practiced, there may be no reduction in yield.[1]

[1] Okla. Bul. 65, p. 6.

CHAPTER V.

HARVESTING

EARLY CUSTOMS

Risks and Customs of the Harvest Period.—Man has a regulative control that is sufficient to insure a crop over so few of the essential conditions of wheat growing that there is always a very large element of risk involved from the time the wheat is sown until it is harvested. Increasing control due to accumulating knowledge acquired from past experience continually diminishes the risk. When a balance of all these things has been struck, however, the fact remains that the modern wheat grower is playing with many factors, anyone of which may cause a partial or complete crop failure. Extremes of heat and cold; drought; superabundance of rainfall; destructive hail or wind storms; floods; parasitic plants, such as smut and rust; predatory insects, birds and rodents; fire; various diseases and other unfavorable conditions may defeat all means to success at the farmer's command. Thus wheat raising, like most of the extractive industries, has a large aleatory element which cannot be eliminated, though it may be reduced to a constant factor by means of the insurance principle, to which we will give subsequent attention.

This risk which is involved reaches its maximum at the harvest period. In most regions wheat must be promptly harvested when it is ripe. If not then attended to, not only is the period of risk prolonged when there is no possibility of further gain, but an actual loss is sustained under the most favorable conditions, and the grain is also more susceptible to the destructive influences of its environment. The grain will now be shelled or lodged by wind previously harmless, many birds seek food in the ripened fields, and the rain causes the seed to lose color and to sprout. The ripening grain must be closely watched, for the determining change in the heads may occur between one day and the next. The field must usually be harvested within two weeks.,

In wheat raising the whole year's toil meets with no reward before the harvest. If this is lost, the fruits of all previous

73

labors go with it. Before the advent of modern machinery, harvesting was the most burdensome and exacting operation on the farm. It could not be delayed. The completion of the harvest gave relief from this season of toil and anxiety, and replenished bountifully the stores of grain which had become scant. Secure in this abundance and free from the arduous labors, the early husbandmen enjoyed a time of unusual license during which they dispelled their cares with rounds of uproarious jollification. During these general rejoicings practically all nations celebrated with games and rustic fêtes the final ingathering of the sheaves. In England the close of the season was marked with the "Harvest Home." A procession led by a pipe and tabor marked the bringing home of the last sheaves in the hock-cart. The load was surmounted by a sheaf shaped and dressed to represent the goddess Ceres, or by pretty girls of the reaping band in fantastic attire. The reapers danced about the procession, shouting:

> Harvest-home, harvest-home,
> We have plowed, we have sowed,
> We have reaped, we have mowed,
> We have brought home every load,
> Hip, hip, hip, harvest-home, etc.

In France and western Germany was found the Harvest May, *bouquet de la moisson*. A green sapling or branch was selected at harvest time and adorned with flowers, ornaments, and dainty eatables. It was often set up in the field that was being reaped. When the harvest was made, it was brought home on the last sheaf or load. The farmer received it with a solemn welcome and attached 't to some conspicuous spot on the barn or house, where it remained until replaced by its successor. These harvest festivals of modern Europe are very similar to those of ancient Greece, from which they have descended. There a branch of olive or laurel was used for the *eiresione*, or harvest bush, and it was carried to the temple of Apollo.

The great harvest festival of Rome was the Saturnalia, held late in December at the end of the vintage and harvesting. All classes, even the slaves, devoted themselves to feasting and mirth. Probably one reason why our Christmas was placed at the end of December was that it might supplant the Saturnalia and other heathen festivals. While Christmas is a festive occasion, and once took the place of harvest festivals, Thanksgiving Day is our national harvest festival. It ranks as a legal holiday and is fixed by proclamation. This day was suggested by the Hebrew feast of tabernacles, or the "feast of ingathering at the end of the year." Occasionally in our country there is also an after harvest dance.

Our festivals, however, have lost the rude simplicity and rustic romance characteristic of the past, and they are less immediately connected with the harvest. Modern invention has quite changed the nature of harvesting, rendering it an ordinary process and depriving it of many features which made it important and interesting in the olden times. One feature which has survived is the annual migration of harvest laborers. The novelty, the hardship, and the adventure incident to the travel, and the unusual compensation for the toil, so often performed with emulative zeal, have always lent a peculiar charm and enchantment to this occupation for a certain class of humanity. Every harvest, bands of the Irish used to travel to England, while the Italians and Austrians still go to France and Germany to help reap the grain. Shiploads of Italians regularly go to Argentina for the harvest time, and return to Italy when the season is over. Every year great numbers of agricultural laborers, both men and women, emigrate from the central and western provinces of Russia to the steppes of the east and southeast.

Nowhere else has this feature of harvesting evolved to such an extent as in the United States. The characteristic attractions are here found in an unusual degree, especially upon the bonanza farms of the northwest. In this district there is no farming in the usual sense of the word, for wheat raising has become a business interest differentiated from all others. The hard and practical business atmosphere of our age is everywhere prevalent, an atmosphere that would soon chill the simple home customs of our fathers. Not even home life is found here, for the year around the bulk of the work is done by transient laborers who live at the division dormitory, or in quarters far out on the fields. Nor is there the association of the factory, for men working on different parts of the same farm will often not see each other a single time from one year's end to another. But for the harvester the fascination of magnitude is always present, for magnitude is characteristic of every phenomenon and of every operation. The mere sight of a field of swaying, rippling wheat, with its green and gold, and with wave upon wave rolling away beyond the observer's horizon, surpasses description.

The Harvest Laborer.—In the United States, the wheat harvest begins in earnest by June. It is September before the last harvester passes northward out of the Red river valley, and during this time the merry click of the reapers is heard from sun to sun. This harvest-time succession has developed its own typical harvester. He first appears in Oklahoma. As the wheat ripens, he travels northward. Before Kansas and Nebraska are left behind, his possessions include a little money, a blanket, and perhaps a sooted tin tea pail. He is now one of an army of many thousands, a great number of whom follow the harvest through the Dakotas and beyond the Canadian border. The typical men of this class rarely pay railroad fare. Many of them ride into the bonanza district on the "blind-baggage" of passenger trains. Perhaps most of them ride on freight trains, at times over a hundred on one train. As a rule, the men of this class are not "hoboes," though now and then a tramp does work. The tramp element helps some, especially when laborers are scarce, but they are poor and unsatisfactory workmen, and are avoided when possible.

Perhaps a large majority of the men required to harvest the wheat of the middle west do not follow the harvest northward, but merely work through the season in one locality. Tempted by low railroad fares and large wages, they come from nearby cities, and from the states east and south of the wheat district. Many of them are farmers and farmers' sons. A large per cent are foreigners, especially Scandinavians. The personality of the men varies much. Among them the writer has found the city banker again seeking in the harvest fields during a brief vacation the health and pleasures experienced in younger years; the refined college youth earning the means with which to finish his course in the east; the western pioneer making a desperate effort to keep the wolf from the door of the shanty that sheltered his family, and to save the homestead by paying the interest on the mortagage which drought and frontier misfortunes had placed upon it; the dreamy faced wanderer who merely drifts with his environment; and the coarse, hard-featured criminal and ex-convict. It has been estimated that 15 men for every 1,000 acres of wheat migrate annually to the wheat districts. The number recruited from other sections for the harvest of Kansas alone in 1903 is claimed to have been 28,000, a force half as large as the standing army of the United States. Employment agencies in adjacent states sent men into Kansas in companies of 100 and 200. Some farmers used all the guile and promises at their command to induce men to stop with them instead of journeying farther. Some men were actually kidnapped, it is claimed, from the platforms of the trains, and held by force till their train had gone. "In Saline and Cloud counties, when the harvest started and there was a shortage of hands, the farmers' daughters went into the fields while the thermometer was close to the one hundred mark and did the work of men." Many of these harvesters remain over for the threshing, which often lasts until the snow flies in December. Doubtless a majority of the men go as they came, on special railway excursions, for which fares are frequently one cent a mile. The itinerant harvesters disappear so gradually that no one knows where they have gone. Some of them find their way to the mines of the Rocky Mountains. Many of them go to the logging camps of Minnesota and Wisconsin.

In Argentine there is a succession of harvest-times similar to that in the United States. It begins in the northern provinces in November and continues to move southward until February. The succession of wheat-harvesting seasons in different countries of the world is given below:[1]

January.—Australia, New Zealand, Chile.

February and March.—Upper Egypt, India.

April.—Lower Egypt, India, Syria, Cyprus, Persia, Asia Minor, Mexico, Cuba.

May.—Texas, Algeria, Central Asia, China, Japan, Morocco.

June.—California, Oregon, Mississippi, Alabama, Georgia, North Carolina, South Carolina, Tennessee, Virginia, Kentucky, Kansas, Arkansas, Utah, Colorado, Missouri, Turkey, Greece, Italy, Spain, Portugal, South of France.

July.—New England, New York, Pennsylvania, Ohio, Indiana, Michigan, Illinois, Iowa, Wisconsin, Southern Minnesota, Nebraska, Upper Canada, Roumania, Bulgaria, Austria, Hungary, Southern Russia, Germany, Switzerland, South of England.

August.—Central and Northern Minnesota, Dakotas, Manitoba, Lower Canada, Columbia, Belgium, Holland, Great Britain, Denmark, Poland, Central Russia.

September and October.—Scotland, Norway, Northern Russia.

November.—Peru, South Africa, Northern Argentina.

December.—Argentina, Burmah, New South Wales.

Proper Stage of Maturity for Harvesting.—In most of the wheat growing countries it is a very general practice to begin harvesting before the wheat is quite ripe. This lessens the danger from loss on account of over-ripeness, and if the grain is properly cared for, it does not seem to diminish the yield. Ordinarily, cutting should begin as soon as the straw turns yellow and the grain is in the dough. A good test is that the kernel "should be soft enough to be easily indented with the thumb nail and hard enough not to be easily crushed between the fingers."[1] In a climate like that of California, wheat may stand without injury for over two months after it is ripe. There is no danger from rain, and the only loss occurring results from an occasional sandstorm.

HARVESTING IMPLEMENTS.

Machinery for Harvesting.—The development of agricultural machinery is a very important factor in the world's economic progress. Growth in this direction has been very marked in recent years, and in no class of agricultural implements has it been more so than in that for reaping grain. The primitive method of harvesting wheat doubtless consisted in merely pulling up the plants by the roots and stripping the heads from the stalks by means of a comb or hackle, but long before written history crude implements were devised to assist the hand in pulling or breaking off the straw. From these rude beginnings to the modern combined harvester and thresher is a far cry, and the wheat grower who sacks his thousands of bushels of wheat from over 100 acres in a single day has little conception of the amount of painful study and experimentation, and of the numerous inventions it has required to evolve from the ancient sickle the perfected machine with which he so easily gathers his grain.

The Sickle.—Flint implements resembling a rude form of sickle or reaping hook are found among the remains of the later stone age in Europe. The remains of the early European habitations contain bronze sickles. The earliest records of Egypt contain accounts of reaping by means of crudely constructed implements similar to the modern sickle in form.

[1] Hunt. Cereals in America (1904), p. 103.

Greece, receiving the art of agriculture as a heritage from Egypt, had similar forms, as did also the Jewish nation. Since ancient times, the Chinese and Japanese have reaped with an implement resembling the sickle.

All sickles were used with one hand only. The grain was not

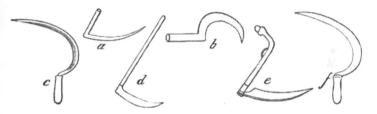

DIFFERENT FORMS OF EARLY SICKLES AND SCYTHES

As lettered above: *a.* Egyptian sickle; *b.* sickle of the middle ages; *f.* smooth-edged sickle; *c.* toothed sickle; *d.* early form of scythe; *e.* Hainault scythe and hook.

always bound in sheaves. One man could bind what six reapers cut, using "corn" for binding. A reaper cut an average of one acre per day.[1] Brewer, however, states that in England in 1844 seven persons usually cut one to one and one-half acres in ten hours.[2] Besides being still widely used in China and Japan, the sickle is also a common implement among the Russian peasants, and in Sicily. The first wheat raised in the Red river valley in America was cut with sickles and bound with willow withes by women and children.

The Scythes and Cradles are all used with both hands. They evolved from the sickle and form the second class of reaping appliances. The Hainault scythe, a Flemish implement, was a form intermediate between the sickle and scythe. It had a wide blade about 2 feet long. The handle, about a foot in length, was held in the right hand, and had a leather loop into which the forefinger was inserted. The handle also had a flat part which projected against the wrist, and served to keep the blade in a horizontal position. The left hand, aided by a hook, gathered the grain. The early scythes were clumsy and heavy. They had straight handles, and were used for cutting grass

[1] Rogers, Hist. Agr. and Prices, Eng., 5:53.
[2] First Cen. of Repub., p. 176.

only. As the scythe evolved, the blade became lighter and the handle passed through many forms before it permanently assumed the crooked wooden pattern. When fingers were fastened to the snath to assist in collecting the grain into bunches or gavels, the scythe became a cradle. The latter implement was perfected in America during the last quarter of the eighteenth century.

The scythe seems to have appeared first among the ancient Romans. Before 1850, the scythe or cradle and the sickle were the implements almost universally used in harvesting grain. The perfected American cradle spread rapidly to other coun-

AN EARLY CRADLE

A MODERN CRADLE

tries, but not without opposition. In England such violent opposition developed at Essex that the farmers were "deterred from the practice." The scythe and cradle are still frequently found in use in Russia and in various other parts of Europe. They are also found in America under conditions which render other implements impracticable. Within fifty miles of New York City are farms on which the grain is still reaped with the cradle. Brewer gives 1½ acres a day as the amount of grain cradled in this country by one man. It required two others to rake, bind and "stook" it. Others say 4 acres a day could be cradled by a good worker while another raked and bound it.

HARVESTING MACHINERY

The Header.—All reaping devices thus far considered have aimed at mechanical advantages alone. All of those subsequently discussed endeavor, not only to extend and improve the mechanism of the machine so that it will perform perfectly each and every operation connected with harvesting, but also to apply a power that will operate the machine. Under headers are included all machines that are designed to gather only the heads of the wheat, leaving the straw in the field. Such machines are of two kinds; stripping and cutting headers. The former has the distinction of being the first grain gathering

THE GALLIC HEADER, DESCRIBED BY PLINY A. D. 70

machine mentioned in history. It was used by the farmers of Gaul as early as the time of Christ. Pliny described it. A series of lance-shaped knives was fastened into one end of a large-bodied, two-wheeled cart. An animal yoked behind the cart pushed it through the grain. After the heads of the wheat were stripped from the stalks by the knives or teeth, they were raked into the box-like frame by an attendant. Palladius gives a similar account of the machine in the fourth century.

After being used during hundreds of years, the Gallic header disappeared, and it seems to have been completely forgotten for several centuries. Only through literature did it escape the fate of permanent oblivion and become a heritage for the modern world. The published descriptions of the machine by Pliny and Palladius furnished the impulse in which modern harvesting inventions originated. Its distinctive features are retained in several modern inventions of this class, machines

THE WHEAT HEADER IN OPERATION ATTENDED BY STACKING WAGONS

which have a practical use and value under conditions similar to those which existed on the plains of Gaul. Toward the close of the eighteenth century, the social, economic and agricultural conditions in England, on account of increasing competition and the higher value of labor, were ripe for the movement of invention that was heralded by the printed account of the Gallic header. The first header was constructed by William Pitt in 1786. It was an attempted improvement on the ancient machine in that the stripping teeth were placed in a cylinder which was revolved by power transmitted from the wheels. This "rippling cylinder" carried the heads of wheat into the box of the machine, and gradually evolved into the present day reel.

Nearly all of the principles involved in the header seem to have been developed mainly in connection with other machines, such as the reaper and combined harvester, in connection with which they will be discussed. Before 1823 only four inventors of harvesting machinery placed the power in front of the machine. This involves either a side cut or driving the power through the grain. On account of the great width of cut in the header the side cut would give great side draft, and as there is nothing to counterbalance this, all headers are propelled in front of the power. Omitting minor details, the evolution of the header was completed in Haines' celebrated machine of 1849, which was widely known as the "Haines Illinois harvester." It was thoroughly successful, and was practically the same as the machine of today.

The modern header has a cutter with a reciprocating and advancing rectilinear motion; the reel brings the grain upon a traveling canvas apron which delivers it to an elevating apron on one side, and this in turn discharges it into the header-box placed upon a wagon driven along with the machine; it has a swiveled steering wheel, operated by a suitable tiller; and an evener, to which the four or six animals are hitched, is pivoted forward of the steering wheel. The header ordinarily clips the stalks a few inches below the heads of the grain, but it can be run very low for lodged or short grain. It saves binding and shocking, but it is essential for the wheat to be dry before it is cut, as it must immediately be either threshed or stacked. If slightly damp, green, or weedy, it cannot be threshed at once, and may stack-burn if stacked. This confines the use of the header largely to the western part of the United States, where peculiar conditions exist which make it possible to let wheat ripen completely without much danger of loss, though the machine is used to some extent in the Mississippi valley. Some wheat growers cut with binders until the grain is ripe, and then use the header. It cuts from 12 to 20 feet in width, and from 15 to 50 acres a day. In Washington three headers and one threshing machine usually work together. From 50 to 75 acres a day are thus harvested. Three header-boxes, or barges, are usually used with one header. These are often unloaded at the stack or machine by horse power. A peculiarly arranged netting is laid in the box, and by means of ropes and a derrick the whole load is hoisted to the stack or feeder.

The header was used very extensively on the Pacific coast before the combined harvester came in use. Sixteen-foot headers drawn by six mules were used. The grain was usually threshed as fast as it was headed. The ordinary crew for a 44-inch cylinder thresher and 26-horse-power engine was as follows: Seven headers operated by 42 animals and 14 men; 21 header-boxes, requiring 42 animals and 21 men; and at the machine there were 11 animals and 32 men; this made a total of 95 animals and 67 men. In 1880 such an outfit averaged 3,800 bushels per day in California. Many headers are in use in South America, and a machine similar to an American header is also being used in Russia. The stripping header is still used in Australia. About 20 per cent of the headers manufactured in the United States are sold in foreign countries.

The Reaper.—Under the reaper are included all machines designed to cut the grain and gather it in bunches, gavels, or rows. While the header was the first harvesting machine that was invented, it was not the subject of so many improvements, nor did it have, in modern times, such wide and early practical utility as the reaper. The ingenuity of man is well shown by the numerous devices that were invented to accomplish the two objects of the reaper. Nearly all of these inventions were made in England. Two forms of motion were utilized in cutting the grain, circular and rectilinear. Both forms shared the continuous advancing motion of the machine to which they were fastened. The type now universally used, except in stripping headers, is a reciprocating rectilinear motion. As perfected, this type involves the principles of both the saw and the shears.

In Pitt's "rippling cylinder" were combined the first use of the circular motion, the first forerunner of the reel, and the first utilization of the principle involved in transmitting power from the wheels of the machine to operate some of its parts. The latter principle has been utilized in practically all harvesting machines ever built, excepting some of the combined harvesters constructed since 1903. Some form of the reel is also found on every harvesting machine which has had any success. In consideration of these facts, Pitt's name holds high rank among inventors of harvesting machinery.

The first patent on a reaping machine was granted in England to Joseph Boyce in 1799. Its only title to fame is priority.

AN EARLY ENGLISH REAPER INVENTED BY BELL, 1823

A year later an unsuccessful attempt was made to adopt shears as a cutting apparatus. This machine was unique in being operated by human power. Outside and inside dividers to separate the swath from the grain left standing, now found on all harvesting machines, were apparently first used in 1805. With Gladstone's machine (1806), the first to be drawn instead of pushed, appeared the side cut and the platform upon which the severed grain falls. Salmon (1808) first utilized the reciprocating cutter combined with the advancing motion of the machine. His reaper was also the first to have a self-delivering apparatus for the grain. Dobbs, a theatrical genius, invented a reaper (1841) and introduced it to the public in a play adapted to this purpose. The stage was planted with wheat which was harvested by the machine during the course of the play.

While English genius invented the essential contrivances of the reaper, American ingenuity must in the main be accredited with the rapid perfection of the machine for practical use. The first patent issued in the United States on an invention in this line was in 1803. The inventions of Hussey and McCormick came before 1835. McCormick's machine (patented 1834) was first used in the harvest of 1831. It was drawn by one horse, and seems to have possessed in crude form all of the essentials

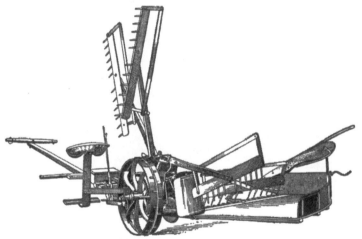

A MODERN SELF-RAKE REAPER

of a modern reaper. The grain was raked from the platform by a man walking behind the machine. Developing the reaper of today consisted solely in perfecting contrivances for utilizing the principles already discovered. The devices for automatically removing the grain from the platform were many, and they varied greatly in principle and crudeness. The revolving vane, the first form of which was invented by Hoffhein (1852), finally became established as the most advantageous method.

The reaper was virtually perfected by 1865, but in the United States other forms of harvesting machines soon entirely supplanted it in cutting wheat. It is widely used in Europe at the present time, especially in Russia and France, and nearly all machines of this kind manufactured in the United States are sold abroad. A reaping attachment is often used with a binder to drop the grain in bunches, and it is also widely used with a mower by small farmers in Europe.

The Self-Binding Harvester.—All machines which deliver the grain bound in sheaves, whether it is bound automatically or otherwise, are considered as binders. The reaper cut and collected the grain. This is only a part of the harvesting problem, and before this part was fairly solved, inventions began to appear seeking by means of an automatic binder to do away

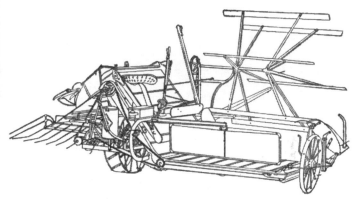

A MODERN SELF-BINDING HARVESTER

with the slow and laborious process of hand binding. In the case of the binder, discovery and invention must both be credited to the United States. Better economic and social conditions, dearer and scarcer labor, and more level and extensive grain fields were the conditions that made all agricultural machinery very profitable in the United States, and caused this country to outstrip England in the development of harvesting machinery.

Binders have been divided into two classes: Those in which the binding device is attached to a machine of the self-rake pattern, called the "low-down" class; and those in which the grain is elevated to the binder. Straw, metal strips, wire and twine were the four types of band with which experiments were made. Some machines carried an attendant to do the binding; others required an attendant to aid in this; others were automatic, but the power had to be furnished; while still others were automatic and received their power from the machine. The first effort to bind grain by machinery was made by John E. Heath of Ohio. His patent (1850) was on a twine or cord binder of the low-down type. Next (1850-1851) appeared the first machine with men riding on it to bind the grain as it was cut. It had a box for carrying the sheaves, the first forerunner of the bundle-carrier. Other contrivances that now appeared were: The cord knotter (1853); the wire twister (1856); the straw braid twister (1857); the automatic trip regulating the action of the binder and the canvas to elevate the grain over the drive wheel (1858); and the knotting bill and revolving cord holder (1864).

The Marsh machine began its successful career on the market in 1864, and from this date the "low-down" type of machine had a minor popularity. There is, however, still a successful binder of this type on the market which is unique and very popular for certain classes of grain harvesting on smaller farms. The Lake wire binder (about 1873) was perhaps the first commercially successful automatic binding machine brought out. There were, however, serious objections to wire binders, for pieces of wire were carried into threshing machines, and even into flour mills, where they occasioned fires by coming in contact with rapidly moving machinery.

The name best known among persons interested in harvesting machines is that of John F. Appleby. He had the genius to combine the advantages of preceding inventions with some of his own inventions in such a manner as to attain success. The Appleby binder on the Marsh frame was an irresistible combination that outstripped all competitors, and at once sprang into such popular favor that it swept over the world with overwhelming rapidity. The problem, at the solution of which many inventors had aimed in hundreds of patents during 30 years, was solved.[1]

[1] Ardrey, Amer. Agr. Implements, pp. 64-77; Miller, Evolution Reaping Machines, pp. 34-37.

The standard binder combines the cutter and draft of the reaper with the reel and traveling canvases of the header, and adds the automatic device for binding the grain in sheaves, and the bundle carrier for collecting them in piles. The operator can adjust the reel at will while the machine is in motion. An endless canvas on the platform of the machine conveys the cut grain to two similar canvases, between which the grain is elevated to the opposite side of the drive wheel. It is there received and packed into a bundle by the binding device. As the size of the bundle increases, the resulting pressure trips the binder, which binds automatically as often as it is tripped. The pressure required for this, and consequently the size of the bundle, can be regulated. While in operation, the entire machine can be adjusted to variations in the grain and in the levelness of the field. The most usual width of cut is 6 feet, but machines cutting different widths are made. One man with three horses will harvest from 10 to 20 acres per day with the binder, and it requires two other men to shock what is cut. A bonanza farmer expects such an outfit to cut 250 acres in a season. On the Dalrymple farms of Dakota, binders with 7-foot cut are used, and about 15 are run in one crew. Each crew or gang has its overseer. A wagon follows with water, twine and other articles, while a gang of shockers set up the wheat as fast as it is cut. In the United States the binder is used in every state which raises wheat, while abroad it is used quite extensively in England, Russia, Germany, France and parts of South America, and to a less extent in other countries where wheat is grown.

The Header-Binder is the most recent development in binders, and is, as the name suggests, merely a binder attached to the header. It has the wide cut of the header and the grain can be cut in the same condition as with a binder or reaper. These machines have found quite extensive favor in the Dakotas, Kansas, Oklahoma, on the Pacific coast, and in Argentina.

In binding wheat, a 10,000-acre farm uses two carloads of twine in a single harvest, an amount that would lay a line around the whole coast of England, Ireland and Scotland. It is estimated that the United States consumes annually from 110,000 to 120,000 tons of binder twine.

Shocking.—We found in a previous chapter that the ripening process in wheat involves a transfer of material from the straw to the grain. If the grain is cut before it is dead ripe, as is usually the case, this transferring process is not completed at harvest. Under these circumstances the completion of the ripening process is greatly aided by prompt shocking and capping, and loss will result if the grain is not thus protected from the hot sun and wind. This purpose is best accomplished by round shocks with caps. If the sheaves are large, or if the grain is green or weedy, it is customary to put 12 bundles in a shock. Their disposition is as follows: Three pairs are placed in a row; two bundles are then placed on each side of the row; the eleventh bundle is placed on top of the shock, and the twelfth, after its ends have been spread fan-shape, is placed crosswise of the eleventh. In a shock of 16 bundles, the disposition is the same, only that four pairs are placed in the row, and three in each side. A method of shocking that is quicker and more advantageous when the grain is practically ripe at cutting consists of placing any convenient number of pairs of bundles in a row. In any method, efficiency and economy of time demand that two sheaves be handled at once.

Combined Harvesters include all combinations of machines designed to leave both straw and chaff in the field and to deliver the wheat cleaned ready for market. The combined harvester is the culmination of the modern movement of discoveries and inventions pertaining to harvesting machinery. With this machine the wheat is cut, gathered, threshed, cleaned, and even sacked without a single touch from the human hand. On one side the grain is cut, and on the other side it is dropped at regular intervals in piles of filled and tied sacks, ready for the market. Every operation, except sewing up the sacks, is mechanically and automatically performed by the application of horse or steam power. In economy, in capacity and thoroughness of work, in perfection of mechanical construction, and in ease of operation, there is apparently little more to be attained. The combined harvester can be used advantageously in a dry climate only, where there is little fear of rain, and no great dews, which should be off before the middle of the forenoon. It also cannot be used where the grain is moistened by the damp breath of the ocean, as in western Oregon.

Ridley, an Englishman residing in Australia, invented a combined harvester in 1845 which employed the principle of the ancient machine of Gaul and attracted considerable attention. This type of combined harvester, commonly known as the "stripper," is still used in Australia, and is especially adapted to the dry harvest seasons prevalent in that country. Strippers have been manufactured in Canada and in the United States. They have been tried in California and Washington, but the atmospheric conditions did not seem suited to them. In Argentina, however, their introduction seems quite successful.

This machine strips the heads from the stalks of standing wheat by means of a comb resembling the ordinary sickle guard in appearance. Directly above the rear of the comb is a drum about 18 inches in diameter in which works a rapidly revolving beater which aids the comb in the decapitating process and furnishes a draft which carries the heads up into the threshing cylinder. This consists of teeth revolving within stationary teeth, and the threshing is more of a rubbing than a battering process. From the cylinder the grain and straw pass to the sieves over a vibrating metal table. Imperfectly threshed grains are returned to the cylinder. The straw and chaff is discharged at the rear of the machine, and the winnowed grain is carried to the top of the machine by a belt and cup elevator. Here the grain is screened. The screenings and the perfect grain pass to separate bins, from which they are bagged. A receiving box drops the bags in piles of four or five. Some of the machines discharge the straw and chaff under the middle of the machine, and fill the bags automatically.

The stripper can be used only in wheat that is ripe, dry, and free from weeds, for otherwise the grain will not thresh clean and the machine will clog. It is suited only to non-shattering wheat, which is not lost in the operation of harvesting. The expense of harvesting in this manner is estimated to be from one-fifth to one-half that of binding and threshing wheat. With a boy to ride the lead horse, one man can operate the machine, and from four to seven horses can easily draw it. A machine taking a five-foot swath will cover from 6 to 10 acres per day. In 1902 the price of these machines was $750 gold in Argentina, but it has since been reduced.

The combined harvesters used in the United States are restricted by climate to the Pacific coast, and may be divided into two classes on the basis of the power used, whether animal or steam. In the work and operation of these two classes of machines, there is, in the main, only a difference in capacity. The standard horse-power machine cuts a swath from 16 to 20 feet wide; is drawn by 24 to 40 horses; harvests from 25 to 45 acres of wheat per day; and requires four men to operate it. It requires a machine man to regulate the cutting bar and look after the machine in general; a steersman, a man to manipulate the sacks and tie them, and a driver. This is the most advantageous harvester to use on the smaller farms, those having less than 3,000 acres. It was used successfully before 1880, but its sale and manufacture in a commercial way did not begin until 1885.

The Steam Harvester has a cutting bar from 24 to 42 feet long, requires eight men to operate it, and harvests from 75 to 125 acres per day at a cost of from 30 to 50 cents per acre, which is the cost of the mere twine with which the sheaves are bound when the wheat is harvested with a binder. The traction engine or motive power is independent of the harvester proper. An auxiliary engine is mounted on the frame of the harvester. Steam conveyed to this engine from the boiler of the traction engine constitutes the driving power for running the cylinder, separator, header, and recleaner, ''the effect being a steady and uniform motion of all parts at all times and in all conditions of the grain and at any speed at which the harvester may be traveling.''

The traction engine is 110 horse-power, has double engines, and nine to 12-inch cylinders. The driving or carrying wheels are eight feet in diameter, and have a width of 32, 40 or 60 inches, according to the nature of the ground on which the machine is to be operated. This style of outfit is used very largely on the reclaimed tule lands. The separator has a cylinder from 26 to 40 inches in length. The mechanism of the machine is so perfected that the feeder, cylinder, grain carrier, shoe, and all cleaning devices remain in a level position upon uneven land and no matter how the machine is set. Thus under all conditions the machine does substantially the same work as upon a dead level.

The greatest width machine that was ever put out was an experimental one of 52 feet. It was built by a farmer, and was not a success on account of its construction. While successful machines with a width of 40 and 42 feet have been turned out, there are two standard large size machines, both smaller. One cuts a width of 25 feet, while the other consists of a 22-foot header with a 12-foot extension, making 34 feet in all. The machines of a greater width can scarcely be considered as a single machine. They consist of a regular cut of about 16 feet, with an addition of about 12 feet, making 28 feet for the machine proper. Then an independent header pushed by horses delivers to the outer canvas, thus making the 42 feet. Such an outfit is used only in the very lightest crop, and its exceptional cut is of advantage, not only in covering more ground, but also in keeping the thresher and cleaner sufficiently supplied with grain to insure the best work. The manufacturers claim that ''the steam harvester can handle grain in almost any condition, whether it is standing, lodged, tangled or overgrown with weeds.''

A Complete Outfit for thus harvesting grain consists of traction engine, auxiliary engine, thresher, header, water-tank wagon and cook-house. The average price of such an outfit is about $7,500. The great expense and capacity of these machines make them suitable only for the larger farms, those containing from 3,000 to 20,000 acres of land. The steam combined harvester was put on the market in a commercial way in 1892. The average life of the machine is from 8 to 15 years. The great advantages of this machine are economy in time and power on account of combining so many operations in one, the rapidity with which grain may be marketed after it is ripe, the small amount of human labor required, the diminution of risk from fire, and the waste of grain which is avoided.

It is a Pacific coast production and its sale is at present confined almost exclusively to that section of the world. It is the typical machine of the ''Inland Empire,'' a name applied to all of the Pacific northwest east of the Cascades and Sierras. At least two-thirds of the wheat of California is reaped with the combined harvester. It is a novel, interesting and picturesque valley scene to see this ponderous harvester sweeping through miles upon miles of ripened wheat, devouring swaths from 16 to

42 feet in width, raising its cloud of yellow dust, and leaving behind a long train of sacked grain, ready to be hauled to the warehouse, railroad, or mill. It is estimated that 3.000 combined harvesters were operated on the Pacific coast in 1903.

THRESHING.

Threshing is the operation of separating the grain from the chaff and straw. It is perhaps an entirely safe proposition to say that this has been accomplished in every imaginable manner. Perels states that the oldest method of threshing was by utilizing animals in tramping out the grain, but the flail, according to the same author, was known in grayest antiquity in a form similar to that of the present day. Both methods have been used in modern times. It is more probable that the first grain was shelled by hand, and that the first advance was to an auxiliary implement, a staff or rod with which the heads were pounded. The heads were also whipped across sticks or poles. The flail was early invented by attaching a club to the staff. The wind was the first fanning mill, the grain being thrown up so that the chaff would be blown away. The same forces, gravity and a current of air, are still utilized. The only improvements have been in the manner in which they are applied and in the addition of the screen.

Horses were used to tramp out the grain in early times in the United States, or a great roller with large wooden pins was dragged over the grain. These methods were still used in this country in 1835 or 1840. From 23 to 30 bushels per day for three horses, a man and a boy were the usual results. This method is still often used in Russia, where, in cleaning the wheat, the "shovel and wind" plan is utilized for the chaff, and a sieve 3 or 4 feet in diameter is used for removing weed seeds and grading the grain. In Spain and Syria, the threshing is also frequently accomplished by driving oxen or horses over the grain. The same method is occasionally found in remote parts of Argentina. Even in New Mexico one could find grain reaped with the sickle and threshed by the trampling of goats as late as 1899.

The Flail.—Where this implement was used, threshing was the chief farm work of winter. The flail was not rare in the United States as late as 1830, was common in Great Britain until 1850, and was still used in Germany in 1872. It is used now in parts of Europe where the holdings are very small or the peasants poor, notably in Russia. From 8 to 12 bushels of wheat was considered a good average day's work.

The Second Method of Applying Animal Power to threshing was by drawing over the grain an implement made rough on the bottom. It has been used in Egypt from ancient to present times, and consists of a wooden frame with three cross bars or axles on which are fixed circular iron plates. In ancient times the grain was usually at the circumference of the circle over which the machine was drawn, but now it is stacked in the center. It was called the *noreg*, and another form was known as the *charatz*. The *moreg* of the Hebrews was a similar device, and the old Roman devices corresponding to these inventions were the *traka* and *tribula*. Italy and some of the eastern countries still use substantially the same implement. A knifeboard construction known as the *trilla* is used in Spain.

The Evolution of Modern Threshing Machines.—During the eighteenth century three Scotchmen made separate inventions that led up to the modern threshing machine. Michael Menzies came first (1732). He contrived to drive a large number of flails by water power. It was called a "wonderful invention," "capable of giving 1,320 strokes per minute, as many as 33 men threshing briskly," and as "moved by a great water wheel and triddles." Its only contribution was to demonstrate the impracticability of the flail motion. About 1758 a Scotch farmer named Lackie invented a rotary machine which could thresh dry oats, but in wheat it merely knocked off the heads. Its value lay in showing the superiority of the rotary motion, and it was the first suggestion of the modern cylinder. The first machine of the modern type was invented by Andrew Meikle in 1786, patented in 1788, and completed in 1800 by the addition of a fanning mill. This was the first machine to thresh, clean, and deliver the grain in one operation.

The early machines were driven by water, or worked by horses, though wind power was also used. "Cider mill" horsepowers were most frequently used at first. Tread or railway powers came next, and soon afterward, the sweep powers. All of the early threshers were stationary. The first threshing by steam was in 1803. The first machines to be successfully placed upon the market were open-cylinder threshers, known under various names, as "chaff-pilers," "bob-tails," "ground-hogs," and "bull-threshers." They simply threshed the grain and did not clean or separate it. H. A. Pitts (1834) success-fully combined the "ground-hog" with the common fanning mill in portable form. He and his brother patented (1837) the original of the great type of "endless apron" or "great belt" separators.

Threshing machines were first brought into general use in Great Britain. Many were introduced from 1810 to 1820. In the southern counties of England, the machines were the ob-ject of popular attack, and in many districts the farmers were obliged to abandon such as had been erected. Pusey wrote in 1851: "Open air threshing may appear visionary; but it is quite common with the new machinery." The coal burnt by the best engines per horse power per hour was 28 pounds in 1847. Four years later it was less than one-fourth as much. Steam was soon universally used for threshing in England. The first "bull-threshers" were used in the United States about 1825. They spread rapidly until 1835, when separating devices had been added. Five years later little threshing was done by other means. Horse power was used exclusively, and it was not until about 1876 that steam power began to come into use.

In Germany there were many lever "hand threshing ma-chines" in use in 1850. Two men worked the lever, and a third fed the grain, but these three laborers could thresh more grain with less labor by using the flail, while the machine also cut up the straw and wheat.[1] By 1872 steam threshing had well begun to drive out other methods of threshing in Germany. In 1854 a steam engine of three-horse power threshed 160 bushels of wheat in a day. Similar engines up to nine-horse power ex-isted, and they threshed more grain. An American machine threshed 25 bushels per hour in the early sixties. In 1876 a steam thresher operated by 18 hands threshed well 2,000 bushels of wheat in one day. The bulk of the grain was always quite easily threshed from the straw. The great difficulty was to save the little that was usually left. It was estimated that from 5 to 10 per cent of the wheat was left in the straw by hand threshing.

Practically all threshing in the United States is now done by steam. The musical hum of the machines, which could be heard for miles, and which possessed a peculiar fascination that always charmed the threshermen, accompanied the sweep powers with gearing and tumbling rods to their oblivion. The side gear driving the cylinder of the separator made most of the noise. When this gear was cut off to give place to the belt pulley, the noise was reduced to a minimum, although the hum of the cylinder is still maintained. A few farmers own their own machines, but generally the threshing is done for a stated price per bushel by the itinerant outfit. In some sections the farmers still exchange work in the threshing, while in others the whole crew travels with the outfit. The farmer then simply takes care of the grain. On the smaller farms, 500 to 1000 bushels are threshed per day.

On the large farms, whether the grain is bound or headed, the last day of harvesting is the first day of threshing. If bound, the grain is not stacked, as it generally is on the smaller farms, but is threshed from the field. It is usually considered fit to thresh after it has cured in the shock for about ten days. When wheat is stacked, it begins to ''sweat'' about three days after stacking, and the process is over in about three or four weeks. It has been claimed that this is beneficial to the wheat in that it is fed from the straw, and that the berries thus become plumper and heavier and also acquire a better color. English writers seem to say nothing concerning this process of sweating. The northwestern wheat growers of the United States claimed that the wheat would sweat in the bin if this process had not taken place in the stack before threshing. When it is dried by seasonable cutting and threshing, however, it is very questionable if it can sweat or heat in the bin.[1]

The Modern Threshing Machine has a self-feeder, a band-cutter, and an automatic straw-stacker. There are also automatic weighing attachments. The grain is pitched upon the self-feeder, and the machine performs all the other operations. There are two forms of automatic stackers, the swinging stacker with rake to elevate the straw, and the wind stacker, in which the straw is forced through a long air-tight chute by a blast

[1] U. S. Dept. Agr., Spec. Rept. No. 40, p. 30; Hunt, Cereals in America (1904), p.107.

from a fan within the machine. But even with "blowers," as the latter are called, the straw pile often becomes awkwardly high, and the machine is moved from it. Sometimes the straw is also dragged away by horses hitched to a large rack, an operation which is called "bucking the straw." The cleaned grain is delivered from the machine through a spout. On the bonanza farms it is run into grain tanks holding about 150 bushels, which are hauled to the elevators or railroads, by four-horse teams. About 30 men are employed with each machine, and they thresh and haul away from 2,000 to 3,000 bushels per day; 1,300 acres is the minimum capacity of one machine. Ordinarily it will thresh 2,400 acres, 2,500 acres require two machines, and 6,500 acres require three. Straw is usually burned in the engine. During the season of 1903 one of the

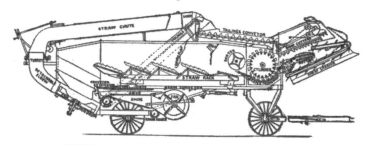

SECTION OF A MODERN THRESHING MACHINE

largest threshers in Kansas turned out 3,500 bushels of wheat in 9 hours and 45 minutes. This seems to be the usual maximum. Only 4 men are required to operate this machine. It takes 18 men and 10 two-horse wagons to bring the shocked wheat to the thresher. The largest amount of wheat which the writer has found recorded as being threshed in one day is 6,183 bushels in 1879. The work was done under the most favorable circumstances by a steam thresher having a 48-inch cylinder. [1]

A complete threshing outfit consists of a traction engine (which also hauls the whole outfit from place to place), a separator, a straw or coal wagon, a water wagon, a "cook-shack," and a sleeping tent. The cook-shack, a product of the west, is a small house on wheels which serves as a kitchen

[1] 10th U. S. Census, 3:457.

and dining room. In the early fall before it is too cold, the men often sleep upon the straw in the open air.

Distribution and Manufacture of Machinery.—The figures of the following table pertain to the United States only. A summary of patents on machinery which does not include machines used exclusively in industries other than that of wheat is not available. Over 2,000 patents were on wheat harvesters and over 3,000 on wheat threshers. The figures on the sales are to a certain extent approximations.

	Patents granted before 1902[1]	Average sale per year[2]	Per cent of machines sold abroad[2]
Plows	11,625	?	?
Harrows and Diggers	5,774	?	?
Seeders and Planters	8,566	?	?
Reapers		35,000 to 40,000	75
Binders	11,258	150,000 to 225,000	15
Headers and Header Binders		4,000	20
Combined Harvesters (horse power)		200	?
Combined Harvesters (steam)		25	?
Threshers	4,951	?	?

Little attention has been given to the export trade of the combined harvester, principally because the capacity of the manufacturers has been taxed to the utmost to fill home orders. Machines have been shipped, however, to Australia, Argentina and Spain, and though they work fairly well, the people do not take kindly to them. They lack the proper amount of intelligence to operate the machines with the best results, a difficulty not experienced to any great degree in the United States. Argentina, Paraguay and Uruguay have taken most of the machines that have been exported to South America, about one-fourth of the total exports. Another one-fourth has gone to the colonies of Australia and New Zealand, and the others have gone mainly to European countries. Many also go to Canada; 718,113 binders were sent there during the 9 months ending March 31, 1903. Over two-thirds of the exports are mowers and reapers. As many as 9,000 tons of machines have been shipped abroad in a single steamer.

[1] Census Bul. 200, 1902, p. 17. [2] Letters by competent observers.

CHAPTER VI.

YIELD AND COST OF PRODUCTION

Factors Increasing Yield.—As farming methods are improved, the yield of wheat per acre is being increased. Some of the main factors causing the increase are: (1) The use of drills in seeding results in greater immunity against drought and winter-killing, especially if press drills are used; (2) crop rotation; (3) improved methods in plowing and cultivation; (4) improvement of seed by natural and artificial selection, and by hybridization; (5) fertilizing; (6) irrigation; and (7) tile drainage.

Factors Decreasing Yield.—Nearly all of the factors just mentioned are inoperative in a new country, for their product gives intensive cultivation, while extensive cultivation is always characteristic of a new country under ordinary conditions. The yield is always low under extensive methods of farming. Such methods lower the fertility of the soil and a further decrease in yield results. The rapid improvement in farm machinery has favored extensive cultivation. It has also cheapened the cost of production, so that comparatively poor grades of land which it was previously unprofitable to work can now be farmed at a fair rate of profit. The operation of these factors is perhaps best shown by the wheat statistics of Australia.

From 1873 to 1898 the acreage of all the provinces of Australia except that of Tasmania increased, in some very greatly, while in every province (except Tasmania, where there was a decrease in acreage until the last eight years of the period), the yield decreased, in some cases over one-third. During the ninth decade in New South Wales the increase in acreage was slight and the decrease in yield insignificant, but in the next eight years the acreage increased nearly fourfold, while the yield fell off about one-third. The apparent lack of correlation between increase in acreage and decrease in yield in one or two of the provinces is doubtless due to some other factors.

The yield of wheat per acre in different countries is shown in the following table.[1] Figures in parentheses show limits to which acreage had increased or decreased by end of decade, in round thousands. The bushel is the unit in this table.[1]

	1860	1870	1880	1890	1900	1904	Av. 1894 to 1898	Av. 1899 to 1904
	(15,000)	(18,000)	(37,000)	(36,000)	(42,000)	(47,854)[3]	......	......
United States......	9.9[2]	11.9	12.4	11.8	13.3	14.5	14.0	13.7
United Kingdom.	......	(3,831)	(3,065)	(2,483)	(2,158)			
Australia[4]...........	......	......	25.4	28.5	32.6	27.7	32.9	32.4
		(349)	(977)	(1,145)	(2,154)			
Victoria.....			12.7	10.1	8.0			
		(166)	(252)	(333)	(1,319)			
New S. Wales....			14.5	14.3	10.6			
		(3)	(10)	(10)	(46)			
Queensland			15.0	11.0	14.9			
		(25)	(27)	(33)	(74)			
West'n Aust'a..			12.4	11.5	10.5			
		(784)	(1,733)	(1,673)	(1,788)			
South Aust'a....			8.3	6.4	4.6			
		(58)	(50)	(39)	(85)			
Tasmania..........			17.9	18.0	18.8			
		(132)	(324)	(301)	(339)			
New Zealand....			27.0	24.5	24.2			
France..................	16.0	16.3	16.4	17.9	18.1	18.6	19.8	20.8
European Rus'a..	......	......	7.9[5]	6.5	8.0	11.4	9.3	9.7
Germany..............	......	......	18.2[5]	21.5	27.9	29.4	25.7	28.1
Hungary..............	......	......	14.1[5]	19.8	17.3	16.3	17.6	17.6
Italy......................	......	......	12.1[5]	12.1	11.5	11.8		
Spain....................			......	14.7	10.5	10.6		
Canada................			13.7[5]	15.5	14.3	16.8		
Argentina............			......	......	12.7	12.1		
India....................			......	9.2	10.7	10.1		

France is a good example of an older country where the yield is being increased by intensive cultivation. In 1840 the yield was 14.6 bushels per acre, and in 1850, 15.6. The constant and regular rise in the yield per acre for nearly three quarters of a century in France is remarkable. The acreage rose gradually from about 12,500,000 acres in 1831 to about 17,500,000 acres in 1898. If poorer wheat land was brought under cultivation, the advance in methods of culture more than counterbalanced its effect. It is very interesting to compare the United Kingdom with France. The data cover the years from 1871 to 1898 inclusive. In this period the acreage of the

[1] Data taken from Yearbook U. S. Dept. Agr., 1902, 1905. Mo. Summary of Commerce and Finance, Jan., 1900, pp. 2039-2065. U. S. Dept. Agr., Bu. of Sta., Bul. 42, 1906, p. 26.
[2] 1866.
[3] 1905.
[4] All dates for Australia begin in 1873 and end in 1898.
[5] 1883.

United Kingdom decreased over one-third while the yield increased one-sixth. In the same period the acreage of France increased about one-eleventh, while the yield increased less than one-ninth. Presumably both countries made approximately the same advance in the arts, that is, in methods of production. It does not appear that there ever was a case where an advance in the arts supplanted with wheat a crop more profitable than wheat was before the advance. In increasing her acreage France had to utilize lands of lower yield, thus reducing the average yield of all, while the United Kingdom raised the average by exactly the opposite process, namely, by reducing her acreage in ceasing to sow to wheat those lands of such a low yield as to be unprofitable.

In the United States the causes and effects cannot be traced easily or clearly. We see that the greatest increase in acreage was in the eighth decade, but this acreage was located in the Mississippi and Red river valleys. It consisted of some of the most fertile land of our country, and proved to be better wheat land than any which had previously been sown in that grain. Consequently, it was but natural that the yield should rise, especially as there had been but little intensive farming. The rise in yield would doubtless have been constant since that date, had it not been for abnormal natural conditions which seem to have decreased the actual yield slightly in the ninth decade, although the potential yield has increased uninterruptedly. Since the ninth decade the increase in acreage has been comparatively rapid, doubtless largely due to great improvements in machinery, but the arts have advanced rapidly enough to more than counteract these results. The average yield from 1866 to 1886 was 12.2 bushels per acre, while that from 1886 to 1906 was 13.7. There is such a great annual variation in yield that statistics are not conclusive unless they are averages extending over at least a decade.

Columella gives 19.5 to 27 bushels as the amount of wheat that the Romans raised per acre. From 1200 to 1500 England raised 4 to 8 bushels per acre, while she raises about 30 now. The testimony of a contemporary observer shows the yield of wheat near Philadelphia in 1791 to have averaged less than 8 bushels per acre. It is now more than twice that amount. The greatest yield of wheat in the United States seems to be in the

Pacific northwest. This is perhaps partly due to the ideal weather prevailing there. A long, wet winter with little frost; a cool, wet spring, gradually fading away into the warmer summer; only light rains after blossoming; abundant sunshine and rather dry air toward harvest; and dry weather for harvest seem to be the most favorable weather conditions for the maximum yield. Sixty to 70 bushels per acre were harvested, even in the sixties. A volunteer crop may give 25 to 30 bushels.

COST OF PRODUCTION.

The itemized cost of raising an acre of wheat in different localities and years is given in the table below:

	Argent'a[1]	Kan.[2]	Wash.[3]	Kan.[4]	Red Riv.[5]	Minn.[6]	Minn.[7]	Japan[8]	Eng.[9]	France[10]
Date	1904	1903	1903	1902	1902	90-00	1880	1873	1601	1250
Cost of raising per bu.	$0.52	$0.26	$0.24	$0.37	$0.46	$0.64	$0.35		$0.92	$0.42	......
Preparing land: *Misc.*	0.18	......		1.24		1.15			3.19	
Plowing	0.64	1.00		0.96	0.70	1.00	0.31		1.74	0.33
Harrowing	0.18			0.28		0.15			1.45	0.03
Seeding:				0.92	1.00	1.05				0.22
Drilling or Seeding	0.09	0.25				0.30	0.03			
Seed	0.89	0.45	0.80			0.75	1.00		2.17	
Harvesting & Thresh.			0.40							
Harvesting				1.36	0.60	1.90	0.65		0.72	0.09
Cutting and Twine	2.12	1.25				0.75		3.60		
Twine										
Shocking						0.15				
Stacking						1.00				0.01
Thresh. & Market						0.90			0.20	
Thresh. & put in bin	2.20			1.60	1.50					
Threshing		1.20				0.60	1.25			
Fertilizing								4.00		
Int. on land: (*ins.*)	0.39		1.80	1.90	1.95	1.40		4.00		
Int. on mch'ry, etc	0.89		0.20	0.29						
Interest		3.60	2.00		2.28				2.17	
Tax	0.71				0.25	0.30				
Total Cost	$8.29	$7.80	$7.20	$7.31	$8.28	$6.40	$7.00	$11.60	$8.45	

[1] U. S. Dept. Agr., Bu. of Sta., Bul. 27, pp. 56-59.
[2] Eastman, Rev. of Revs., 28:198.
[3] Letter, H. Haynes; loc. cit., & Spokesman-Review, Wash., Oct. 1, 1903.
[4] Rept. Kan. State Bd. Agr., p. 12, 1902.
[5] Ency. Brit., 10th ed., 1:217.
[6] Indus. Com., 10:707. Also U. S. Dept. Agr., Bu. of Sta., Bul. 48, 1906, p. 54.
[7] Atlantic Mo., 45:34-35.
[8] U. S. Agr. Rept., p. 369, 1873.
[9] Hartlib, Legacy of Husb.; Rogers. Hist. Agr. & Prices, Eng., 4:493.
[10] Henley, Walter L., quoted in Rogers, loc. cit., 1:270.

SIDE VIEW OF A COMBINED HARVESTER AND THRESHER, CAPACITY 125 ACRES PER DAY

The variations in these accounts suggest the difficulties incident to obtaining reliable figures. The cost of production varies at different times and on different farms. In most cases it is impossible to give a reliable average, for the statistics are wanting. According to the table, Argentina can raise an acre of wheat at the same cost as that of the Red river valley in the United States. The average cost of raising an acre of wheat in Russia is about $8.

There are also many accounts of the cost of raising wheat which are not itemized, and consequently still less reliable. In the United States cost varies greatly in different sections. In Washington it is from 20 to 35 cents a bushel. In Oregon 20 cents is recorded. It is not likely that this price can include interest on capital, in any section. In North Dakota 50 to 54 cents is the cost; $5.72 per acre is also given for this state, not including interest on land. The running expenses averaged $3.77 in South Dakota from 1894 to 1900 inclusive. The total expense in Minnesota is $6.40.[1] In the early nineties the expense of raising an acre of wheat was $7.50 in Arkansas, from $6.13 to $10.32 in Nebraska, and $10.38 in experiments in Wyoming. Where wheat was the sole crop, $10 was given as a total average cost per acre in the United States on a farm of 160 acres in 1882. It is claimed that the shores of the Great Lakes could raise wheat at 15 cents per bushel before 1850, while the river counties of Illinois raised wheat for 30 cents, including hire of land and all expense.

The cost of raising a bushel of wheat in England was given as $1.76 in 1821 and $1.45 in 1885. In the black-earth region of Russia the cost of producing wheat, including rent, was said to range from 35 to 73 cents per bushel during the last part of the nineteenth century. In the first quarter of that century the cost, exclusive of rent, was given as 97 cents. The average cost in Russia during the years 1899 to 1903 inclusive, not including expense of rent and seed, varied from 34 to 48 cents per bushel for spring wheat.[2] Poggi says that the cost of a bushel of wheat in Italy is 69 cents, its production being at a loss. He criticises others who state its cost as only 44 cents, and who say that it can be profitably produced.[3] In Hungary

[1] Indus. Com., 10:ccxv.
[2] U. S. Dept. Agr., Bureau of Sta., Bul. 42 (1906), pp. 85-6.
[3] Atti del Instituto Veneto, etc., Tomo lvi, 7th s., T. ix. p. 723.

the cost is from 52 to 63 cents per bushel, or from $10.58 to $12.79 per acre, not including land rent. In Germany the cost is 95 cents per bushel. It costs 65 cents a bushel to raise wheat in India, but according to rather extensive data collected by the department of agriculture of that country the cost was exactly half this amount in 1884. The average annual cost of cultivating an acre of land in England rose from about $17.45 in 1790 to about $34.90 in 1813.[1] Threshing wheat by flail in that country cost about 8 cents a bushel. By the old system of horse-power machines, it cost about 5 cents, and by steam this was reduced to 2 cents.

The cost of raising wheat in the United States has not been reduced so greatly in the older wheat states as in the new states of the west, where the level and extensive farms give the greatest opportunity for the use of labor-saving machinery. For example, the combined harvester saves from 3.6 to 5.4 cents a bushel on the cost of harvesting with the header alone.

The Profit on Raising Wheat usually is not large, and it has often been denied that there is any profit at all.[2] Under the most favorable average conditions the bonanza farmers of the Red river valley do not make a net profit of over $3.32 per acre, or 8 per cent on the capital invested.[3] In England before the plague of 1332-1333 a lord possessing feudal rights over all the land in a manor made a profit of about 18 per cent on agricultural operations. After the plague, 1350-1351, profits were very low, at the best less than 4 per cent on the capital invested in the estate. Hartlib gives the profit on an acre of wheat in the middle of the seventeenth century as about $9. In order that there may be any profit in raising wheat in Argentina it is said that the yield must exceed 10 or 12 bushels per acre.

Amount of Labor Required.—About 1775 in the United States it was 3 days' work to cut 100 bushels of wheat, to bind and "stook" it took 4 days, while threshing and cleaning required 5 days more. In all, it required about 15 days of hard manual labor to get 100 bushels through these processes.

[1] Lowe Pres. State of Eng., p. 153.
[2] Indus. Com., Vol. 10.
[3] Ency. Brit., 10th ed., 1:217.

Thus it took about 1 hour and 45 minutes of human labor to harvest and thresh each bushel. These figures of Brewer are too small, however, as compared with those given by the department of labor for 1830. According to the latter figures it required 2 hours and 32 minutes at that time for the same operations. In 1896, by the use of the combined harvester, this time had been reduced to 5.6 minutes. The cost of human labor per bushel had declined from 15 cents to 2.2 cents. The entire time of human labor necessary to produce one bushel of wheat, including sowing, reaping and threshing, fell from 3 hours and 3 minutes in 1830 to 10 minutes in 1896. In the same period of time the cost of human labor per bushel fell from $17\frac{3}{4}$ cents to 3 1-3 cents. The cost of both animal and human labor fell from 20 cents to less than 10 cents. The greatest saving has been in harvesting. The human labor which does remain is quite light compared to that of 1830. This reduction in cost of production represented a saving of about $91,000,000 for the United States on the wheat crop of 1907.

CHAPTER VII.

CROP ROTATION AND IRRIGATION

CROP ROTATION.

The Effects of Continuous Cropping.—Different crops remove from and contribute to the soil elements of different kinds or in different proportions. The availability of plant food is also influenced. Continuously raising one crop tends to exhaust the soil of the food elements available for that crop. In a rotation of crops these effects are not so manifest. Some crops also contribute to the soil elements needed by others, as, for example, leguminous plants fix nitrogen which becomes available for wheat the next year. A rotation involves different methods of cultivation, which are often very effective in eradicating certain weeds. Continuous cropping and cultivation change the physical condition of the soil. This often results, particularly in prairie regions, in the soil blowing and drifting. Rotation of crops, especially when grass is introduced, will soon return the soil to its proper physical condition and prevent blowing. There is little profit in using commercial fertilizers unless rotation of crops is practiced.

Comparative Utility of Crop Rotation.—As a rule the pioneer farmer in a new country never practices much rotation of crops. This is one of the factors of high and intensive farming, which is never found on the frontier. The main reason for this is that land, being plentiful, is cheap, while all other forms of capital, as well as human labor, are comparatively scarce and high. It is but natural for the pioneer to endeavor to diminish those elements entering into the cost of production which are most expensive by substituting others less expensive. Land is the cheapest factor, so he uses this more lavishly, not to say recklessly, and saves the labor and other capital required to farm intensively, which is to cultivate more carefully, to rotate and diversify crops, to keep stock, to fertilize, to irrigate, and to follow many other practices requiring additional labor and capital. This fundamental advantage of extensive farming due to the cheapness and abundance of land is augmented by the fact that the pioneer usually is farming a soil of such virgin fertility that for a number of years it will produce large crops in spite of extensive culture. Often, as has been the case in the United States from the very beginning, when the soil has lost its fertility so that it will no longer yield standard crops, the farmer leaves the solution of the problem of its further profitable culture to others than himself by removing away from it to settle again upon virgin soil, and to repeat there his previous operations. While labor and all capital except land are higher in price in a new farming country, so little capital is required that its cost is usually below the cost of that required in the older country. In 1860 the United States was a half century behind England in intensive methods of farming, yet the cost of production was much lower for the American farmers than for the most scientific farmer of England, even if the latter paid nothing for the use of his land.[1]

If most of the members of a community are engaged in agriculture, the supply of agricultural products is not apt to fall below the home demand. Where such a large proportion of the people have an opportunity of producing at cost, home demand is not apt to raise the price greatly above the cost of production of older countries, and exportation is possible. Exportation involves the cost of transportation. Under normal conditions then, prices must always be lower at home than abroad before it will be profitable to export. As long as these conditions obtain, it will be impossible for prices of agricultural products in a new country (generally an exporting country) to be as high as those of an older country. High farming involves more expense than extensive farming, and consequently a larger capital is essential. But as prices cannot be so high in the newer community, and as capital is not so abundant, it follows directly and imperatively that farming cannot be of such a high and intensive grade. Unfortunately, however, as is so frequently the case with the recklessness of plenty, the most loose and careless methods of farming come in vogue, methods that are certain to exhaust the soil to such a degree within a limited number of years as to necessitate either improved methods of culture or its abandonment. While there may be extenuating circumstances in pioneer times which will

[1] 8th U. S. Census, Agriculture, p. viii.

excuse extensive methods of farming when the future must be forgotten because of present necessities, when many of the advantages of an older society are wanting, and when the burden of public improvements perhaps falls comparatively more heavily, nevertheless such a course long pursued is not only short sighted and suicidal from the standpoint of the individual, but it is also unjust to the future.

When extensive methods of farming have once become customary, changes take place slowly, unless they are necessitated by the growth of population and the exhaustion of the land. These conditions continually repeat themselves in history, for the ancients were already well acquainted with intensive methods of farming.

Summer Fallows.—When land does not produce the usual crops, there is a wide practice of letting it rest one year. No crop is planted, but the land is generally cultivated. This almost invariably results in an increase of yield during succeeding years. It has been claimed that this gain is at the expense of heavy loss in humus matter and available plant food.[1] Fallowing encourages the development of nitrates. One of its greatest advantages is that it enables the soil to store up moisture for the wheat crop of the following year.

Historical.—The farmers of ancient Egypt rotated crops. The same practice was followed in the time of Virgil, as well as the fallowing of land. The three-field system was not new in England in Norman times. It consisted of wheat the first year, barley or oats the second year, and fallow the third year. According to Gibbins crop rotation was not practiced in England in the beginning of the sixteenth century, but the triennial fallow was usual in the first half of the eighteenth century. It was known as the "Virgilian" way of farming. Clover and lucern were introduced in the eighteenth century, and brought a new rotation of crops that saved the wasted year during which land used to lie fallow. In the middle of the nineteenth century, rotations were practiced which brought a wheat crop every fourth or fifth year, or twice in 6 years. The Japanese sowed the wheat in rows, and cultivated vegetables between the rows at the same time, in addition to raising other crops before or after the wheat crop on the same ground during the same year.

[1] N. D. Bul. 24, p. 73.

Before the twentieth century, American agriculture consisted mainly in raising cheap crops, and little attention was given to resulting effects upon the soil. After the soil was robbed of its fertility, various devices were resorted to in order to get a paying crop. The most common of these was to seek new land, or to give the land a rest from production. Reports from thousands of correspondents show that little systematic crop rotation was practiced in the United States even as recently as 1902.[1] At the close of the eighteenth century the deterioration of the soil became apparent, particularly in Virginia and Maryland, and as early as 1882 it was noticed that the yield of wheat was declining on account of continually cropping this grain on the same land. The most skilled farmers were unanimous in recommending rotation of crops. The most generally advised rotation gave one wheat crop in three years. Under the stress of hard conditions a true conception of the necessity of rotating crops gained a foothold and expanded into farm practice. As would be expected, the longer the occupation, the more developed is the crop rotation. In passing from the east to the west, the degree of rotation begins to diminish in Ohio, and by the time Kansas is reached, it has practically disappeared entirely. One-crop or two-crop production was characteristic of the first agriculture of the north central states.

On the Dalrymple farm of North Dakota wheat was grown continuously for about eighteen years, by which time the soil had been so impoverished that a system of crop rotation and summer fallow became necessary. Generally corn and barley are sown and cut early so that the land may be plowed in July before the wheat harvest. Considerable land is also barren summer-fallowed, in which case it is plowed twice during the summer. In Canada, experience with continuous cropping has been much the same as in the United States. Large areas in different parts of the early settled portions which once yielded fine crops of wheat have been abandoned to pasture and other purposes.

Experimentation.—In experiments in North Dakota, the plots which had been rotated with corn or potatoes yielded about twice as much as the best continuous wheat plot. Good cultivation alone was not sufficient to produce the best crops, and

[1] Yearbook U. S. Dept. Agr., 1902, p. 520.

TYPICAL WHEAT FIELD ON A FARM WHERE CROP ROTATION IS FOLLOWED

other crops gave a poorer yield on land that had been continuously sown in wheat. "Land which produced three crops of wheat and one cultivated crop in a period of four years, gave almost as much wheat and more profitable returns than did the land which produced four crops of wheat in succession."[1] Experiments have been made in the continual culture of wheat on a certain piece of ground, there being no fertilizing of any kind, as, for example, the "experimental acre" in Kansas. This trial was begun in 1880, and by 1896 the yield was falling off. Permanent spots of diminished fertility had then appeared. Though they may have been due in part to surface-washing, partial exhausting was undoubtedly a factor.[2]

Historic experiments in growing wheat continuously without fertilizing have been carried on in England for over 50 years. "The yield has fallen to about 12 or 13 bushels to the acre, but for the past 20 years there has been little or no difference in the yield, except slight fluctuations due to seasonal conditions. So far as is known, the soil will produce 12 or 13 bushels to the acre annually for hundreds of years."[3]

The Crop Rotations of the United States now generally practiced in some typical counties of states leading or prominent in their geographical divisions, are given below:

Pennsylvania.—Corn, wheat two years, grass two years (York, Franklin, etc.). Corn, oats, wheat, grass three years (Chester, Westmoreland).
Minnesota.—Wheat two years, oats, wheat, flax (Marshall). Corn, wheat two years, oats (Lac qui Parle). Corn, wheat two years, grass two years (Ottertail, Todd, etc.).
Washington.—Wheat, rest (Adams).
California.—Wheat, rest (Solano, San Joaquin, etc.).
Maryland.—Corn, wheat two years, grass two years (Montgomery, Frederick, Talbot, etc.). The rotation on dairy and stock farms includes wheat for only one year.
Oklahoma.—Wheat without rotation (Grant, Garfield, Kingfisher, etc.). Wheat, corn, (Dewey). Wheat three years, oats (Kay).

No crop, nor even any one class of crops, such as the cereals, should be continuously grown on a soil that will produce a variety of crops. On ordinary soils, cereal crops should be rotated every two to four years with a leguminous crop, such as clover or alfalfa. The North Dakota experiment station finds that wheat should have a good place in the rotation because it is a particular crop, and that the average yield of

[1] N. D. Bul. 48, p. 735: Bul. 39, p. 458.
[2] Kan. Bul. 59 (1896), p. 90.
[3] Indus. Com., 10:clxxxviii.

wheat is greatest when the crop follows either corn or potatoes. After these crops, placed in the order that they merit for preparing the soil for wheat, come summer fallow, millet, vetch, peas, wheat and oats. The more dry and unfavorable the season, the more important it was to introduce a cultivated crop into the rotation. The best rotations included a perennial grass, for which purpose brome grass is well adapted to North Dakota. The rotations vary greatly in different states, and soil, climate, and economic causes must determine which rotations are most advantageous for any locality. Summer fallowing is widely practiced on the Pacific coast, largely because there is practically no rotation feasible.

Crop Rotations in Foreign Countries.—In Canada, summer fallowing is rapidly becoming general throughout the territories, where the profitable corn crops of the United States cannot be grown on account of the latitude. The system of agriculture most prevalent in Russia is the three-field system, which is universally practiced in the center of the Russian wheat belt. The usual sequence of crops is winter rye, spring wheat and fallow. The arable land is divided into three corresponding parts. At a given time each part is in a different stage of the system. Other crops are being introduced, and this is lessening the area of fallow land. Among the private land owners this signifies progress in agricultural methods. Among the peasants it frequently signifies a harmful overworking of the land, the penalty of which is the drastic retribution of greatly reduced yields. Another system, still more primitive than the three-field one, is also found in Russia, especially in the steppes of the southeast, where the greatest extension of the wheat area is taking place. By this system the land is tilled until it becomes exhausted. It is then allowed to lie fallow in order to recover its fertility. This may require 10, 15, or even 30 years. In Archangel, Olonetz, Vologda, Viatka and Perm, the forest must be cleared to prepare the new land for cultivation, but in the southeastern provinces of Orenburg and Astrakhan, in New Russia, Kherson and northern Caucasia, all that is required is to plow the land. As population grows, this wasteful method of farming is being replaced by the three-field system.

Impoverishment of the land by continuous wheat cropping

has been the custom in Argentina. Sixty per cent of the wheat is raised under the renting system. The colonist owns nothing which grim necessity does not compel him to own, and he practices his ruinous methods of farming until the land is completely exhausted. Then he fastens the bullocks and horses to the carts, packed with his many children and his few miserable pots, boxes, beds and implements, and travels until he finds new fields. Mixed farming as known in the United States is little understood or practiced in Argentina, and the farmer is generally either a wheat grower or a maize grower. There is complaint of the methods of farming in all parts of the Republic, however, and a practice of rotating crops is already beginning, by alternating wheat and maize, or by planting the land with alfalfa after three or four years of wheat cropping.[1]

For the best crops of wheat in Egypt, it is sown every fifth year, the rotation being (1) cotton; (2) "birsen" (clover) or "full" (beans); (3) wheat; (4) dura (maize); (5) "birsen." A commercial success has been made of growing wheat and alfalfa together on the dry uplands of North Africa. In Algeria two rows of wheat are sown 4 inches apart. A space of 40 inches is left between the double rows, and in this space the alfalfa is sown. Wheat is sown only every other year. This is of interest, as alfalfa is now the greatest American fodder crop, especially in the arid southwest where durum wheat is being more extensively grown.

Experiments with Mixed Crops have been made, chiefly in Canada and North Dakota. Results seem to be in favor of unmixed grain, although wheat and flax have an advantage under certain conditions, as when wheat is apt to lodge, or when there is a superabundance of moisture. In the latter case flax has increased the yield of wheat as much as 6.5 bushels per acre, in addition to giving 1.2 bushels of flax per acre.

IRRIGATION.

Historical.—Irrigation is of prehistoric origin. Water, as was shown in a former chapter, is one of the greatest essentials of all plant growth, and it is also one of the most variable quantities involved. Since the effects of these variations upon vegetation appear quickly, they must have been noticed at an

[1] U. S. Dept. Agr., Bu. of Statistics, Bul. 27 (1904), pp. 41-42.

early date, and then it was only another step to supply artificially the needed water. Irrigation was a condition that was
indispensable to the settlement of large portions of western
America, Australia and South Africa. In meeting these problems during the nineteenth century, the Anglo-Saxon race had
its first experience with extensive irrigation. Throughout all
the centuries of previous history, the art of irrigation was quite
exclusively the possession of Indian, Latin and Mongolian races.
It was used extensively by the ancient Chinese, Egyptians,
Persians and by the people of India. The Homeric Greeks
used small canals in irrigating. In Italy, it was probably as old
as the Etruscans. The Romans borrowed the system from the
east, and brought it to their country and southern France. The
ancient Peruvians also practiced it, and in Spain it dates back
to the Iberian life existing under the Roman conquerors.

Modern Irrigation in Foreign Countries.—Irrigation is more
or less extensively practiced by all of the great nations of the
globe, even in subhumid and humid regions. As a rule, however, the wheat crop is not extensively irrigated, for irrigation
is more profitable with other crops. The total area watered
runs into millions of acres in most of the European nations.
Wheat is frequently irrigated in the Po valley. In Mexico,
Argentina and Australia, wheat is irrigated to some extent.
Both streams and wells furnish the water. Extensive systems
have been planned for Australia, and over 1,000,000 acres could
be irrigated in New South Wales alone. Argentina contains
large areas which are irreclaimable except by irrigation. The
lower valley of the Nile with its delta comprises another great
irrigation system, 6,000,000 acres being under cultivation.
Egypt is so arid that dry farming is impossible. In 1902
British enterprise completed a dam across the Nile at Assuan.
It is built of granite, and is 70 feet high, 23 feet wide at the
top, 82 feet wide at the bottom, and $1\frac{1}{4}$ miles long. It is the
largest irrigation dam in existence, and the reservoir has a
storage capacity of over thirty billion cubic feet. The largest
increase in irrigated area in recent years has been made in
British India, where about 30,000,000 acres have been reclaimed or made secure for cultivation by constructing new supply works. It has been estimated that 80,000,000 acres more
can be reclaimed in India. In 1892 over $150,000,000 had been

invested, and yielded a large profit, though it was often obtained indirectly. India has the largest reservoir in the world. It covers an area of 21 square miles, and it was constructed for irrigating in Rajputana. It is known as the great tank of Dhebar.

Irrigation in the United States.—In America, the town-building Pueblo Indian tribes practiced irrigation perhaps a thousand or more years ago. Their ditches and canals can still be traced in the little valleys near the mesas of southwestern Colorado and adjacent portions of Utah, Arizona and New Mexico, where the cliff dwellings are found, as well as across the border valleys through which are scattered numerous ruins of community dwellings. Their knowledge of engineering is evident, and remarkable. Careful levels have been run over several miles of their canals. The grade was found to be fairly uniform and suited to a canal of such dimensions, as well as in accord with present day knowledge of hydraulics, safe velocities and coefficients of friction. While these well defined remains of ancient irrigation works have long outlived the civilization to which they belonged, there are cases where they have been utilized in modern works. The ditches at Las Cruces, New Mexico, have been used uninterruptedly for over 300 years. Some 70 years before the settlement of Jamestown, the Spaniards irrigated on the Rio Grande. Adventurous mission fathers pushed on to California, carrying the art of irrigation with them.

The beginnings of irrigation by English-speaking people in this country were in the Salt Lake valley of Utah, in July, 1847. The Mormon pioneers, driven out from Illinois and Missouri, stopped from necessity on the shores of the Great Salt Lake. They diverted the waters of the little canyon streams upon the present site of Salt Lake City, so that they might raise a crop from the very last of their stock of potatoes and save the band from starvation. At about the same time water for irrigation was drawn from the ditches used for placer mining by the gold miners of California. After the stoppage of hydraulic mining by the passage of anti-débris laws, the ditches were either abandoned or used exclusively for irrigation. Many were enlarged and are still used.

The Extent of Wheat Irrigation in 1899 is shown in the table below:[1]

	Acreage			Production		
	Total	Irrigated	% Irri-gated	Total	Irrigated	% Irri-gated
Arizona..........	24,377	24.137	99.0	440,252	436,582	99.2
California.......	2,683,405	161,086	6.0	36,534,407	1,649,455	4.5
Colorado........	294,949	247,644	84.0	5,587,770	5,309,350	95.0
Idaho..............	266,305	82,708	31.1	5,340 180	1,799,028	33.7
Montana........	92,132	37,710	40.9	1,899,683	843,143	44.4
Nevada..........	18,537	18,246	98.4	450,812	448,802	99.6
New Mexico ..	37,907	36,638	96.7	603,303	589,185	97.7
Oregon...........	491,258	16,092	3.3	7,280,443	387,201	5.3
Utah..............	189 235	108,630	57.4	3 413,470	2,554,248	74.8
Washington...	1,073,827	14,204	1.3	20,817,753	328,958	1.6
Wyoming.......	19,416	14,753	76.0	348,890	288,180	82.6
Nebraska.......	....	14,143	....	....	185,481	....
Total..........	5,391,348	761,848	14.1	82,716,963	14,634,132	17.7

While considerable wheat is irrigated in some states, practically all that is grown in them, yet the average per cent of irrigated wheat in all the irrigating states is relatively small, only 14 per cent. Excluding California and Washington, where much wheat is raised and little irrigated, this rises to 36.5 per cent; 17.7 per cent of the wheat produced is irrigated, compared to 14.1 per cent of the acreage. On this basis which, however, takes no account of differences in soil, rainfall and climate, the yield in these states would be increased over 25 per cent if all the wheat were irrigated.

The Problems of Irrigation in our country are, and have been, along two general lines: Agricultural and engineering; and legal and social. Of these two lines, the latter has presented the greatest difficulties. Litigation and controversy have been a menace and a source of loss to many communities because no institutions existed for adequately defining, limiting and protecting water rights. The claims of navigation came into conflict with those of irrigation. When streams flowed through more than one state, interstate difficulties arose. Some of these are the basis of a suit by the state of Kansas against the state of Colorado.

Work at the solution of either class of problems has been immensely handicapped by a most lamentable lack of knowledge

[1] 12th U. S. Census, 6:825-870.

of certain essential facts and conditions. Among these are existing water supply, quantity required to grow crops, losses from seepage and evaporation in distribution, character of the control over streams already vested, and measures of administration requisite for an equitable and effective division of water supply among a multitude of users. Such unforeseen results as alkali lands and seepage waters, formerly secondary considerations, are now often the most primary problems. Such irrigation as could easily be accomplished with simple means independent of co-operative institutions has largely been effected. As the work extended, greater problems arose, claims became hopelessly conflicting and united effort under institutional administration became an imperative condition of advantageous development.

Water Supply.—There are two sources of water for irrigation: Surface waters, such as streams and lakes, and subterranean waters. The former supply over 90 per cent of the irrigated land. There are three ways of obtaining underground waters: By pumping from wells; by driving tunnels into the sides of hills and mountains; and by using flowing wells. Artesian areas are widely scattered, and individually they are of small size, except in the Dakotas and California. In 1889, 51,896 acres, or 1.4 per cent of the irrigated land, were irrigated from wells. In 14 irrigating states there were 8,097 wells, nearly half of which were used in irrigation. Each well supplied on an average 13 acres, had a depth of 210 feet and discharged 54 gallons per minute; 169,644 acres were irrigated from wells in 1899. Underground waters seem to be present very generally. It is claimed that there is not a farm of 160 acres upon the great plains region without the requisite moisture absolutely needed for from 10 to 30 acres of tillable ground.[1] The average depth of water applied to crops in 1899 was 4.35 feet, and in 1900, 4.13 feet.

Application to Crops.—The two principal methods of irrigation are by flooding and through furrows. The former is generally used in growing grain. There are two methods of flooding, the check system and by wild flooding. By the latter process a level field is completely submerged. When the ground is not level enough for this, the field is divided into compart-

[1] Hinton, Rept. on Irriga., Cong. serial No. 2899, part I, p. 8.

ments by ridges. The highest compartment is flooded to the
top of the ridge, which is then opened on the lower side. The
water thus passes into the next compartment, and this procedure is continued until all the compartments are irrigated.
If the land is properly prepared and irrigated before the
wheat is sown, two subsequent irrigations will make a good
crop. When the soil is thus used as a storage reservoir, in
parts of Kansas and California no irrigation is needed between
planting and harvesting.

Alkali.—Arid region soils are usually rich in mineral ingredients. This is because such soils originated in the de-

THE FURROW METHOD OF IRRIGATING

composition of rocks in regions where the rainfall is too scanty
to wash out the soluble elements as in humid regions. The
soluble salts are naturally distributed throughout the soil, and
are not harmful until the application of irrigation water. They
are then leached out of the higher grounds and concentrated
in the lower lands. Evaporation tends to bring them to the
surface. Many irrigation waters also contain much salt in solution, which results in a further deposition of salt. The result
of these factors is often ruinous to vegetation. Many thousands of acres have been thus rendered unfit for cultivation in
the United States, and the agricultural industries of 59 villages in India were wholly or partly destroyed by the rise of

alkali previous to 1864. Water containing over 1,000 parts of salt in a million has been used without injury. Most of the artesian wells of Dakota have a salt content much higher than this, and the effects of irrigating three or four years with this water rendered wheat lands of the Red river valley almost wholly unsuited to raise current crops.[1] The most effective method of removing alkali from land is by underdrainage and flooding.

The Cost of Irrigation in the United States as shown by the eleventh and twelfth census is as follows:

	Average values per acre	
	1889	1899
Irrigated land	$83.28	$42.53
Water right	26.00	—
Annual cost	1.07	0.38
First cost of water rights	8.15	7.80

A rise in values would be expected, instead of a fall, as good lands with water supply were scarce in 1899, and those lands were first irrigated which required least labor and capital. It has been estimated that a perpetual water right in a grain country is worth from $25 to $50 per acre. The cost of irrigation from many of the original ditches was as low as $2 to $5 per acre.[2]

The Semi-Arid Region of the United States.—There are men still living who knew the Mississippi valley as a wilderness. For several generations a popular American slogan has been ''westward the course of empire takes its way,'' and the rapidity with which the fertile lands of the great river valleys were brought under cultivation has been almost incredible. As this huge wave of immigration swept across the prairie to the great plains, it encountered the subhumid belt as a buffer between the humid and the arid regions. Gradually the settlements proceeded westward from the abundantly watered Mississippi and lower Missouri valleys, and pushed into the well defined subhumid slope which rises progressively toward the Rockies. These virgin lands, bordering upon the greatest wheat raising region of the world, and fully as fertile, since they were not washed by frequent rains, were a continual temptation to

[1] Mon. U. S. Geol. Sur., 25:546-547.
[2] Indus. Com., 10:xxxii.

carry the "empire" yet farther west. The "Great American Desert" disappeared from the maps. During a series of years in which the rainfall was more adequate than usual, the agricultural areas leaped forward to the west from county to county. The first general advance was in 1883. Within five years, western Kansas and Nebraska and eastern Colorado were largely settled. To the east of the arid region is a strip of territory embracing portions of Kansas, Nebraska, the Dakotas and northwestern Texas, which has been designated as the "rain belt." Its name resulted from the theory that the humid region was gradually extending itself toward the west as a consequence of the breaking of the prairie sod, the laying of railroad and telegraph, and the advent of civilization. There was supposed to be a progressive movement of the "rain belt" as civilization advanced. While thorough cultivation undoubtedly makes a material modification in the effects of a given degree of aridity, it has been declared that the probability of a perceptible change in climate does not merit serious discussion.

The theory received a serious setback from the periodical exodus which occurred when succeeding years brought a rainfall at or below the normal. There were years when the average rainfall (10 to 20 inches) decreased by almost half; there were months without a cloud; there were days in the southwest when the winds were so dry and hot that green corn was turned into dry and rattling stalks. When crops shriveled and died on millions of acres, men lost hope and means, and they were forced to abandon the homes that represented the earnings of a lifetime. Whole counties were nearly depopulated. These vicissitudes caused the tide of migration to ebb and flow, and continually wore out its resources. The desert had been removed from the maps. The supplications of the devout and the dynamite of the "rainmaker," a suggestion of the Indian medicine men who had held sway on the plains less than a century before, had vainly implored the heavens for the rain which alone was wanting for the production of profitable crops. Yet the blunt fact remained, and still remains, that many millions of acres were dead, vacant, and profitless simply because of their aridity. This land has little value now, for in many places a whole section does not yield enough to keep a fleet-footed sheep from starving.

South of Yellowstone park in the Wind river mountains of Wyoming rises Mount Union in majestic grandeur. Three streams take their course from this peak—the Missouri, the Columbia and the Colorado. Embraced in the branching arms of these streams is the industrial future of a region greater in extent than any European nation save Russia. Could this vast district be reclaimed for settlement, it would be a task second to none in the realm of social economics, for here millions of people could find homes. Within this region is contained practically all that remains of the public domain. The only element lacking to make the land valuable is moisture. New influences are at work to remedy this, the bitter failures of 20 years ago have been largely forgotten, and a second wave of settlement is sweeping over the plains. Rather slowly and unwillingly public attention became fixed upon irrigation. While the water supply is sufficient to irrigate only a small fraction of the arid domain, approximately three-fourths of a billion acres, several million acres are already under irrigation, and there is a good prospect that many more millions will be irrigated in the future. At present this area forms potentially the best part of our national heritage. Although most of the land would be typical for raising wheat, and the completion of the irrigation works which the government now has under way will add millions of bushels to the annual production of wheat, the better adaptability of other crops to intensive cultivation under irrigation will doubtless soon render it unprofitable to irrigate wheat extensively. The introduction of irrigation will make possible the growing of diversified crops in some sections where wheat alone can now be profitably raised. Where the supply of water is insufficient for irrigation, the only remedy is the development of drought resistant crops for dry farming. One of the greatest of these is durum wheat. If there is water enough to irrigate but one acre of ground on the dry farm, this will make a green oasis with shade and foliage for the farmer's home, a pleasant contrast to the monotony of the gray and dusty summer plains with their shimmering waves of heat.

CHAPTER VIII.

FERTILIZERS

Fertilizing consists in the physical application to the soil of elements which are immediately or mediately available for plant food, or which aid in changing from unavailable to available forms of plant food any elements already existing in the soil. It is meant, of course, to exclude water, the contribution of which is irrigation, but any elements held in suspension or solution by irrigation waters, and falling under the conditions of the definition, are fertilizers.

Historical—The Homeric Greeks were familiar with the use of manure as a fertilizer. Cato mentions irrigation, frequent tillage and manuring as means of fertilizing the soil. To these Virgil adds ashes. The ancient Peruvians were skillful in the application of manure, a practice that has existed in parts of Russia from time immemorial. The earliest records on agriculture show that the value of fertilizing had already been taught by experience. The degree to which intensive cultivation had developed, the natural fertility of the soil, and the incidental occurrence of materials that could be used as fertilizers have always been, in general, the factors determining the extent of the practice.

NATURAL FERTILITY.

Soil Composition and its Relation to Plant Life.—From a physical point of view the soil of the field may be analyzed as follows: (1) The soil proper, consisting of various sizes and arrangements of grains made up of insoluble or imperfectly soluble minerals; (2) humus, more or less decomposed organic matter derived from the decay of former animal and plant life; (3) the soil moisture, covering the soil grains, and containing in solution a varying amount of the soluble soil constituents; (4) the soil atmosphere, differing from air in composition to some extent, and usually saturated with water vapor; and (5) soil ferments, or bacteria, which so permeate the soil that it should be considered as a living mass and not as dead, inert matter. Indeed, the inanimate parts of the soil have their highest significance as the environment of the bacteria which they contain, and in part nourish.

To understand the effect and value of fertilizers, a knowledge of the chemical and physical composition of soils, and of the relation of their composition to plant growth is essential. These things must be clearly understood, because fertilizers act upon the plant indirectly through their influence upon the composition of the soil.

At the beginning of the nineteenth century Sir Humphrey Davy said that the substances which constitute the soil ''are certain compounds of the earths, silica, lime, alumina, magnesia and of the oxides of iron and magnesium; animal and vegetable matters in a decomposing state, and saline, acid or alkaline combinations.''[1]

He also fully understood that the soil furnished nourishment for the plants, and that different plants flourish best in different soils. While he described the soil elements, often with surprising accuracy, and was the most expert chemist of his time, he did not adequately appreciate the plant foods contained by the soil, and his conception of the functions of the elements which he described was often extremely vague. For example, he held that the silica which plants contain imparts to them their rigidity. He recognized in a general way, however, that phosphoric acid, potash and lime enter into the composition of plants, and he successfully combated many unscientific notions. The derivation of soils from rocks was also known in his time.

Mineral or artificial manures were first studied systematically by Liebig, whose views found their way into the United States before the middle of the century. The publication of his work in 1840 marked a new era in agricultural chemistry. Before his time it was very generally held that organic substances were the chief food of plants. This has been called the humus theory. It was rejected by Liebig, who went to the opposite extreme and held that organic matter has no part in plant life. Practical knowledge of the use of manures, wood ashes, slaughterhouse refuse, gypsum, lime and plaster as fertilizers was widely diffused and acted upon before the time of Liebig, but

[1] Yearbook U. S. Dept. Agr., 1899, p. 203.

it required his work to bring about a full appreciation of plant requirements and of the important office of the soil. Through the vehement discussions of his work, Boussingault, Lawes, Gilbert and others were led to a critical study of these problems. The exact needs of plants for mineral nutrients were carefully investigated by means of experiments of water-culture and sand-culture. This work was carried on by the foreign experiment stations between 1865 and 1873, and its results contributed very materially to the subsequent development of the enormous industry of manufacturing and selling commercial fertilizers.

With prophetic vision Liebig said: "Manufactories of manure will be established in which the farmer can obtain the most efficacious manure for all varieties of soils and plants."[1] Systematic work in the chemical analysis of soils in the United States began in 1850, when D. D. Owens made an extensive chemical examination of the soils of Kentucky in connection with its geological survey. The most recent developments seem to show that the amount and proportion of the elements contained by the soil are of less importance than was formerly supposed. It is of far greater importance that such elements as are present should be in a form available for plant food. Just what form an element must assume to be most available seems to be in a large measure an unsolved problem yet, but evidently the texture and the structure of the soil are fully as important as the chemical condition of its elements. By texture is meant the relative sizes of soil grains, and by structure the arrangement of these grains under field conditions. After exhaustive investigations on many types of soil, the conclusion has been reached "that on the average farm the great controlling factor in the yield of crops is not the amount of plant food in the soil, but is a physical factor, the exact nature of which is yet to be determined."[2]

Most of the fertilizing which has been done has been according to the theory that the soil is a lifeless mass composed of so many elements, and that some elements were absent, or not present in sufficiently large proportions, it being the object to contribute in the form of fertilizer the elements which were needed. While the benefits of fertilizers have been unquestioned

[1] Yearbook U. S. Dept. Agr., 1899, p. 340.
[2] U. S. Dept. Agr., Bu. of Soils, Bul. 22 (1903), p. 63.

for over a century, it is, nevertheless, doubtful whether quite the right path has been followed by investigations which endeavored to determine just how those benefits arose. Air and soil are the media through which the growing plant receives its nourishment, but this is more than a mere mechanical process. In some cases at least there must be some sort of digestion or decomposition of foods before there can be assimilation. Silica, highly insoluble and apparently the least suited of all the mineral constituents of the earth to enter the vital organism of the plant, however finds its way into the plant tissues. Phosphorus, one of the most important mineral foods of plants, exists in the soil, or is applied in fertilizers, almost exclusively in the form of mineral phosphates, but appears in the plant largely in organic combination, while the mineral phosphates which do appear are not those which pre-existed in the soil, such as those of lime, iron and alumina, but chiefly those of potash. It is also found that soils of different composition, texture and structure supply different quantities of water to the plant, irrespective of the percentage of water actually present in the soil. As water conveys the nutritive solutions to the plant, when the supply of water is inadequate, there may also be a deficiency of nutrient materials. It is probable, then, that fertilizers, by temporarily increasing the concentration of the solution, increase the food supply. Such fertilizers seldom permanently affect the nature of the solution, and the concentration with respect to the mineral plant food constituents per unit of solution is considered approximately constant. In the same and in different soils, however, the water content varies widely, and usually the greater the water content, the more diluted is the solution.

In 1902 such exceedingly delicate and sensitive methods for analyzing soils in the field were devised that "the amounts of nitrates, phosphates, sulphates and the like, which may be present, as indicated by water solutions, can be determined to within four or five pounds per acre one foot deep." Fertilizers applied in the spring can be traced from the place of application down through the different depths of the soil which they invade. Much progress has been made toward determining by analysis the fertilizers needed by a particular soil.

A BATTERY OF TWENTY SELF-BINDING HARVESTERS AT WORK ON AN IOWA FARM

Humus.—Opinion as to the value of humus to plants has, pendulum like, swung to extremes. According to the early alchemists, decaying animal and vegetable substances yielded their spirits to the new plants. Many of the earlier chemists believed that the larger part of the materials entering the growing crop was supplied by humus. The net result of the combined labors of DeSaussure, Boussingault, Dumas and Liebig on this problem was to demonstrate that plants obtain most of their food from the air, and particularly that part which was supposed to be furnished by humus. Subsequently to this, humus was supposed to have a low value, but it is now known to perform many functions of the greatest consequence in plant growth. A certain amount of humus is essential to the proper physical condition of the soil. Besides influencing tilth, permeability and weight of soils, it facilitates drainage and prevents baking and cracking. Humus increases the power of the soil to absorb and retain moisture and renders it more friable and mellow. It supplies nitrogenous plant food and aids in making mineral plant foods and fertilizers more available and effective. It also lessens the danger of the winterkilling of wheat, and it furnishes food for the myriads of bacteria which live in all fertile, aerated, moist and warm soils.

The best method of keeping an adequate humus supply in the soil is to grow clovers and grasses in the crop rotation and plow under all plant refuse. Leguminous inter- or cover-crops can sometimes be grown advantageously after the main crop of the year is gathered, such crops being plowed under in the fall or spring for the purpose of supplying humus for the next crop. , These inter-crops also tend to prevent plant food from leaching out of the soil between crops. Among the best humus furnishing crops to be grown thus are soy beans and cowpeas, but even rye may be used.

As to the amount of humus contained in dried soils, those from the Red river valley contained 4.82 per cent, those between the Snake and Pelouse rivers, 6.4 per cent, those near Walla Walla, Washington, 4.8 per cent, and those of Missouri 4 per cent. Many of the soils of the south are deficient in organic matter. The native prairie soil of the Red river valley was very rich in humus, but the amount has been greatly depleted by continuous wheat growing. The soil humus can be increased by diversified farming. To keep the soils in the best physical and chemical condition, such a system of rotation should be practiced as will include both humus producing and humus consuming crops. Leguminous crops seem to have a marked effect in increasing the organic matter in the soil.

Soil Moisture.—A considerable amount of work was done on every type of soil during 1902 by the bureau of soils. As a result, soil moisture is now looked upon as a great nutritive solution which has approximately the same composition everywhere, and can vary only within narrow limits if plant development is to be successful. It is thought that the kind of crop adapted to a soil is largely determined by its physical characteristics, while yield is more influenced by chemical characteristics. The dissolved salt content of soils seems to be only a minor factor in determining the yield and quality of crops, the wide differences observed on different soils being mainly due to other factors. "It appears, further, that practically all soils contain sufficient plant food for a good crop yield, that this supply will be indefinitely maintained, and that this actual yield of plants adapted to the soil depends mainly, under favorable climatic conditions, upon the cultural methods and suitable crop rotation, a conclusion strictly in accord with the experience of good farm practice in all countries." It seems that a chemical analysis of a soil, even if made by extremely delicate and sensitive methods, will in itself give no indication of soil fertility. If the probable yield of a crop can be determined at all, it is likely to be by physical methods.[1]

BACTERIA.

Bacteria and Nitrification.—For cereal crops a previous leguminous crop is practically equivalent to the application of a nitrogenous fertilizer. In effect this was known by the Romans 2,000 years ago. Many theories were advanced to explain the beneficial effects of a leguminous crop, but the true explanation was not found until 1886, when Hellriegel convinced the entire scientific world that bacteria cause and inhabit the root nodules of leguminous plants, and that the symbiotic relation between

[1] U. S. Dept. Agr., Bu. of Soils, Bul. 22 (1903), p. 64.

these bacteria and the plants enables the latter to feed indirectly upon the limitless and costless store of free atmospheric nitrogen.

Nitrogen is one of the most costly and important of all plant foods and most crops remove large quantities of it from the soil. This applies with especial force to the wheat crop. The commercial supply is so limited that a "nitrogen famine" had already been predicted, but the discovery of nitrogen-gathering bacteria seems destined to lead to the utilization of air nitrogen at a nominal cost. Plants normally obtain through their roots nitrogen in some highly organized form. All non-leguminous plants placed in soil entirely destitute of nitrogen will wither and die. Bacteria alone have this power of fixing nitrogen. Not only have these bacteria increased the nitrogen content of soils planted with leguminous crops, but it is now claimed that for many centuries they have been continually fixing atmospheric nitrogen in certain regions of Chile and Peru, thus creating the extensive deposits of nitrate of soda there found in a natural state.

Men realized the increasing importance of the nitrogen problem as the supply decreased, and it was but natural for scientists to turn to atmospheric nitrogen in an endeavor to replenish the stores being so rapidly depleted, especially when they remembered that nearly eight-tenths of the air is nitrogen, and that plants are able to obtain all their carbon from a gas constituting only 0.1 per cent of the air. During the last quarter of a century bacteriologists have made numerous experiments which have thrown much light upon the subjects of nitrification, denitrification, and the fixation of free nitrogen in the soils. It has been found that the soil is alive with countless micro-organisms. The activity of some of these ferments favors, while that of others retards, plant growth. One group carries on the process of nitrification, and another that of denitrification. In the latter process the nitrates are broken down, deprived of their oxygen, and reduced to ammonia or nitrogen gas. Nitrates thus lose their availability for plant food. Great losses in manures may often occur from this source. It is the part of scientific agriculture to determine how to minimize the activity of inimical ferments.

Bacteria providing nitrogenous food for plants seem to be of three classes. One of these works on the nitrogen contained by the soil humus, and comprises three genera, each of which has an essential function in reducing nitrogen to a form available as plant food. Another class develops symbiotically with the growing plants, swarming in colonies upon the rootlets. Their vital activity oxidizes atmospheric nitrogen. The third class apparently secures the same result without symbiosis.

Efforts were made to inoculate soils with artificial pure cultures of the third class and thus increase the nitrogen content without the aid of manure or mineral fertilizer. While some very successful experiments were made, the percentage of failures was too great for practical purposes. The root tubercle bacteria seem to give the greatest promise of success. All of those which have yet had any practical importance were found exclusively on the roots of legumes. Some cultures of these organisms, known as nitragin, were placed upon the market a few years ago by German experimenters. They were adapted to specific crops only, for it was claimed that each kind of leguminous plant had a special germ which was more successful upon it than any other form. There were so many failures that the manufacture of nitragin was abandoned.

Previous to 1902 the United States department of agriculture inaugurated extensive practical experiments in an effort to find improved methods of soil inoculation. The reasons for the failure of the German pure culture method were worked out. Improved ways of handling and preserving pure cultures were discovered, as well as means of rapidly and enormously increasing them after they were received by the farmer. Great progress was made toward developing an organism effective for all legumes, and the virility of the bacteria was so increased that they fixed over five times as much nitrogen as formerly. When the department could send in perfect condition to any part of the United States "a dry culture, similar to a yeast cake and no larger in size," the nitrogen-fixing bacteria of which could be "multiplied sufficiently to inoculate at least an acre of land," the prospects of an early and complete solution of the nitrogen problem seemed to have a rosy hue indeed.

In spite of the great progress that has been made, however, there has been little to encourage the hope of directly increasing the nitrogen supply of the soil for the wheat crop by means of bacteria. The more practical solution seems to be the indirect one of growing in rotation with wheat leguminous crops aided by artificial cultures. Success in this has been pronounced and practical. In 1904 the United States department of agriculture made a very extensive experiment with artificial inoculation of leguminous crops. About 12,500 tests were made under all sorts of conditions and in almost all of the states in the union; 74 per cent of the tests properly made proved successful.[1] Not only was nitrogen thus fixed in available form for subsequent grain crops, but the leguminous crops sometimes yielded five times as much as non-inoculated ones grown under similar conditions, the usual increase ranging from 15 to 35 per cent. One result of the success of the experiment was a demand for cultures far beyond what the department could supply. A great improvement was made in 1906 by abandoning dry cultures for pure liquid cultures hermetically sealed in glass tubes.

Perhaps the best method of distributing and applying the organisms is by inoculating the seed of the legumes used. This way is thoroughly effective and costs but a few cents per bushel of seed treated. One gallon of liquid culture will inoculate 2 bushels of seed. Soil may be inoculated and then distributed as fertilizer would be, or earth may be transferred from a field containing the bacteria. Both of these methods are expensive, less certain of success, and weeds or pests may be transferred with the soil.

The nitrifying bacteria are parasitic plants that penetrate the roots of legumes to obtain food carbohydrates. After the roots are from 2 to 4 weeks old, the bacteria are unable to enter them. It is now known that tubercle formation is not essential to successful inoculation, and that the bacteria may be present in an efficient state in the absence of tubercles.[2] Humid soil and a temperature of 60 to 80° F. are most favorable to the growth of soil bacteria, 35° F. being the lowest, and 98°F. the highest temperature at which growth is possible.

[1] Bu. of Plant Industry, Bul. 70, p. 41.
[2] Yearbook U. S. Dept. Agr., 1904, p. 49.

Denitrifying organisms thrive best in a soil at least slightly organic, and so packed as to exclude the oxygen of the air. The nitrifying bacteria are unable to develop in organic matter, but its presence to some extent is not fatal to them. The presence of nitrogenous substances has a deleterious effect upon the cultures of nitrifying bacteria, which seem to fix atmospheric nitrogen only in the absence of plenty of nitrogen in the soil, consequently little benefit is to be obtained from inoculating soils containing a good supply of nitrogen. Most of the nitrifying germs seem to exist in the first foot of soil, while few, if any, exist at a greater depth than 18 inches. Other bacteria, such as those which change the sulphur and the iron compounds, also exist in the soil.

APPLIED FERTILIZERS.

Need, Time and Application of Fertilizers.—Nearly all wheat land that is under continual cultivation, even if crop rotation is practiced, yields larger returns when fertilizers are properly used. In each individual case local conditions and the economic position of the wheat grower must determine to what extent it is advantageous to fertilize. Each farmer must, in a large measure, learn by experience whether the application of a certain fertilizer is profitable under his circumstances. It is now well known that yield does not increase in proportion to the amount of plant food applied, and that the increase in straw is greater than that in grain. In determining the value of such application, it must, of course, be remembered that more than one crop is benefited. The wheat crop may be increased either by direct fertilizing or by the residual effect of fertilizers applied to other crops in the rotation. The composition and condition of the soil determine the relative importance of different fertilizing constituents. Phosphoric acid used alone generally increases the yield of wheat grown anywhere on the glacial drift area of the United States.[1] Either nitrogen or potash applied alone does not seem to increase the yield greatly, while the application of both with phosphoric acid gives the greatest gain. A fertilizer that can usually be found on the market is one containing 4 per cent each of ammonia and potash

[1] Hunt, Cereals in Amer. (1904), p. 75

and 12 per cent of available phosphoric acid. By applying from 250 to 500 pounds of this commercial fertilizer per acre the best general results are obtained. If land has been quite exhausted by continuous wheat growing, the proportion of nitrogen and potash should be greater. Commercial fertilizers are best applied by means of an apparatus made for this purpose and attached to the wheat drill. They may also be broadcast just in front of the drill. In the case of winter wheat, most of the nitrogen is often applied early in the spring so as to prevent loss through drainage during the winter.

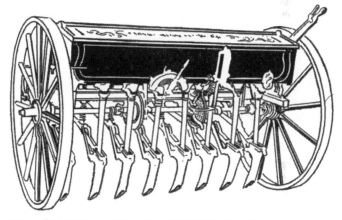

THE COMBINED GRAIN AND COMMERCIAL FERTILIZER DRILL

Kinds of Fertilizers.—These naturally fall into two classes, barnyard manure and commercial fertilizers. It is only on farms where the supply of manure is not adequate to preserve a high state of soil fertility that commercial fertilizers are economical. In general farming the former has the greatest relative value on account of individual or combined physical, chemical and bacterial influences not yet fully understood. Considering equal weights, however, most commercial fertilizers contain more plant food than manure does. It is also claimed that when applied to wheat they will produce larger returns. Nevertheless, the lower cost of farm manure always makes its use more profitable than that of other fertilizers. Where both are used, it is most profitable to apply the manure to crops grown in the rotation, such as maize, and to apply the commercial fertilizers directly to the wheat crop.

Manure.—It is claimed that the soils of China have been in continuous cultivation for more than 4,000 years without falling off in productiveness, and that the continued soil fertility is due to the utilization of all animal manures and of sewage. During the eleventh century in France, stable manure was unknown as a fertilizer, though flocks of sheep were used for this purpose. Stable manures were utilized in the medieval husbandry of England, and they have been used to great advantage in France and Germany for over a century.

In America manure has always been utilized as a fertilizer by progressive farmers, but it has also been looked upon as a farm nuisance. It has been charged with producing dog fennel and various other weeds, and with "poisoning" the soil. In parts of Oregon and South Dakota it has been burned, sometimes for fuel. It has been hauled into ravines in California, into the creek in Oklahoma, into a hole in the ground or to the side of the field in Kansas, to the roadside in Missouri, to great piles in North Dakota and Idaho, and to the river in the Mohawk valley.[1] It is estimated that the farmers of the United States annually lose over $7,000,000 by permitting barnyard manure to go to waste. As the fertilizing value of the manure annually produced by the farm animals of the United States is calculated at over two billion dollars, it must, however, be very generally utilized, a fact which does not excuse the foolish and useless waste. The fertilizing value of the average amount produced yearly is estimated for each horse at $27, for each head of cattle $19, for each hog $12, and for each sheep $2. The amounts of fertilizing constituents in the manure stand in direct relation to those in the food of the animal, and have a ratio to them varying in value from one-half to unity.

Experiments have shown that equal weights of fresh and of rotted manure have equal crop-producing powers. As 60 per cent of the weight is lost in the rotting process, manure should be used in fresh condition. "Barnyard manure contains all the

[1] Yearbook U. S. Dept. Agr., 1902, p. 529; Industrial Commission 10:clxxxviii.

fertilizing elements required by plants in forms that insure plentiful crops and permanent fertility to the soil. It not only enriches the soil with the nitrogen, phosphoric acid and potash, which it contains, but it also renders the stored-up materials of the soil more available, improves the mechanical condition of the soil, makes it warmer, and enables it to retain more moisture or to draw it up from below."[1] It has a forcing effect when fresh.

Barnyard manure rapidly undergoes change and deterioration. The latter results mainly from two causes: (1) Fermentation, and (2) weathering or leaching. Losses from leaching may be prevented by storage under cover or in pits, while proper absorbents and preservatives, such as gypsum, superphosphate and kainit, will almost entirely prevent destructive fermentation. The manure should be kept moist and compact. The loss is less in deep stalls than in covered heaps. The fertilizing constituents of well rotted manure are more quickly available to plants than those of fresh manure, and the former should be used when prompt action is desired. In the wheat lands of California manure is more or less visible for four or five years after its application to the land, and in the semi-arid region it must be used cautiously on unirrigated land. The light soils of the Pacific coast lack the moisture requisite for the absorption of wheat straw plowed under, and consequently it must be burned. This wastes the nitrogen element of the straw, but saves the ash ingredients for the land. Land treated with stable manure for 6 years gave an increase of 60 per cent in the yield of wheat. Ten tons has been given as a reasonable amount to apply to one acre of wheat land.

Guano.—The first shipload of Peruvian guano was imported by England in 1840. Two years later a company was organized to trade regularly in this substance. From 1.5 to 2 cwt. per acre of wheat was harrowed in with the seed. In the United States it quickly gained in popular favor. By 1876 the trade was regulated by national treaties, and millions of dollars were engaged in its transportation. Peruvian guano was used chiefly for its ammonia. The later guanos of the West Indies were rich in phosphates, and of greater advantage to some crops than the Peruvian. Guano was also one of the principal

[1] Yearbook U. S. Dept. Agr., 1895, p. 570.

sources of nitrogen. Most of the guano beds are now completely exhausted.

Phosphoric Acid.—The four main sources of this are:
(1) Bones; (2) phosphatic deposits like those of South Carolina, Florida, Tennessee, North Carolina and Virginia in the
United States, or the keys of the Carribean sea; (3) accumulation of fossil and excrementitious material; and (4) Thomas
slag, a by-product of the smelting of iron ores. Bones were
used to a limited extent in England before 1810. They were
ground until Liebig made the discovery of preparing superphosphate of lime by dissolving bones in sulphuric acid. One
bushel of bone dust dissolved by one-third its weight of the
acid is superior as a manure to four bushels of bone dust. The
lime of the bones is converted into gypsum, and the phosphoric
acid is reduced to a state more easily soluble and assimilable.
Formerly bones were also often burned.

In 1817 superphosphates were first manufactured in England, and the first phosphate mined commercially in South
Carolina was in 1867, six tons. The earliest form of mineral
phosphate used for fertilizer was apatite. In Canada 10 years'
experience has shown that finely ground, untreated mineral
phosphate has no value as a fertilizer. The Bessemer process
of manufacturing steel gives a by-product, rich in phosphoric
acid, ''produced by the union of the phosphorus of the iron
with the lime of the flux employed.'' This is reduced to a
fine powder and applied, without treatment, to the soil. It
contains from 15 to 20 per cent of the acid. The yield of
wheat seems to be little affected by the carrier or source of
phosphoric acid if the material used is finely ground. Some
30 to 60 pounds of the acid should be applied to an acre of
land.

Lime.—It has been claimed that this is one of the first
mineral elements to show depletion. Sourness of soil often
results. To correct this and supply lime, 20 to 40 bushels of
lime per acre may be used. Lime has a tendency to work
down, and should never be deeply plowed under. It can be
applied to the soil most advantageously prior to the planting
of maize in the rotation, for wheat does not seem to be directly
benefited by it. If it is applied directly to wheat, it should be
scattered over the plowed field a few days before seeding and

at once harrowed in. The ground should be stirred again before seeding. The application of one ton per acre every 4 to 6 years is advised for Illinois uplands.

Marl.—This consists essentially of carbonate of lime. Often considerable amounts of organic matter, sand and clay are also present. It originated in the breaking down of fresh water shells. Its action is more slow and "mild" than that of lime. It has been "regarded rather as an amendment than a fertilizer." Its chief functions are in improving the tilth, neutralizing acidity, and promoting nitrification, besides supplying lime. The marls of New Jersey also contain potash and phosphoric acid.

Nitrate of Soda.—Trade in this as a fertilizer began between 1830 and 1840. The supply is limited. In 1860 all estimates showed that it would last nearly 1,500 years. By 1900 these estimates had fallen to less than 50 years, and the world's markets were annually consuming nearly 1,500,000 tons, the United States requiring about 15 per cent of this amount. It is by far the most expensive fertilizer in use, and it is the best carrier of nitrogen.

Potash.—The main sources are wood ashes, and, since 1860, the products furnished by the potash industry at Stassfurt, Germany. Several forms are imported from Germany, each containing a different but correctly warranted quantity of this fertilizer. Nearly 500,000 tons are imported annually by the United States. This is over half the product. Since 1860, the Stassfurt salts have been almost the only source of concentrated potash. Good wood ashes contain perhaps 10 per cent of potash. They were long used without the real reason of their value being known. Besides potash, ashes often contain considerable lime. Hartlib gave 23 fertilizers and means of fertilizing the ground, and among them were included lime, marl, ashes and chalk.

Gypsum or Land Plaster.—Those who are experimenting with this material report varied results. It has long been used, however, and the most reliable conclusions seem to be: (1) That gypsum has undoubted fertilizing value on many soils; (2) that its chief value depends on three processes: (a) Preservation of ammonia and perhaps other nitrogenous forms;

(b) decomposing potash and phosphorus-bearing silicates, liberating these two elements for plant-food; (c) affecting soils physically, making them granulated, or loose and mellow; (3) and that it decomposes sodium carbonate and thus breaks up the so-called "black-alkali."

Common Salt.—This has also been used as a fertilizer for at least several decades. In the eighties it was a common practice in England to sow salt in the early spring on wheat land that was too rich, the idea being that a larger deposit of silica in the stalk would result, enabling the wheat to stand better. While it has been found a valuable agent for increasing the yield of barley, it is of less importance in raising wheat.

Miscellaneous Fertilizers.—A great many other materials have been used to a greater or less extent as fertilizers. Among them are: Animal products, as wool waste and the refuse of modern slaughterhouses, blood, bone, hair, horn, hoof, etc., which with fish, manure and sulphate of ammonia from the gas works, are still the main sources of nitrogen applied to crops; swamp muck, marsh mud, sea-weed, sludge, poudrette, potassium, cottonseed meal, rape-cake, burnt clay, charred peat, soot and green manuring crops. The latter are simply plowed under, a practice widely followed in the United States, especially with alfalfa and other legumes. Where stock can be raised, green crops and cottonseed meal have nearly as great a value for fertilizer after feeding as before, and yield the additional intermediate product of milk or beef.

It is interesting to note that the aborigines taught the early settlers of New England the value of fish as a fertilizer. Fish or fish waste should be composted. Quicklime is used in France. Fish compost readily yields its elements to growing crops, consequently it should be applied in the spring, and not deeply covered. Sludge is the precipitant of sewage, and poudrette is the same reduced to a dry powder. A part of their value lies in the germs of nitric ferment which they contain. Some 40 tons of wheat straw leached and burned on the soil contribute to it 8 pounds of phosphorus and 680 pounds of potassium, besides the nitrogen leached into the soil before the straw was burned. This immensely increases the yield of wheat. Mulching with straw does not seem to be of any benefit to wheat, whether applied for fertilizing or for winter protection.

Fertilizing by Irrigation.—To show the fertilizing value of irrigation waters, some analyses are given below.

COMPOSITION OF IRRIGATION WATER.

Results expressed in parts per 1,000,000

Description	Total Solids	Organic Matter	Silica	Ferric Oxide and Alumina	Lime	Magnesia	Phosphoric Acid	Potash	Soda	Sulphuric Acid
Middle Creek[1]	77.0	24.0	15.8	0.9	22.6	6.1	0.41	0.8	3.7	0.34
Yellowstone River[1]	96.0	32.0	11.2	1.1	21.0	8.3	0.41	1.9	22.1	17.0
Missouri River[1]	221.0	64.0	18.9	3.1	52.8	18.1	0.22	1.9	30.0	21.9
Shields River[1]	186.0	34.0	6.8	1.3	74.5	15.0	0.35	0.14	18.0	20.0
Bozeman Creek[1]	117.0	31.0	20.2	1.5	32.2	12.8	0.44	0.9	6.3	3.4
Nile River[2]	1700.0						3.4	10.2		

The waters of the Nile seem to have the largest amount of nitrogen, 1.7 per cent, all the others having merely a trace. Some 24 acre-inches of Rio Grande water add to the soil about 1,075 pounds of potash, 116 pounds of phosphoric acid, and 107 pounds of nitrogen. The same amount of Delaware river water contains 741.08 pounds of materials, while the Mississippi, St. Lawrence, Amazon and La Plata rivers average 655.6 pounds of solid matter for every 24 acre-inches. As a rule sewage waters from the cities have the highest value for irrigation, and muddy river waters stand next. Waters containing sulphate of iron are positively injurious when applied to land. They usually issue from peaty or boggy swamps. While the fertilizing value of sewage waters is unquestioned, and while their use has been almost universally favored, objections have been made to them on other grounds. To say the least, they undoubtedly contain a hidden danger, and if used at all, it should be with the greatest of care. It has been claimed that "the use of sewage for fertilizing purposes is not to be commended because of the danger of contaminating the soil with

[1] Rept. Mont. Exp. Sta., 1902, p. 62.
[2] Exp. Sta. Record, V. 14, No. 11, p. 1057.

pathogenic ferments, which may subsequently infect the health of man and beast. These ferments may attach themselves to vegetables and thus enter the animal organism, or they may remain with a suspended vitality for an indefinite period in the soil and awaken to pernicious activity when a favorable environment is secured.''

Vast stores of fertilizing materials are continually being washed from the earth by floods, and carried away by streams and rivers. The Seine river thus annually carries two million tons of silt, a greater weight than the merchandise which its waters transport. The Var carries seaward yearly 23,000 tons of nitrogen, and one cubic meter of water per second from this stream could be made to produce crops valued at 35,000 francs each year. The river Durance, an Alpine stream, annually carries silt, the fertilizing power of which is equal to 100,000 tons of stable compost or excellent guano. It would take 119,000 acres of forest trees to yield the carbon that this volume of silt contains.

FERTILIZER EFFECTS

Effect on Germination.—In general, fertilizers never seem to aid in the germination of seeds, and may be harmful if used in large quantities. One per cent of muriate of potash, or of sodium nitrate, is very detrimental, whether applied directly, or mixed with the soil. Phosphoric acid and lime are much less injurious, and may be harmless if not used in excess. It is safest not to bring commercial fertilizers into immediate contact with germinating seeds, and the effect of chemicals applied to seeds before they are planted is no index of their action in this respect when used as fertilizers. When injury does result, it is chiefly to the young sprouts during the time between when they leave the seed coat and when they emerge from the soil, the seed being affected but slightly, if at all. Salts injurious to wheat seedlings have been given in the following order: Magnesium sulphate, magnesium chlorid, sodium carbonate, sodium sulphate and sodium chlorid. Different varieties of wheat vary in their ability to resist the same toxic salt, as does also one variety in different salts.

Effect on Yield and the Supply of Plant Food.—There seems to be a certain minimum yield of wheat which a soil will give under continuous cropping and ordinary cultivation, and this yield can be increased by rotation of crops, and still more by improved methods of cultivation. Fertilizing is also a factor which generally increases the yield, whether utilized by itself, or in conjunction with other factors. The use of commercial fertilizers must, however, be accompanied by intensive methods of cultivation in order to be profitable, and now and then the returns seem to diminish with continued use. Mr. Whitney, chief of the bureau of soils in the United States department of agriculture, maintains that he never saw a case of soil exhaustion which was probably due to the actual removal of plant food. He considers the so-called worn-out soils of the United States due to conditions which make the plant food unavailable, and holds that the primary object of fertilizing is the adaptation of soils to any desired crop or crops. Fertilizing can also be practiced to force growth, even on rich soil. Texture and drainage of soils can be improved, the ratio of soil constituents balanced, and acidity neutralized. Attention is called to the facts that "the soils of India, which tradition says have been cultivated for 2,000 years, under primitive methods, without artificial fertilizing, still give fair returns. In Egypt, lands which have been cultivated since history began are as fertile as ever. In Europe there are records of cultivation of soils for 500 years." [1]

Tradition is not always scientific, however, and soil is not greatly taxed by such primitive methods of culture as have existed in India for 2,000 years. The sediment which is deposited by the waters of the Nile at every annual overflow is entirely adequate to maintain the fertility of the cultivated lands of Egypt, while fertilizing, improved methods of cultivation, and crop rotation have greatly increased the yield of European soils. On some of the fields in France 28 bushels of wheat are raised per acre where 17 bushels were raised 50 years ago. The soil of France is more fertile today than it was in the time of Cæsar. The fertility of the soil in Germany has increased proportionately. In England, land on which wheat was grown continuously for 50 years without fertilization yielded 12 to 13 bushels per acre, while adjacent plots to which fertilizers were applied averaged about 30 bushels per acre. Mr.

[1] Industrial Commission, 10: clxxxviii, cxcii.

Whitney takes the position that if the soils of eastern and southern United States have any less plant food than when first cultivated, they at least have all the ingredients essential for crop production. This position is certainly supported by statistics that have been given on the amount of plant food contained by soils. An acre of very fertile soil contains about 70,000 pounds, or 2 per cent, of potash in the first foot of ground. A crop of wheat removes about 15 pounds of potash from each acre. It has been estimated that the first eight inches of soil contain on an average enough nitrogen to last 90 years, enough phosphoric acid to last 500 years, and enough potash to last 1,000 years.[1] This supply is materially increased when we consider the great depth penetrated by the roots of wheat. It must also be borne in mind that the loss of plant food is often much greater than that removed by crops; for example, it has been given as three to five times as much in the case of nitrogen.

Extensive farming, nevertheless, soon reduced the productivity of our first cultivated soils, and with the opening of the large and level western wheat fields of fully as great fertility as was ever possessed by any soils of the United States, many of the older lands were abandoned. Now, however, most of the farm lands of the west have been occupied, the standard or farming is being raised, and conditions have so changed as to make it seem profitable again to resume the cultivation of these abandoned lands. But they must be farmed by intensive methods, of which fertilizing is a valuable part. Some lands have already been restored to fertility and are being cultivated with profit.

Missouri soils are still rich in plant food, yet their productivity is much less than it was 50 years ago. Commercial fertilizers had been profitably used in wheat raising in Ohio over a decade ago. Growing a leguminous crop on light sandy soil deficient in humus increased the yield of the following crop of wheat over 50 per cent in Arkansas. When 400 pounds of a complete fertilizer were used in addition, the following 2 years the wheat crop averaged over 70 per cent more than on soil not thus treated. Manure treatment and the application of phosphorus is found very profitable in Illinois.

[1] Rept. Mich. Board Agr., 1905, p. 147.

As a rule, the land in the Red river valley is not fertilized, and produces less than 15 bushels per acre, but the application of fertilizers has given 26 bushels per acre. Rotation of crops is already widely practiced in the northwest, and as the soil becomes more exhausted and the prices of land and produce rise, fertilizers will be used there, just as they have in every other country where similar conditions arose. Even the largest bonanza farmers are looking forward to the time when they must fertilize. Stock will also be raised, and farming will become more diversified. This will give opportunity to utilize many of the products of wheat on the farms where they are produced, and the need for commercial fertilizers may ultimately be removed altogether. One-fourth of the nitrogen and nearly all of the phosphoric acid and potash which enter into a crop of wheat are contained in bran, screenings and middlings. Most of these can be returned to the soil by raising stock. These principles are not mere theories, for their practical working has been demonstrated in Michigan and Illinois, in portions of which the land has been continually growing more fertile under cultivation without the use of a pound of commercial fertilizer.

The average Kansas wheat grower has given little thought yet to fertilizing, but "his methods will change with the years and the necessities." In Minnesota, "results already reached warrant the statement that the average yields per acre of wheat can be increased 25 to 50 per cent by so rotating the crops and manuring and cultivating the fields (as) best to prepare the soil for this grain." [1] The necessity of fertilizing is little felt in Canada as yet.

Fertilizer Laws.—A majority of the United States, including nearly all the states east of the Mississippi, have statutes, most of them rather stringent, governing the sale of commercial fertilizers. That dealers were cheating farmers was first shown by the Massachusetts experiment station. This station was instrumental in the passage of the Massachusetts fertilizer law, which was more or less closely followed by other states. The department of agriculture has made efforts for a more uniform system of laws, and to regulate interstate trade. No general fraud is now practiced. Thousands of official analyses are

[1] Minn. Bul. 12, p. 321.

made yearly, and these very largely eliminate fraud and extravagant claims.

Fertilizer Statistics.—The annual sales of fertilizers in the United States exceed $50,000,000, and perhaps over 2,000,000 tons are consumed. During 1896 over 375,000 tons were imported, valued at about $19 a ton, and over 514,000 tons, valued at about $8.50 a ton, were exported. During the first six months of 1905 the importations were valued at nearly $2,000,000. The annual import of nitrate of soda is nearly 200,000 tons, having an average value of about $30 a ton. The first guano sold for about $95 a ton, but later fell to half that amount. A ton of cottonseed meal has a fertilizing value of over $20. A ton of fertilizer, costing $25, is applied to an acre of wheat land in New York. The phosphate mined in the United States in 1899 amounted to 541,645 tons. The average cost of phosphates at the quarry was $4.42 per ton in 1893. In Illinois rock phosphate could be procured at about $8 per ton in 1906, and bone phosphate at $25 per ton. California expended six times as much for fertilizers in 1900 as in 1890. About 1890 the farmers of Ohio were expending a million dollars annually for commercial fertilizers used in the production of wheat.

It is claimed that a trust caused prices of fertilizers to advance from 15 to 25 per cent in 1900. Below is a schedule of prices given for the different fertilizing substances per pound:[1]

Nitrogen ....$16\frac{1}{4}$ cents
Potash .... 5 cents
Water-Soluble Phosphoric Acid .... $5\frac{1}{2}$ cents
Citrate-Soluble Phosphoric Acid .... 5 cents
Phosphoric Acid in fine bone ....: 3 cents
Phosphoric Acid in medium and coarse bone........$2\frac{1}{2}$ cents

In Illinois the annual cost of fertilizing an acre of land is about $1.70. Tubercle organisms enable leguminous crops to add from $8 to $10 per acre in nitrogenous fertilizer. In the early nineties over 75 per cent of the fertilizers sold were equal to or above the guaranty under which they were sold. Most of the others were much less than 1 per cent below their guaranty. It has been estimated that with the use of all barn manures and proper cultivation, the soils of the Red river

[1] Miss. Bul. 77 (1902), p. 3.

valley should yield good crops without commercial fertilizers for over a thousand years. "From 7 to 135 pounds of nitrogen, from 3 to 55 pounds of phosphoric acid, and from 3 to 36 pounds of potash are sold with every ton of produce leaving the farm."[1] Eighty billion pounds of nitrogenous material entered into the creation of one harvest in the United States in the early nineties. The cereals annually took from the earth nearly 3 billion pounds of phosphoric acid, and the loss of potash was not less than 4 billion pounds.

[1] Yearbook U. S. Dept. Agr., 1897, p. 301.

CHAPTER IX

DISEASES OF WHEAT

Introductory.—Studies in plant pathology of any great practical bearing or importance are decidedly and characteristically modern and recent. In 1885 there were three institutions in the United States besides the department of agriculture which were making systematic efforts in experimental work with plant diseases, and in disseminating such knowledge as then existed in these lines. Ten years later over one hundred special investigators were devoting their time to this work, and 50 colleges and stations were endeavoring to solve its practical problems. The science of plant pathology has had its highest appreciation during the last decade, for some of the outlying problems have been solved, a working foundation has been laid from the facts that were acquired, and a body of institutions with specialists and resources has been developed for scientifically prosecuting the work.

The Classification here followed is practical and inclusive, rather than scientific and exclusive. There is nothing pathological in the sudden destruction of a field of wheat by floods or locusts, but excessive moisture or the presence of a parasite may each bring about diseased conditions, and every gradation of phenomena between these two types must be considered.

Sources of Injury and Weakness are of three kinds: (1) Unfavorable inanimate environment; (2) unfavorable animate environment; and (3) poor seed wheat. When these three sets of factors occur in combination, as they frequently do, their relations and inter-relations are so intimate and intricate as to be inseparable. Only in a general way can they be individually studied.

WEATHER AND SOIL INFLUENCES

Unfavorable Inanimate Environment.—Drought; Hail, Wind and Rain Storms; Floods; Fire. The operation of these destructive influences needs no further elucidation than a mere mention. Very little specific information can be found as to the extent of damage caused. Such scattering data as have been collected can be most advantageously presented under the subject of insurance.

FROSTS.—The most usual injury by frost is the winterkilling of fall wheat. This may occur whenever the ground freezes to any appreciable depth, and in two ways. The plants either freeze to death, or are lifted out of the soil by alternate freezing and thawing. A good covering of snow is very protective. Seeding with a press drill lessens the danger. Frost may also injure wheat when it is filling, or it may cause the stems to burst after they have jointed.

HOT WAVES OR WINDS are most liable to occur during a period of drought. It is thought that these waves can be forecasted for a period of about four days. At such times the eastward circulation of the atmosphere is practically suspended, and radiation is at a minimum. A hot wave is defined as a period of three or more consecutive days with a maximum temperature reaching or passing 90° F. In years when hot waves are unusually severe, the harvest returns are decreased by one-fourth in quantity, and the quality is quite inferior. The heat seems to mellow the ground, however, and to put it in such ideal condition as to increase the crop of the following season. Hot winds have a velocity of 20 to 30 miles per hour, a temperature often ranging from 100 to 106° F., and 20 to 30 per cent of relative humidity. The roots cannot supply moisture, even if it is present in abundance, as fast as it is evaporated from the leaves by this great blast of hot, dry air. The cells are completely desiccated, and the whole structure of the plant collapses. A hot wind is most destructive immediately after a rain, which temporarily checks and lessens the transpiration of which the plant is capable. In the United States these winds are most apt to occur in the central prairie regions. In Argentina, a similar dry, hot wind known as the *pampero* comes up suddenly, destroys all vegetation "and even cracks furniture and timber in buildings." A wind-break of trees, or anything else that tends to lessen the movement of the air, has a remedial effect.

EXCESSIVE MOISTURE.—This may be injurious in a number of ways. If too much water is present, the wheat may be "drowned." It also tends to develop the straw indefinitely, and at the expense of the grain. Rains during heading are apt to prevent filling, and are by far the most common cause of blight. In a very wet harvest wheat is apt to germinate before it can be threshed. One-third of the wheat crop of lower Canada was lost in 1855 by the grain germinating in the straw.

UNFAVORABLE SOIL.—The soil texture may be such as to deprive the roots of the proper air supply. Certain elements, as in the case of alkali, for example, may be present in such abundance that their chemical action upon the wheat plants causes disease. Some of the essential plant foods may be absent, or present in improper proportion.

All types of disease mentioned thus far arise from physiological variations due to abnormal variations in the growth factors, are not transmissible, and consequently never spread from plant to plant or field to field, as do the infectious wheat diseases.

PLANT INFLUENCES

Unfavorable Animate Environment.—WEEDS—Plants out of place are called weeds. They deprive wheat of its nutriment and ordinarily give very little in return, except in the case of certain legumes. Weeds once introduced into a region spread rapidly. With runners, rootstocks, running roots and apparatus for throwing seeds, they effect a dispersion of their kind independently of any external agencies. Wind, water and animals are the natural agencies that aid in the dispersion, but rarely carry seeds long distances. Man aids weed migration more than all natural means combined, and consequently its general direction is in the wake of the progress of cultivation. Commerce in wheat makes some weeds cosmopolites. These plants have a wide range of adaptability, which grows wider under conditions of cultivation. Some seeds, especially those of cockle, have a tendency to approximate the wheat grain on account of selective influences arising from cleaning seed wheat. Those which differ most from wheat are the ones removed. Sowing the remaining ones develops a strain more closely resembling the wheat grain.

Weeds injurious to wheat may be divided into three classes, based on the point of incidence of the damage caused: (1)

Those which choke the crop, preventing its growth; (2) those which interfere with harvesting and curing; and (3) those whose seeds injure the commercial value of the grain by mingling with it. Deterioration in the quality of the grain by the third class is perhaps the greatest damage resulting from weeds.

Kinds of Weeds: CHESS OR CHEAT (*Bromus secalinus* L.)—This is an annual grass that will not produce seed unless sown in the fall, and consequently it is not found in spring wheat. It is less vigorous than wheat, but more prolific, and also more resistant against cold and insects. One pound of seed has been known to multiply 99-fold in one generation, and one seed 3,000-fold. The ordinary observer cannot distinguish the young chess plant from wheat. Chess injures flour and must be cleaned from the wheat before grinding. Seed wheat properly cleaned by a fanning mill is quite free from chess. If wheat is treated for smut by stirring it in a solution, the chess seeds will rise to the surface and can be skimmed off. A pound of chess and a bushel of wheat have about the same number of seeds.

Russian Thistle or Cactus (*Salsola kali tragus* L.).—This weed is neither a thistle nor a cactus, but a saltwort, closely related to the tumbleweed, lamb's quarters, and pigweed. In parts of Russia, where it has been known over 150 years, extending now to northern Russia and central Siberia, it is known as Tartar or Hector weed. It was first introduced into the United States in 1873 or 1874, being sown in South Dakota with Russian flax seed. As the weed prefers a dry climate, it could not have found a more congenial habitat. When the plants were uprooted in the fall they rolled across the prairie with the speed of the wind, scattering seeds at every bound, and stopping only when they were worn to pieces, or when the wind ceased, for in the early Dakotas there were few fences, forests, or streams to stop their course.

Thus they covered an advance of 5 or 10 miles in a season, though stray plants went much farther. As a rapid traveler thoroughly covering territory it surpassed any other weed known in America, and very few cultivated plants intentionally distributed have such a record for rapidity. Within 20 years it infested a continuous area of about 35,000 square miles, and caused at least $1,600,000 damage to wheat every year.

By 1894 it had crossed to the west side of the Missouri river, and was spreading in Minnesota, Iowa and Nebraska. In 1895 the injuries which it caused extended from Michigan to Colorado, Idaho and California, but the greatest damage resulted in the Dakotas and Nebraska.

The Russian thistle is an annual with a dense, yet light growth of circular or hemispherical form. The average plant is 2 to 3 feet in diameter, weighs 2 to 3 pounds when matured and dry, and is estimated to bear 20,000 to 30,000 seeds. Single plants have been found 6 feet in diameter, weighing about 20 pounds when thoroughly dry, and estimated to bear 200,000 seeds. It is ideally fitted to be carried by the fall winds, which easily break off or pull out its slender roots. A severe frost kills the plant at any time, but it produces seeds abundantly as far north as the Canadian boundary.

In Russia no effectual method of exterminating the weed is known. It is continually growing worse, and as a consequence the cultivation of crops has been abandoned over large areas in some of the provinces near the Caspian sea. Laws for its eradication were passed in South Dakota in 1890, in Iowa in 1893, and in Kansas, North Dakota, Minnesota and Wisconsin in 1895. In 1892 and again in 1893, the department of agriculture sent an assistant botanist to the Dakotas to work out the whole life history of the plant, and the method of dealing with it was established chiefly on the basis of the knowledge thus obtained. It is claimed that if this had been done before 1885, a saving would have been effected for the wheat growers of the northwest "sufficient to pay the cost of maintaining the whole department of agriculture for many years to come."[1] Wetter seasons, more intensive farming, the building of fences, and the planting of trees reduced the Russian thistle to the ranks of comparatively unimportant weeds in the Dakotas, but in Nebraska stringent measures were necessary. Farmers co-operated in the work, and the weed was "hunted almost as strenuously as game would be," so that for some years it has not been an important factor in wheat growing in Nebraska.

Darnel (*Lolium temulentum* L.) has its widest distribution in Europe. It also occurs in the wheat fields of California, where

[1] Yearbook U. S. Dept. Agr., 1897, pp. 95-96.

it is mistaken for ·chess. It is an annual grass, and can be cleaned from seed wheat by the chess method.

Cockle (*Agrostemma githago* L.) is a member of the pink family and a widely distributed weed of the wheat fields since ancient times. In size and weight its seeds resemble wheat grains so closely as to be removed with difficulty. They are easily seen in the grain, however, and are injurious to flour, consequently they render wheat less marketable. As cockle is usually not abundant, grows over a foot high, and is conspicuous because of its large pink blossoms, it can easily be pulled from the growing wheat. The seeds have great vitality and will germinate even if they have lain in the ground several years.

Wild Garlic (*Allium vineale* L.) is most troublesome to wheat in eastern United States. It grows about two feet in height. The flour is spoiled when the bulblets of the plant are ground with the wheat. These can be removed by careful screening. Badly infested land should be put into cultivated crops for a few years.

Wheat-Thief (*Lithospermum arvense* L.).—Other names by which wheat-thief is known are bastard alkanet, corn gromwell, redroot and pigeonweed. Its greatest damage is to meadows following wheat in rotation. Cultivated crops are the best remedy.

Wild Mustard or Charlock (*Brassica sinapistrum* L.) is so uniformly found in spring wheat that flouring mills make a by-product of its seed. When not very abundant it is easily pulled in the field, for it grows nearly as high as the grain and has conspicuous yellow flowers. In small wheat fields where it is very abundant it can be killed by spraying the field with a 3 per cent solution of copper sulphate, using about 50 gallons of solution to the acre. It is claimed that the wheat is not injured.[1] If wild mustard seed is covered with at least five inches of soil it will not grow, but thus buried it will retain good germinating power for over 56 months. It comes up most abundantly through one inch of soil.

Thistles are of two varieties: Canada thistle (*Cnicus arvensis*), and common or sow thistle (*C. lanceolatus*). The latter is also known as spear, bur, and bull-thistle. It grows 2 to 4 feet high and has the better hold on the land where both

[1] Cornell Bul. 216 (1904), p. 107.

infest the same area. The Canada thistle is a native of Europe and Asia. In Great Britain it is called corn or creeping thistle. Its growth is rather slender, and from one to two feet in height. Its deep-laid, extensively-creeping and sprouting rootstock make it one of the very worst of weeds. Both of these plants are constantly invading new territory, especially such as is continually cropped in wheat. The areas infested become useless for growing small grain. Intensive cultivation with a certain amount of hand work is the best remedy.

General Remedies.—While experiments in killing certain weeds by spraying with chemical solutions have been made, the most practical method is to prevent as far as possible the spread of their seeds. This is accomplished by sowing clean seed wheat, and by killing the weeds when they do succeed in starting. Conditions most favorable to the growth of wheat place the weeds at the greatest disadvantage. Intensive farming always finds effectual methods of dealing with them.

Losses.—The wheat growers lose millions of dollars annually on account of weeds. They are the largest factor in elevator dockage. In North Dakota[1] 47 samples of wheat were found to have an average real dockage of .64 pounds per bushel. This varies each year, and the average dockage for the 1906 crop of wheat was about 2 pounds in the Minneapolis market. If 500,-000,000 bushels of wheat are grown in the United States annually, and the dockage averages only half of one pound per bushel, then over 4,000,000 bushels are not only waste, but they also injure the commercial value of the real wheat. Where large quantities of wheat are cleaned, the cost is about one-fourth of a cent per bushel.

Birds.—In the United States 29 species and subspecies represent the family Icteridæ, a group of birds including those commonly known as bobolinks, meadowlarks, orioles, blackbirds, grackles and cowbirds. Rating the blackbirds in the order of their grain-eating propensities revealed by stomach examinations, and putting first those that eat least, the list reads: Bobolink, redwing, cowbird, rusty blackbird, yellowhead, crow blackbird, boat-tailed grackle, Brewer's blackbird and California redwing. Since the first two are the ones most complained of, the amount of grain actually eaten would not seem to be the

[1] N. D. An. Report (1904), p. 45.

only factor to be considered in determining the relative harmfulness of the species. The dove, the sparrow and the crow also eat grain from the fields, as do many other species. None of these birds are, however, entirely harmful; indeed the good that many of them do by destroying injurious insects and seeds more than counterbalances the damage occasioned by the eating of grain. At least 50 different birds act as weed destroyers and help to eradicate nearly 100 species of noxious plants. The number of weed seeds eaten is enormous, one bird eating a thousand seeds of some kinds for a single breakfast. The insects eaten by these birds are also generally noxious. There is usually an equilibrium of organisms in nature, and birds become harmful only when they disturb this proper balance by increasing out of proportion to their environment.

"**Yellow Berry.**"—This occurs in hard winter wheat. Some of the wheat berries are often lighter in color and weight than the hard red ones, and also have a lower gluten content. Overripeness and failure to stack the sheaves have been given as causes, but opinions seem to differ as to this. It is claimed that the annual loss in Nebraska is from one-half to one million dollars.[1]

FUNGOUS ENEMIES

"**Glume Spot.**"—This fungus receives its name from the dark spots that it causes upon the glumes of wheat. It has been studied but little, and no remedy is known.

Wheat Scab (*Fusarium culmorum*) also attacks the glumes and causes lead-brown-colored sections in the spike, or even destroys the spike entirely. The loss is usually light, but may reach a maximum of about 15 per cent. The only remedy suggested is the burning of the stubble.

Smut.—Two kinds of smut attack wheat, stinking smut, or bunt (*Tilletia tritici* (Bjerk.) Wint.), and loose smut (*Ustilago tritici* (Pers.) Jens.). The enormous damage resulting from this disease attracted attention in ancient Greece and Rome. Hartlib called attention to the fact that smutty seed produces smutty grain, and he was perhaps first to record a remedy (1655). His three remedies for smut in wheat were liming the

[1] Neb. Bul. 89 (1905), p. 50.

THREE THRESHING OUTFITS IN OPERATION ON AN OKLAHOMA RANCH

field, liming the seed, and soaking the seed over night in common salt lye. Though the efficacy of the remedies was doubted, there must have been some beneficial results, for it is claimed that the seed was invariably steeped. By the middle of the nineteenth century there were many methods of steeping, liming and brining wheat to prevent smut. It was found that the use of a solution of arsenic gave a clean crop from smutty seed.

Just how fully the nature of smut was understood by these early writers does not seem to be clear, but they must have had considerable knowledge of the disease in order to pursue such correct principles in endeavoring to effect a cure. The black dust frequently filling the kernels of ripening wheat consists of thousands of germs of the parasitic smut fungus. These germs, or spores, have a great capacity for spreading over the fields, but the only real danger seems to be from those spores which lodge on healthy kernels, generally in the hairy ends. Chances are slight of clean seed being infected by being sown on ground containing smut spores. Spores have the same function as seeds of higher plants, and, the infected kernels of wheat being sown, the spores germinate at the same time as the wheat. The slender filaments penetrate the tissues of the wheat plant before the first leaf is put forth. From this point its growth is within the wheat plant, both plants growing together. It seems to be still undecided how these threads proceed below the upper two joints of the mature wheat plant, whether they pass upward through the pithy region of the stalks, or whether they follow the surface tissues, but it is probable that the method of smut growth is uniform throughout the entire plant. The fungus seems to die as it passes upward, and leaves few traces of its path.

In the mature wheat plant smut seems to be found only in the chlorophyl bearing parenchymatic tissues. Nearly 30 rows of breathing pores in the skin covering of the straw run lengthwise with the stalk. Under these rows of pores are layers of succulent cells which produce the food eventually used in forming the wheat grains. The smut filaments remain close to the open pores, absorbing and taking the place of this cellular structure with its chlorophyl and protoplasm. No other cellular tissues are disturbed, and until the heads develop the presence of the smut can scarcely be detected without the aid of a microscope. A mass of smut threads then absorbs all the nourishment, fills the flower or grain, and soon converts it into a mass of spores. As the parasite lives at the expense of its host, the latter is weakened and stunted in proportion to the amount of smut. This may be great enough to dwarf the plants so as to prevent the formation of heads, or even to cause the stalks to die back to the ground, or it may be so little that the heads are never reached, simply the straw being infected. Much of the straw may be thus infected, greatly reducing the yield, even though apparently uninjured heads are formed. Smut filaments have also been found in grains which had formed starch.[1] In general, smut and wheat seem to de-

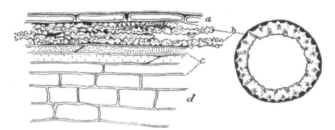

SECTIONS OF SMUTTED WHEAT STRAW

At the left is a longitudinal section of a wheat straw and at the right a cross section. a, epidermal cells; b, smut filaments; c, fibrous cells; d, internal tissues.

mand about the same meteorological conditions for their best growth. Smut will successfully pass the winter, even upon the open ground in North Dakota. Germs two years old have not lost their power of producing smut in a crop.[2]

Grains of wheat affected by stinking smut are slightly larger and more irregular than healthy ones. Such kernels, the so-called "smut-balls," are easily broken open, and the dark-brown powder with which they are filled has a very disagreeable and penetrating odor that pervades the whole bin of wheat, even if only a small per cent of the kernels are smutted. On this account they differ from all other grain smuts in that their presence can be easily recognized. Sometimes

[1] Bolley, Proc. Tri-State Grain Growers' Ass'n., 1900. p. 86.
[2] Rept N. D. Agr. Exp Sta., 1901, p. 34.

STINKING SMUT

50 to 75 per cent of the heads in a field are smutted, and the remainder of the grain is so contaminated by the fetid spores as to be of very little value as flour and worse than useless for seed. Unchecked, these smuts increase from year to year, and they occur more or less abundantly in all wheat raising countries.

Loose Smut of Wheat differs from the stinking smut in these respects: At germination its spores develop a chain of cells instead of an undivided tube; it has no fetid odor; it attacks both kernel and chaff; and it ripens when the healthy wheat is in flour. By harvest time the spores have all taken wings upon the wind, leaving a naked stalk in place of the head. It is known to occur in Europe, North America, northern Africa, central Asia and the West Indies. There are many localities in the United States where it is rare or entirely absent. A loss of 10 per cent or more of the crop is often occasioned by loose smut, and even as high as 50 per cent, but usually it is not as destructive as the stinking smuts. It seems to be more difficult to prevent, however, so that when once introduced into a field, it is more apt to remain.

Remedies.—Any means that destroy the vitality of the smut spores adhering to seed wheat but leave the latter unimpaired in its power of germinating are a safe preventive of smutted wheat. There are several ways of accomplishing this easily and perfectly with the stinking smut. In all treatments by immersion in solution, the seed should first be stirred in water in order to skim off the smut balls. Slaked lime will hasten drying, but is not essential. If the seed is sown without drying, the drill must be set accordingly.

LOOSE SMUT

CORROSIVE SUBLIMATE (*Mercuric Chloride, Hg* Cl^2.)—This may be used at the rate of 1 lb. to 50 gallons of water ($2\frac{1}{2}$ parts to 1,000). The wheat is piled upon a floor or canvas, and constantly shoveled while it is being sprayed or sprinkled, until every grain is wet over its entire surface. The use of more of the solution than is necessary to do this is injurious. The seed should then be dried.

COPPER SULPHATE (*Cu* SO^4).—One pound of crystallized (not powdered) commercial copper sulphate or bluestone is used to every 25 gallons of water. The grain is soaked 12 hours in this solution, being stirred occasionally. Then, to avoid injuring the power of germination, it is immersed for a few minutes in limewater made by adding ten gallons of water to one pound of good slaked lime.

FORMALIN.—One pound of formalin (the trade name for a 40 per cent solution of formaldehyde) is diluted with 50 gallons of water. The grain is treated as in using corrosive sublimate. Each bushel of grain requires about one gallon of the solution. The grain is left in a pile for 2 or 3 hours, and is then spread out to dry. This method is not successful with formalin that is not a 40 per cent solution. Formalin rapidly loses its strength unless kept tightly corked, and careless or unscrupulous dealers sell a solution that is too dilute, or under weight on account of the bottles in which it is sold being below standard size. Formaldehyde vapor has also been found effective in destroying stinking smut.

HOT WATER OR JENSEN TREATMENT.—It is claimed that this process was discovered by J. L. Jensen of Denmark, in 1887. Hot water and quicklime were used several years before this date. In this method the seed is placed loosely in a coarsely woven gunny sack or wire-covered basket, and then dipped in water having a temperature between 132 and 133° F. The volume of water must be 6 or 8 times that of the seed treated at any one time. Lifting out and draining the grain 4 or 5 times during the treatment insures its coming in contact with water at the proper temperature. The treatment requires 10 minutes. The grain should then be dried at once, or dipped in cold water and set aside until it can be dried.

A MODIFIED HOT WATER METHOD is used in treating for loose smut, for this is not destroyed by any of the cures mentioned above. The grain is first soaked four hours in cold water, and set away in wet sacks for four hours more. It is then immersed for five minutes in water at 132° F. Some of the seed is killed by the treatment, and one-half more must be sown per acre. It has been claimed, however, that no sure method of destroying loose smut is known, and that the only available relief at present is to obtain clean seed from a smut-free district.[1]

Results and Expenses.—These remedies seem to be entirely efficacious, and if properly and universally applied, there is no reason why smut should not be practically eradicated from the wheat fields. Before the nature of smut was fully understood, one of the unexpected results of treating it was an increase in yield greater than the result of merely replacing the smutted heads with sound ones. This was explained when it was learned that smut was often present in the straw even though it did not reach the heads. Usually the increase in yield is two or three times as great as the visible smut, but may be six or more times as great. The methods of treatment are inexpensive. In the hot water method the cost is practically only the labor required. In some of the other methods the cost of chemicals is little more than would pay for the labor in the hot water treatment. Liquid formaldehyde, which is used quite extensively in the Northwest, is found very effective, and costs only from three-fourths of one cent to two cents per acre. The treatment by sprinkling and shoveling is cheaper than dipping.

Losses continue in spite of the fact that it has often been demonstrated that smuts are controllable. During the year 1902 wheat smut caused a loss of 2.5 million dollars in the state of Washington alone. In the following year smut destroyed from 10 to 50 per cent of the wheat in parts of Wisconsin. At Winnipeg 3 to 6 per cent of the wheat offered for sale during 1904-5 was rejected on account of smut, and in 1905 as high as 75 per cent of the wheat was destroyed by smut in parts of North Dakota. While seed wheat is very commonly treated for smut, these losses show that there is still need for more educational work. It is necessary to demonstrate repeatedly the efficacy of the treatments in order to secure their adoption by the conservative farming element. This element continues

[1] Freeman, **Minn. Plant Diseases** (1905), p. 297.

to grow smutted wheat because it has always done so, and because the extent of the loss and the ease of its prevention are so little realized. Formerly at least one-fifth of the cereal crops was annually destroyed by smut. Smutty wheat introduces a large element of speculation into the business of elevator men, for it produces a very low grade of flour. Such wheat must be washed, an expensive process which also endangers the quality of the flour. From smutted flour the baker gets a poor, darkened product that finds little market. As a result smutted wheat is justly thrown into a very low, or "rejected" grade.

AECIDIA ON BARBERRY

Rust.—What is popularly termed wheat rust may be the result of one or more of a number of rust fungi, parasitic plants. This disease was mentioned by Virgil. It was known in Britain before 1592. Fontana (1767) is generally accredited with connecting rust of cereals with a specific fungus, which Persoon (1797) investigated more fully, and named. Three kinds of rusts are known to attack wheat. *Puccinia coronata*, the crown rust, is comparatively unimportant. The two distinctive rusts are *Puccinia rubigo-vera*, the early, orange leaf-rust, and *P. graminis*, the late, stem-rust. The former is popularly called "red" and the latter "black" rust. Both species, however, produce first reddish and then black spores, but in the orange leaf-rust the red spores are far more abundant than the black ones, while in the stem-rust the black spores are the more abundant.

Life History.—The wheat rusts belong to that type of fungi which have several stages of development represented by different types of spore formation and separated by two or more rest periods. The life history is the development of the fungus through all its stages, and it is said to be "known" when the experimenter can take one type of spore formation and from this produce artificially all of the other types in turn through the life cycle until a return is made to the type of spore formation with which he began. Usually the types of spores characteristic of the different stages in the life cycle of such a fungus are so different in form and character and so divergent in their modes of development and subsequent habits of growth as to mislead the investigator completely and excite no idea of relationship. Several entirely different types of hosts are frequently utilized in the life cycle. It is because of their complex life history that rusts have so long been shrouded in mystery and confusion, and inadequately understood.[1]

The wheat rusts produce in order, from spring to spring, four different forms of spores. (1) Aecidiospores (injuring spores) are the first spores found in the year. They occur on shrubs, or herbs other than grasses. (2) Uredospores (blight spores) appear in the early summer, and are often called summer spores.

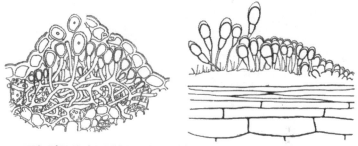

TWO FORMS OF RUST SPORES COMMONLY FOUND ON WHEAT

At the left is shown the uredospores of the red rust, commonly found in early summer. At the right, the two-celled teleutospores or winter spores of the black rust.

These are the red spores that rust the leaf of the wheat. (3) Teleutospores (completion spores) are the last ones of the season. They are also known as resting or winter spores. Their dark color gave rise to the term "black rust." (4) Sporidia are very minute and delicate spore bodies formed in the spring on the germination tubes of the winter spores. The sporidia infect the plant that is host to the æcidium stage. The question of a breeding act in the rust life cycle is still an unsettled one.

When the æcidiospore lodges upon the wheat leaf or stalk in the spring, it remains in a resting condition until a light

[1] The works of Bolley, Carleton and Freeman will be found most useful in a study of rusts.

rain or dew furnishes sufficient moisture for germination. A small thread or filament is then sent out, which requires but an hour or two to pass through a breathing pore of the wheat plant, or, in the absence of a convenient breathing pore, to bore its way into the stem or leaf, within which a mycelium is formed. An ordinary dewdrop may contain hundreds of æcidiospores that have been wafted to it upon the air. The time required for the rust to break out as a spot or pustule, after the germination of the infecting spore, varies from 8 to 14 days, for it is dependent upon atmospheric conditions. This

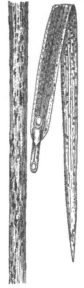

breaking out through the skin of the wheat plant is the result of the great numbers of ovoid spores that are formed, and these red summer spores, the uredospores, are thus enabled to drop off and float away upon the air to other wheat plants. If moisture is present they germinate at once, and the entire above process is repeated. Several generations of the red spores may be formed during one growing season. Countless myriads of spores are thus produced, a pustule 1-16 by 1-64 of an inch in size containing over 3,000 such spores. Under favorable conditions the rustiness of the grain increases with marvelous rapidity. In the meantime the spore beds which produced the first red spores are not inactive, but are producing teleutospores, that is, the black winter or resting spores. These are thick-coated, Indian-club shaped, and two-celled. They may now also appear in new spore beds in which no red spores

BLACK AND RED RUST

have first formed. Over 2,000 of these spores have been counted in a pustule 1-16 by 1-48 of an inch in size. No resting spores have been observed to germinate until late in the following winter. When they sprout a germ tube (pro-mycelium) proceeds from each of their two cells. These germ tubes soon divide into four cells which immediately produce severel minute, delicate cells known as sporidia. If sufficient moisture is present, the sporidia will germinate at once. If not, many of them may dry sufficiently to be carried by the wind, yet not enough to be injured. In the early spring months the damp straw and the ground and surface waters of an old wheat field may be swarming with countless millions of these sporidia, and clouds of them are wafted by the winds to distant points. No evidence has yet been secured, however, that the sporidia directly infect the wheat plant. Those of the stem rust infect the fruit and leaves of the barberry bush, and, as far as known, no other plants. Little cups and clusters of cups of yellow spores are formed. The floors of the cups start palisade-like chains of the spores, which, when mature, again take wings with the wind. These are the wheat infecting æcidiospores with which we began, and the life cycle is thus complete.

At the present writing the life history of all wheat rusts is not perfectly and certainly known in all its phases, but it is quite conclusively known that in some cases certain stages of the life cycle which is given above (and which describes the stem rust more correctly and fully than any other) are not essential. The cluster cup form of rust, found on barberry bushes in black stem rust, forms on common wild plants of the borage family in the case of orange leaf rust. In Europe it forms on hound's-tongue, but this stage of the rust seems to be absent in the United States. In crown rust of wheat the cluster cup stage commonly forms on the buckthorn.

It is now established that the uredospores of a number of the important rusts, including *Puccinia graminis,* can pass the northern winter in viable form. Dried and scattered by the August and July winds, a very large per cent of these rust spores germinated after a dry fall and a North Dakota winter.[1] In the warmer climates the leaf rust not only survives the winter in the red spore stage, but forms new pustules every month of the year. Viable spores of both rusts successfully pass the winter frozen in snow and ice. The very early general infection by rust can hardly be explained by the wintering uredospores, however. Experiments by Bolley show quite conclusively that the infection comes by way of the air and not by way of the soil. It is thought possible that rust filaments passing the winter in green plants, and also broken particles of the mycelium from the crop of the previous year, may aid in

[1] Bolley, Science, N. S. 22:50-1.

the infection. While it has been generally held that seed from rusted wheat will not transmit rust to succeeding crops,[1] the observations of Bolley in 1904 and 1905 proved that seed from badly rusted wheat plants was quite uniformly infected internally, there being spore beds beneath the bran layer, containing both uredospores and teleutospores that subsequently germinated.[2] This demands a new line of investigation, for it has not yet been demonstrated whether or not the internally infected seeds will transmit the infection to the plants grown from them. Variations in the spore forms and in the complicated life cycle of rusts give them great strength in self-perpetuation, the different methods of which present several chances of escaping threatening destructions. Many wild grasses also serve as hosts for the wheat rusts. They may be infected from wheat and wheat from them, which is another resource that aids rust in maintaining itself.

Distribution of Rusts.—Rusts being true parasites able to live only in the tissue of some host, their distribution is co-extensive with that of their native hosts and that of the wheat crop, with one single restriction. Only in countries where no dews set and no rains fall are rusts absent, for moisture is essential to their first growth which causes the infection of the host plant. In irrigated regions where dew and rain are lacking, wheat grows without being rusted. The leaf rust is most regular in its occurrence and is also most widely and universally distributed. It is the most common rust of Australia and India, and in the United States it is most abundant in the Atlantic and southern states. The stem rust is irregular in its occurrence, usually missing one or two years in five or six, especially in some localities. In the United States the severe attacks occur most frequently in the central states, and in parts of Texas and California. This rust is very common in northern Europe, and in some seasons it is also quite abundant in Australia and Tasmania. It seems to be comparatively unimportant in India.[3]

Conditions Favorable to Rust Development.—There must first be such a wind from infected districts as will bring plenty of

---

[1] U. S. Dept. Agr., Farm. Bul. 219 (1905), p. 8; Minn. Press Bul. 24.
[2] N. D. Bul. 68 (1906), p. 646.
[3] Carleton, Cereal rusts of U. S., pp. 21-22, 56-57.

rust spores to the heading wheat. That the spores are thus brought to the wheat is shown by the fact that screened plants are not rusted, and that distilled water exposed to the air will gather great numbers of the spores in the short period of a half hour. The more soft and succulent the wheat straw is, the more open it is to rust infection. "The most effective rust infection weather may be described as muggy, showery, sultry, rather still hot days, with foggy, cool, dewy nights, at about the blossom period. Just following the infection, cool, moist, slow growing, showery weather may result in the most general rust infection, and in the greatest breaking out or rupturing of the straw."[1]

How Damage Results.—Rust deprives the wheat grains of their nourishment. The grains may be only slightly shriveled, or the crop may be completely ruined. It has been claimed that orange leaf rust does little damage to wheat, and that in very wet seasons it may even be of benefit to the grain by preventing superabundant growth of the vegetative parts. This has been denied recently, however, and with good show of reason. It is pointed out that in 1904 and 1905 the leaf rust was so severe that the wheat grains wilted and shriveled before the stem rust was well developed on the straw. The leaf rust may also delay the ripening of the crop until injury from frost results.[2] Since the attack of the stem rust is the more direct, it is unquestionably the more virulent.

The Loss From Wheat Rust in the United States doubtless exceeds that caused by any other fungous or insect pest, and it may be greater than the loss from all other diseases combined. It is often not noticed because it is light. In one or another of the wheat growing sections, great areas are partially to nearly completely destroyed each year. While almost fabulous figures are required to record the estimated annual loss, the probabilities are that this is underestimated, for the slight, unnoticed attacks never enter the computation. If the loss is but 1 per cent of the wheat crop, it approximates $5,000,000 annually in the United States alone. Bolley examined the wheat fields of North Dakota for fourteen seasons, and, leaving out of consideration the years of great destruction, he estimated the

[1] N. D. Bul. 68 (1906), p. 655.
[2] N. D. Bul. 68 (1906), pp. 651-654.

average annual loss at 10 per cent. In 1903 the loss in southern Wisconsin was 50 per cent and in South Carolina 30 per cent. In 1904 the wheat crop of Minnesota and the Dakotas was most promising, but in a few days it was so damaged by rust that experts estimated the loss for the three states at 30,000,000 bushels, and the wheat that was produced in many instances weighed only from 36 to 48 pounds per measured bushel. Many fields were not harvested. Twenty million dollars seems a reasonable estimate for the average yearly loss from wheat rust in the United States. Immense annual losses are suffered by the Australian, Russian and Argentine wheat fields. In some years the loss in England is 50 per cent. Nor is this the only form of loss that must be attributed to rust, for it is one of the chief hindrances that entirely prevent the growing of wheat in parts of certain moist warm countries, such as China and Japan.

Remedies.—Thus far rust has baffled every attempt at a remedy. Fungicides and spraying have been experimented with, but indirect methods are the only ones that have proved of any aid in the combat. All conditions are helpful that tend to mature the wheat crop before the rust becomes abundant. Fields should be properly drained. Good clean seed of a pure variety and of the best germinating powers should be sown in soil properly prepared. The seeding should be early, and the crop should be kept free from smut and weeds. All of these things strengthen the wheat plant and hasten its growth. Rotation of crops is also advantageous. Wild grasses and weeds of the roadsides should be mown, and all barberry shrubs should be killed. The fields should be kept free from volunteer grains. The line of demarcation between the winter and spring wheat belts should be sharply drawn, for the winter wheat, ripening early, develops rust in such abundance that it will greatly injure the later spring wheat. The early maturity of winter varieties generally enables them to escape serious damage.

One of the most hopeful phases of the question is that some varieties of wheat are quite rust resistant. The different rusts are each more easily resisted by certain varieties of wheat. Thousands of varieties have been tested and bred to secure rust resistance. None are absolutely rust proof. "So far as the ordinary wheats are concerned, the resistant varieties are.

as a rule, somewhat dwarfed, are close and compact, and stool but little. The leaves, comparatively few in number, are stiff, narrow and erect, with a more or less tough, dry cuticle, often with a glaucous or waxy surface; heads compact and narrow; and grains hard, red, small and heavy."[1]

Varieties likely to prove considerably resistant to rust in the United States, if they are sown early, are, Kharkof, Turkey, Mennonite, Pringles No. 5, Rieti, Odessa and Pringle's Defiance for winter wheats, and Haynes Blue Stem and Saskatchewan Fife for spring wheats.

Durum wheats are much more resistant than other varieties. During the great rust attack of 1904 in the northwest, the maximum loss for durum wheat seems to have been about 10 per cent while that of ordinary wheats was frequently as great as 50 per cent. The different varieties of durum wheat also vary in their power to resist rust, two of the best being Iumillo and Velvet Don. "Rerraf" is one of the best rust resisters in Australia, but is quite non-resistant in the United States.

Other Diseases.—Leaf blight (*Septosphaeria tritici* Pass.) and Powdery mildew (*Erysiphe graminis* D. C.) occasionally cause slight losses in certain sections.

[1] Carleton, Cereal Rusts of U. S., p. 21.

CHAPTER X.

INSECT ENEMIES OF WHEAT

Species.—It has been estimated that there exist 1,000,000 species of insects of economic importance. About 100 species feed upon growing wheat, and about 50 more are found in granaries. Less than a dozen occasion enough loss to wheat to be of very great importance. Conditions in the United States are most favorable for insects, because the continuous growing of the same grain crops over wide areas, and long, hot summers are very propitious for the multiplication of most species. On account of differences in climatic conditions and in the abundance of parasitic and other enemies, there is a periodicity in the recurrence of grain pests. Since a season favorable to one insect may be unfavorable to another, there is also a more or less marked rotation of different species.

Hessian Fly (*Mayetiola destructor* Say).—Wheat is the natural food plant of this insect, which is also supposed to be native to Asia. It was introduced into America from Europe. The Revolutionary patriots believed that it was contained in some straw brought over by the Hessian troops, hence its name. Some of the ignorant Tory element claimed that General Washington was responsible for its introduction. It was first described technically in 1817.

DISTRIBUTION.—The natural spread of the Hessian fly has been estimated at 20 miles per year. It is now found in nearly all parts of the United States east of the 100th meridian, and on the Pacific coast (since 1884) probably from southern California to British Columbia. In Canada it has been found from Prince Edward Island to Indian Head, Saskatchewan. It also occurs in North Africa, western Asia, Europe and the British Islands. It has been an important grain pest in New Zealand since 1888.

DESCRIPTION AND LIFE HISTORY.—The adult Hessian fly is very fragile, dark-colored, and about ⅛ inch long. It is about half as large as the mosquito, which it resembles. Even when comparatively abundant it will escape the notice of the ordinary observer. It can be caught with a sweeping-net, but is easily confused with other insects taken at the same time. The fly seems to be two brooded in all parts of the United States. In the north the broods follow each other in quick succession, while in the south they are widely separated. The egg of the insect is about 1-50 inch long. The newly hatched larva or maggot is slightly smaller than the egg. The fully developed larva is larger, and on account of its resemblance to a seed of flax it is known as the flaxseed. In fall wheat the fly passes the winter in the young plants, principally in the flaxseed stage, but also in the larval stage, not quite full grown. The flies

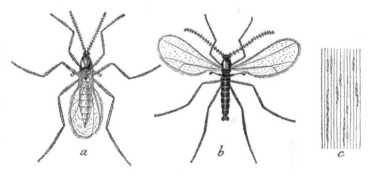

HESSIAN FLY: *a*, FEMALE; *b*, MALE; *c*, EGGS.   ENLARGED

emerge from the flaxseeds when the wheat is about 2 inches high. The time varies from March in Georgia to May in Michigan. Flies from wintering larvæ appear later. The eggs are deposited in the grooves on the upper surface of the wheat leaves, from 100 to 300 by each female fly. They are difficult of perception, even by one who has good eyesight. In a few days the eggs hatch into a pinkish larvæ that soon turn greenish, and descend to just above the roots, or, if the wheat has jointed, to the base of the particular leaves on which they were hatched. Sucking the juices from the growing wheat plant, these larvæ attain the flaxseed stage in about 4 weeks, the time being dependent on the weather. The prolonged southern summer during which there is little food for the larvæ, is passed in the stubble in the flaxseed stage. In Michigan the fall brood appears about the last of August, while in Georgia it appears about 3 months later. The eggs are now deposited on the young fall wheat, and the life cycle begins over again. In regions far north there may be only one brood, and in the south there may be supplemental broods, both in the spring and fall, this being dependent on the weather. Drought prolongs the flaxseed stage.

In the spring wheat regions the insects winter in the flaxseed stage, chiefly in stubble, but also in volunteer wheat. Egg laying begins late in May and continues to October 1st. Eggs are often deposited on grass and weeds, but the larvæ are not known to survive except on wheat, barley and rye. The fly

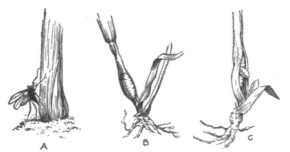

HESSIAN FLY: *a*, ADULT; *b*, PUPA; *c*, LARVA. ENLARGED

is now known to flourish even where spring crops are exclusively grown.

EFFECT OF LARVAE ON WHEAT.—At first the plant seems to be stimulated, and turns a dark green color. Later the infested tillers turn a brownish and then a yellowish color. If the attack comes early, and the plant fails to tiller, death results. If the plant has tillered, some stalks may escape and form the basis for a crop. The larvæ are usually found just above the first joint, but may be found from above the third joint to below the soil. The stalk is usually so weakened that it breaks to the ground, when the wheat is said to be "straw fallen."

LOSSES.—The Hessian fly is the worst insect enemy of growing wheat. It is never entirely absent. The minimum annual damage to wheat is thought to average 10 per cent of the crop, that is, over 50,000,000 bushels. In some localities an injury varying from 50 per cent to total failure is not infrequent. In 1901 the loss in New York was about $3,000,000, and the loss in Ontario was nearly as great. In 1900 it was $16,800,000 in Ohio, and nearly two-thirds of the Indiana wheat was not harvested on account of the fly. The outbreak of the Hessian fly in 1900 was the most notable of recent years. The total loss for the United States was estimated at $100,000,000, and milling operations were seriously hampered in the worst affected region. The damage which the fly does is often laid to rust, drought or other causes. In 1904 there was little complaint of damage from the insect, yet many fields in the Ohio valley were injured to the extent of over 50 per cent.

REMEDIES.—There are a number of natural enemies which attack the Hessian fly in the larval and pupal stages. Some are native, and others are being artifically introduced. While they limit the damage, they are useful mainly where other preventives are neglected. The best remedy for a field of wheat severely attacked is to plow deeply, and plant a spring crop. In case of mild infection, the prompt use of fertilizer may increase the tillering of the wheat so as to produce a partial crop. If the crop has a good growth pasturing or cutting in the fall may be beneficial. When injuries from the fly may be anticipated, moderately late planting of winter wheat is perhaps the best preventive. Seeding for this purpose should be about the middle of September in the northern districts, during the first half of October in Kentucky, and during the first half of November in the extreme south. The rotation of crops should be practiced. Burning or plowing under the stubble is of great advantage. The fly can be starved out almost completely over a district of any size by abandoning for one year the culture of wheat, rye and barley. Volunteer grains should also be destroyed. Early plantings of trap or decoy crops will attract the flies, and, after ovipositing, these crops may be plowed under deeply. While no varieties of wheat are absolutely ''fly proof,'' some tiller more and are less injured than others, such as Underhill, Mediterranean, Red Cap, Red May and Clawson. Preventive measures reduce the annual loss from the Hessian fly by an amount estimated from $100,000,000 to $200,000,000 for the wheat crop alone. The insect shifts so rapidly from place to place that remedies are practically of no avail unless there is concerted action in an infected region.[1]

Chinch Bug (*Blissus leucopterus* Say).—This is a native insect. Its ravages were first noticed toward the close of the eighteenth century, and since that time notable outbreaks and serious losses have been quite constant. It is now found from Nova Scotia and Manitoba southward to the Gulf of Mexico, as well as on the Pacific coast, in Mexico and Central America, and on several of the West Indian Islands. The genus *Blissus* is widely distributed over the Old World. It is a gregarious

CHINCH BUG: VARIOUS STAGES FROM EGG TO ADULT ENLARGED

pest, and its destructiveness is due to this fact rather than to its enormous numbers.

LIFE HISTORY.—Hibernating in grass stools, straw, rubbish or other shelters, the chinch bug begins its life cycle by a spring flight to the wheat fields. The mating occurs at the wheat roots. The eggs are deposited about May 1st, from 100 to 500 by each female, and the egg period is of 2 or 3 weeks' duration. The young hatch in about 2 weeks, and at maturity in July they make a second flight to late corn, millet or other crops. In this country, except in northern regions, a second brood appears after this flight. The second brood is most injurious in August and matures in September and October. It is the first brood that injures wheat, while both broods attack other crops. A short-winged form incapable of flight frequently occurs, especially in maritime districts. There are a number of species of *Hemiptera* that are often mistaken for chinch bugs.

[1] Marlett, Principal Insect Enemies of Growing Wheat; Osborn, Hessian Fly in the United States.

THE CHIEF LOSSES are occasioned in the Ohio and upper Mississippi valleys, and on the Atlantic coast highland. It does more damage to wheat than to any other crop, and the average annual loss is about 5 per cent of the crop.[1] In years when the chinch bugs were unusually severe, the damage to wheat in single states has been estimated to be from ten to twenty million dollars. The losses are great because of the wide distribution of the pest, its prevalence to some extent every year, and its enormous multiplication in favorable seasons.

REMEDIES.—(1) Burning over the land; especially should this be done on waste and grass lands, and all rubbish should be burned. Grass is not injured by being burned over after the ground is frozen. It has been thought that the chinch bug was kept in check by the annual prairie fires in the early years of our country, the hibernating bugs being thus killed. Chinch bugs and other insects injurious to growing grain are practically unknown on the Pacific coast, where the large wheat fields are regularly burned over every year by burning the straw. (2) Trap or decoy crops, such as millet or Hungarian grass; these should be plowed under. When the young insects hatch, they easily reach the surface, but will perish if no crops are near. (3) Rotation; this involves a system disassociating small grains from corn. (4) Plowing; deeply plowing under the bugs collected on the edge of a field is helpful. (5) Spraying; the edge of the field infested may be sprayed with a very strong oily insecticide, even if the crop is killed with the bugs. (6) Protecting furrows. (7) Coal-tar barriers. (8) Artificial spreading of parasitic fungi; considerable work has been done in this line, with the conclusion that it is of little value. The bug is practically exterminated for the season, however, by wet weather and various fungous diseases which this causes. (9) Many bugs are also destroyed by birds, especially quails.[2]

The Wheat Midge (*Diplosis tritici* Kirby) belongs to the same order of insects as the Hessian fly, but in appearance and habit it is entirely distinct. It is believed to be identical with the notorious wheat midge of Europe, and it may also have been in-

[1] Yearbook U. S. Dept. Agr., 1904, p. 466.
[2] Webster, The Chinch Bug; Howard, The Chinch Bug.

troduced into America in straw. It probably appeared first in Quebec, and has now spread throughout the Mississippi valley. The injury is inflicted by its orange-yellow larvæ which extract the milky juice from the embryos forming in the wheat heads, thus causing the grain to shrivel and the heads to blight. In cases of unusual outbreaks the average losses of whole states have been from two-thirds to three-fourths of the entire yield. The wheat midge oviposits directly in the wheat head. The eggs hatch in about a week, and the larvæ enter the kernel at once. They have extraordinary vitality, thrive best in moist weather, and winter in the ground, which they enter about three weeks after hatching. Plowing old wheat

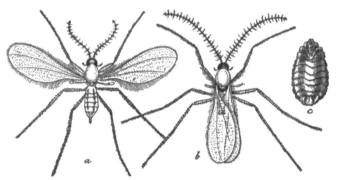

WHEAT MIDGE: *a*, FEMALE; *b*, MALE; *c*, LARVA. ENLARGED

fields deeply, burning the chaff and screenings of wheat from infested fields, and rotating crops are preventives.

The Wheat Plant Lice cause injury by sucking their food from the soft, forming kernels. The yield may be reduced by as much as one-half, but extensive damage rarely occurs. The annual loss is thought to be at least 2 per cent.

Locusts or Grasshoppers.—The locust, formerly present in some years in such overwhelming numbers that large swarms devastated extensive areas of all vegetation, has during the last decade ceased to be of such great economic importance. Locust plagues seem to occur occasionally on all of the continents, and do not seem to be limited to comparatively newly settled regions. From 1889 to 1897 they wrought frightful havoc in Argentina, visiting 347,000,000 acres in the latter year, and destroying 30 per cent of the crops. From 1897 to 1900 the Argentine government spent over $7,000,000 in an attempt to exterminate them. The limit of the invaded region was steadily pushed northward, until in 1901 locusts were entirely absent from the wheat area. They came into Argentina from Bolivia, the territory of the Chaco, and western Brazil. Barcelona, Spain, reported a plague of locusts spreading in 1902. In west central Asia, between Askabad and Krasnovodsk, the cereal and cotton crops are commonly devastated by locusts. In 1903, 50,000 roubles were set aside to be devoted to the destruction of the insects' eggs in trans-Caspia. It is claimed that sacked flour piled on open railway trucks near Krasnovodsk was devoured by clouds of rapacious locusts in an incredibly short time.[1]

In the United States during the early seventies the grass-

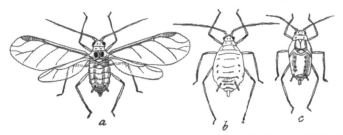

WHEAT PLANT-LOUSE: *a*, WINGED ADULT; *b*, FEMALE; *c*, NYMPH.
ENLARGED

hoppers used to invade Kansas "so they would block railroad trains and destroy all vegetation."[2] In the Red river valley they appeared in great clouds which "cleaned the country quite thoroughly on their flight."[3] These invasions seem to have come mainly from the permanent breeding grounds of the Rocky Mountain locust (*Caloptenus spretus* Uhler). These grounds were located approximately between the meridians of 102 and 112 degrees, and between the 40th and 55th parallels.

[1] Mo. Sum. Commerce and Finance, Feb., 1904, p. 2818.
[2] Industrial Commission, 10:759.
[3] Proc. Tri-State Grain Growers' Ass'n. 1900, p. 184.

East of this territory was a frequently invaded strip about five degrees in width. A great scope of territory farther east, south and west was periodically visited when the natural conditions on the permanent breeding grounds were such as to produce myriads of grasshoppers. They could live only one generation on the lower lands, and then perished. Large portions of their grounds are now cultivated, and this restricts their multiplication. It is thus perhaps impossible for such overwhelming swarms to occur as formerly. Such swarms as do occasionally appear are more localized, and not of such uncontrollable magnitude. They may still be relatively abundant, however. During 1901 in Canada, several hundred insects could be seen "to the yard," and "dead locusts could be gathered up in wagon loads and at times be smelt for half a mile," after poison had been used. In Montana they frequently devastate ranges so that the herds must seek pasture elsewhere.

The Rocky Mountain Locust lays its eggs in almost any kind of soil, preferably in bare, sandy places on high and dry ground. They are laid chiefly in the first inch of soil, and in masses or pods surrounded by a mucous fluid, each pod containing about 30 eggs. The average laying season extends over 6 to 10 weeks, and about 3 egg masses are formed by each female. The time of hatching depends entirely on the climate and latitude. While the young locust is very active, it will remain almost stationary if food is plenty. The migrating propensity is developed only after the first molt, and frequently not until after the second or third. When food becomes scarce the locusts migrate, often in a body a mile wide. From the very first they congregate and display gregarious instincts. They feed as they advance, devouring everything in their path. If they are numerous enough to devastate a region, they are forced to feed upon one another, and immense numbers perish from debility and starvation. They usually move only during the warmer hours of the day, and in no particular

ROCKY MOUNTAIN GRASSHOPPER; *a*, PUPA; *b*, FULL GROWN LARVA; *c*, YOUNG LARVA. NATURAL SIZE.

direction, but purely in search of food. They generally march for one day, however, in the direction begun. If the vanguard does change its course, the new direction seems to be communicated in some way to those in the rear, which follow in wave-like form. There sometimes occurs the singular spectacle of two schools crossing each other, the individuals of each keeping to their own course. Some remarkable records have been made of phenomena resulting from the encountering of obstacles to the march. In Europe Döngingk claims to have seen them cross the Dniester for over one German mile, and in layers 7 or 8 inches thick. "In 1875, near Lane, Kansas, they crossed the Pottawatomie Creek, which is about 4 rods wide, by millions; while the Big and Little Blues, tributaries of the Missouri, near Independence, the one about 100 feet wide at its mouth and the other not so wide, were crossed at numerous places by the moving armies, which would march down to the water's edge and commence jumping in, one upon another, till they would pontoon the stream, so as to effect a crossing. Two of these mighty armies also met, one moving east and the other west, on the river bluff, in the same locality, and each turning their course north and down the bluff, and coming to a perpendicular ledge of rock 25 to 30 feet high, passed over in a sheet apparently 6 or 7 inches thick, and causing a roaring noise similar to a cataract of water." [1]

Their unfledged existence terminates in about 7 weeks. During this time, even without change of direction, they could not travel over 30 miles. The swarms of winged insects will perhaps cover over an average advance of 20 miles a day. They spread most rapidly 4 or 5 days after they become winged, when, with a strong and favorable wind, they may reach a maximum of from 200 to 300 miles a day, and 50 miles per hour. The swarms generally move toward the south and southeast. This locust is single-brooded, dies with the approach of cold weather, and normally hibernates in the egg state. Other kinds of destructive locusts occur, as lesser migratory, non-migratory, red-legged, California devastating, differential, two-striped, pellucid, and American Acridium, but the damage occasioned by these has never been comparable to that caused by the Rocky Mountain species.

[1] U. S. Dept. Agr., Div. of Entomol., Bul. 25, pp. 21-22.

REMEDIES.—Several methods are quite effective in bringing about the destruction of locusts. They have many natural enemies, such as parasitic fungi and insects and birds. These should be protected. Experiments have been made by introducing fungi, especially from South Africa. They were artificially spread, but with little success. Deep fall plowing for the destruction of the eggs is perhaps the best remedy known. In western Colorado "ballooning" used to be practiced. The insects were caught in a large open sack by riding a horse rapidly across the field. A bounty of one cent a pound was paid for the insects, and the rider earned from $5 to $10 per day. Undoubtedly the most effective remedy after the locusts are hatched is to scatter bran or horse droppings poisoned with Paris green around the field before the locusts have entered it.

SPRING GRAIN-APHIS OR GREEN BUG. ENLARGED

In Argentina the best results were attained "by the use of torches dipped in tar." The great abundance of locusts in certain years is doubtless the result of a coincidence of climatic conditions favorable to their development and the absence to a great degree of natural destroyers.

The Spring Grain Aphis (*Toxoptera graminum* Rond).—This species, popularly called the "green bug," was first described in 1852, and 30 years later it was discovered in America. It is found most abundantly in the southwest. This pest can be found in the wheat fields during any year, throughout the infested region, but it is rather erratic in its outbreaks. In ordinary seasons it is held in check by its natural enemies. It is extensively parasitized, and lady beetles devour both young and old. It can withstand a lower temperature than its enemies, however, and outbreaks occur after a mild, open winter followed by a late and wet spring. Such outbreaks occurred in 1890, 1900 and 1907. In the south it may breed all winter, and it has an enormous rate of increase. The eggs are laid among the grain plants in the fields. Wheat and rye are the chief foods, but the insect thrives on the other cereals also, and on orchard grass. Late sowing is a preventive measure.

Other Insect Enemies.—The most important of these are the wheat straw-worms, the wheat bulb worm, the cut-worms, the joint worm, several species of sawflies, and the army worms. The damage caused is local and not great. Most of them can be more or less controlled.

The total loss from insect enemies of growing wheat is estimated to average at least 20 per cent of the crop. That is, in the absence of attacks from these pests, the wheat crop would have a value approximately $100,000,000 greater than it now has.

General Remedies.—Cultivation upsets the equilibrium established by nature. The resulting environment may be so favorable for the development of an insect as to enable it to multiply beyond all previous proportions. The most obvious remedy is to render the conditions unnatural for the insect concerned. Intelligent control presupposes a working knowledge of the insects to be controlled, and frequently the first step to be taken by the American wheat grower is the gaining of this knowledge. Entomological difficulties must be forecast and forestalled. The state agricultural experiment station or the Department of Agriculture can always aid in this, for there is a fairly effective remedy known for every insect of great importance.

Where such large areas are involved as in wheat raising, remedies must be largely preventive and general. Summer fallowing and crop rotation are the most effective. These result fatally for many insects which are not equipped for encountering the sudden destruction of vegetation, or the abrupt displacing of one kind by another. Even if insects are able to migrate from one field to another, disaster from adverse winds, storms, heat or cold may result to the migrants, especially if they are such frail insects as the Hessian fly or the wheat midge. Good seed should always be sown, and in well prepared soil, for a vigorous crop can best withstand attacks.

Insect Enemies of Stored Wheat.—Several species of insects, popularly known as weevils, cause extensive injury to stored wheat. Commerce has distributed them to all quarters of the globe. In warm climates these insects live an outdoor life, while in the colder parts of the temperate zones they pass an artificial or domestic existence. Not only do they occasion loss in weight, but the grain which they infest is unfit for consumption either by man or by most animals, and cannot be used profitably for seed. Three species of insects injuring stored wheat pass their adolescent stages within the kernel and are universally the most injurious forms. They are the rice and granary weevils and the Angoumois grain moth.

The Granary Weevil (*Calandra granaria* L.).—From the earliest times this weevil was known as an enemy to stored grain.

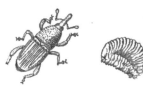

GRANARY WEEVIL, ADULT
AND LARVA. ENLARGED

It became domesticated ages ago, lost the use of its wings, and is now strictly an indoor species. After the grain of wheat is punctured by the snout of the female, an egg is inserted. The resulting larva makes room for its transformations within the kernel by devouring the mealy interior. The grains of most cereals are inhabited by a single larva, but several individuals can thrive in a kernel of maize. The length of the life cycle and the number of generations annually produced depend on season and climate. In southern United States there may be six or more generations per year. One pair is estimated to produce 6,000 descendants in a single year. Besides wheat, they attack all the other grains, and the chick-pea. The greatest damage is caused by the long-lived adults, which gnaw into the kernels for food and shelter.

The Rice Weevil (*Calandra oryza* L.) resembles the granary weevil in structure and habits. It differs from the granary weevil most essentially in having well developed wings, and consequently being often found in the field. It lays its eggs in the standing grain in the tropics, and in the extreme south of the United States, where it is erroneously called "black weevil." It originated in India, was first found in rice, and is now established in most of the grain growing countries of the world.

The Angoumois Grain Moth (*Sitotroga cerealella* Ol.).—Since 1736 the injuries of this moth have been noticed in the province of Angoumois, France, from which it received its name. In the United States it was noticed as early as 1728, and is often incorrectly called "fly weevil." It is widely spread and does incalculable damage in the southern states. It is rapidly spreading, and where it has become established it is more injurious than the weevils, also attacking grain in the field as far north as central Pennsylvania.

The adult insect is often mistaken for a clothes moth. The eggs are deposited in standing grain and in the bin, singly or in clusters of from 20 to 30. It requires at least 4 days for the eggs to hatch. The minute larvæ or caterpillars burrow into the kernels for food, and in 3 weeks or more they are matured. A silken cocoon is then spun within the kernel, the caterpillar transforms to a pupa or chrysalis, and in a few days the moth is again on the wing. In favorable weather the life cycle requires 5 weeks, and about 8 generations are produced annually in the south, where the insect breeds all winter.

GRAIN MOTH, ADULT AND LARVA. ENLARGED

The Mediterranean Flour Moth (*Ephestia kuehniella* Zell.). —The most important of all mill insects, it was comparatively unknown before 1877, when it was discovered in Germany. Its appearance was noticed in England in 1886, in Canada in 1889, in California in 1892, and in New York and Pennsylvania in 1895. While its range is yet limited, it is rapidly becoming distributed throughout the civilized world. The high and equable temperature maintained in modern mills has made the insect a formidable one, for this condition is highly favorable to its development.

Cylindrical silken tubes are formed by the caterpillars. They feed in these until full growth is attained, when a new silken domicile is formed. This becomes a cocoon in which occur the transformations to pupa and imago. In the warmest weather the life cycle is passed in 38 days. It is the habit of web spinning that renders the insect most injurious. Infested flour is soon felted together so as to clog the milling machinery, necessitating prolonged and costly stoppage. Flour or meal is preferred by the larva, but in the absence of these it attacks grain, and it flourishes on bran and all prepared cereal foods, including crackers. In California it lives in the hives of honey bees.

Other Insect Enemies of stored wheat and its products are the Indian-meal moth, the meal snout-moth, the flour beetles, the meal worms and the grain beetles, which all occasion more or less damage. As warm weather favors the rapid breeding of all of these insects, the losses are enormous in the warmer climates. In the single state of Texas, the weevils alone are estimated to cause an annual loss to all grains of over $1,000,000.

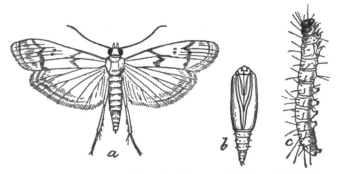

FLOUR MOTH: *a*, ADULT; *b*, PUPA; *c*, LARVA.   ENLARGED

Grain infested by the Angoumois grain moth may lose 40 per cent in weight and 75 per cent in farinaceous matter in 6 months.

Remedies.—Nearly all the insect enemies of stored grain have parasitic or predaceous enemies, or both. Mites and spiders prey on them, and several species of chalcis flies parasitize them. In the field they are preyed upon by nocturnal insects, birds and bats. Preventives and insecticidal remedies are known. Bisulphide of carbon (one pound to one ton of grain or to 1,000 cu. ft. of empty space) is the best insecticide, and naphthaline the most effective deterrent. Hydrocyanic-acid gas is used in fumigating mills to rid them of the Mediterranean flour moth. Perhaps the largest operation of this kind ever made was that of exterminating the moth in a six-story mill and its warehouse, cleaning house, and elevator, a total of over 3,000,000 cubic feet of space. A ton of cyanide of potash and a ton and a half of sulphuric acid were used. Only two living worms and one moth were found, after the operation. It will perhaps require years for the mill again to become so infested as to need another treatment.

There is no weevil-proof wheat, but the small, hard-grained varieties are little troubled by insects. Advantageous practices for prevention are prompt threshing, inspecting, quarantining and disinfecting grain and everything connected with it; scrupulous cleanliness; construction of warehouses and mills to exclude insects; use of improved machinery in mills; and storage in large bulk in a cool, dry, well-ventilated repository.[1]

General Needs and Results.—With a better and increasing knowledge of farm management, of cultural system, and of the natural destroyers of wheat, it should require less than a generation of time to double the yield of wheat per acre. Thirty per cent is certainly a very reasonable figure to represent the average annual loss to wheat from attacks by natural agencies of destruction. One of the greatest immediate needs is to impress the wheat grower with the fact of this loss, for it is often little realized, especially when it results from invisible or unperceived instrumentalities of disease that secretly tread their way through field and plant over great areas of the wheat regions.

To minimize the effects of this great host of natural destroyers of wheat is a task that is profitable, certain of reward, and most imperative in its demands for attention. Since wheat is raised the world around in temperate zone climates, there is little danger of a world famine in wheat on account of their combined effects, for there is always a great probability that large areas will meet with normal conditions. An unusual coincidence of abnormal conditions over wide regions in different parts of the world may, however, raise the price of wheat and greatly change the magnitude and direction of the commercial streams of wheat over the entire world. Such a coincidence occurred in 1897, when the world's wheat crop was greatly reduced by drought in India and Australia, by wholesale destruction from insect pests in Argentina, by a wet harvest in France, and by inundation of the wheat fields of Austria-Hungary.

[1] Chittenden, Some Insects Injurious to Stored Grain.

Insurance.—While scientific and artificial means may lessen losses in many directions, the positive conditions in nature which make possible the operation of natural destroyers of wheat are quite beyond the sphere of man's dominion. In some cases the loss occasioned can be reduced to a small constant factor by means of insurance. This is especially true of loss by storm or fire. Hailstorms occur somewhere every season, but are generally of very limited area, frequently extending over but a few square miles of territory. Consequently a small premium affords protection. Available data concerning wheat insurance are not at all complete or satisfactory. The insurance of wheat in the field is embraced in the more general subject of the insurance of crops against destruction by hail and wind. Companies insuring crops frequently also insure other forms of property. In Scotland, hail insurance existed at least as early as 1780. The first known insurance against hailstorms in Germany is believed to have been in 1797, when the Mecklenburg Hail Insurance Association of Germany was founded. This company was still in existence in 1878. For the first 50 years of its career there was an average rate of 3.8 per cent of the amount insured. In 1812 another company was formed in Germany, having rates from 2.5 to 5 per cent. In 1888 there were 20 mutual and 5 stock hail insurance companies in Germany.

An attempt at hail insurance in France was first made in 1801, by M. Barrau, a philanthropic and enterprising man who was ahead of his time. He lost his fortune in the attempt, for it was thought to be an interference with the dealings of Providence; the government bureaus opposed the plan; and in 1809 the council of state suppressed the society. Permanent hail insurance in France dates from 1823. The average premium received during the subsequent 50 years was 1.05 per cent, while the average loss was 0.81 per cent. The first hail insurance in Austria was written in 1824, and in England in 1842, the latter including 34 acres of wheat at $58.40 per acre. The whole risk was $4304.66, and paid a premium of 1.6 per cent, and also a stamp tax of $5.41. In the United States the first hail insurance was by the Mutual Hail Insurance Company at Milwaukee, Wis., in 1850. It insured on the cash plan with premium notes. In 1878 it was doing business in 5 states.

At least as early as 1880 insurance of crops against fire was becoming popular in some sections of the United States. During the eighth decade 38 hail and tornado insurance companies were transacting business. They were located in 12 different states ranging from Florida to Ohio, Colorado, Michigan and Connecticut. There is apparently no hail in California.

There are at present numerous companies maintained in the north central states for the purpose of insuring crops against destruction by hail and wind. This sort of insurance generally seems to be confined to the co-operative, or assessment plan. The rate varies from 3 to 12 per cent, according to locality. In California several companies insure growing grain against fire at a rate of about 1 per cent per annum, but in the north central states no such insurance is written. In a few of these states there have been excessive losses in some years upon the hail loss branch of insurance. Canada also has a system of hail insurance. Extensive hail storms occur in Argentina and in many districts insurance is generally secured by the colonists.

CHAPTER XI.

THE TRANSPORTATION OF WHEAT

The transportation of wheat has four divisions or aspects: (1) Transportation from the farm to the local market; (2) from the local market to the primary market; (3) from the primary market to the seaboard; and (4) from the seaboard to the foreign market.

Transportation from Farm to Local Market.—On the Pacific coast of the United States all wheat is handled in sacks from the time it is threshed. This is the method of handling wheat in practically all foreign countries. Perhaps the only exceptions to this are in very recent times in certain parts of Russia and in western Germany. In all parts of the United States except the Pacific coast, however, advantage is taken of the flowing quality of wheat by handling it in loose condition as soon as it reaches the elevator, and in some parts, as in the Red river valley, it is never sacked at all, but runs directly from the thresher into the wagon box or grain tank. In transporting wheat from the farm to the local market animal power is well-nigh universally used, whether it is transported on the back of the camel, as in Egypt or India; in the two-wheeled ox-cart, as in Argentina; in the two-horse wagon, as in Ohio; in the four-horse grain tank, as in North Dakota; or on the six-horse, double wagon, as on the Pacific coast.

Transportation from Local Market to Primary Market.—The fact that the production, the internal movement, and the exportation of wheat are greater in bulk and value for the United States than for any other country attaches an unusual interest to a study of the internal transportation of American grain. Those great railway centers into which the wheat of the surplus producing states is concentrated after the first stage of its movement from the producer are designated as the primary grain markets. The ten largest centers are Chicago, Minneapolis, Duluth-Superior, St. Louis, Milwaukee, Toledo, Kansas City, Peoria, Cincinnati and Detroit. An average of from 10,000,000 to 90,000,000 bushels of grain has been annually received by each of these cities. With one exception, all of these primary markets are located at the points where the circumference of an irregular circle intersects the great inland waterways. From each of these centers radiates a fan-shaped net-work of railway lines. In the main, these lines extend to the north, west and south. Sometimes over 25 grain carrying lines come from a single city. Not only do the railroads from any one city compete with each other as carriers of grain, but they also compete with the roads radiating from other cities. The competition is all the more intense because success or failure for certain primary markets in securing the grain often determines whether it goes to the Atlantic or Gulf seaports, and thence to the foreign markets. As a consequence the middle west is well equipped with railway mileage. The net earnings of the railway systems come largely from the grain traffic to the east and south, and from the traffic which this induces in the opposite direction.

The movement of wheat from the local markets of the productive areas to the primary centers for subsequent distribution is almost entirely by rail. There is very little water transportation. In 1899, 50,000,000 bushels of wheat, corn and oats were received in St. Louis. The receipts by wagon were almost equal to those by water, which were little more than a million bushels.

Chicago is the greatest primary grain center in the world, but on account of the great quantity of flour manufactured at Minneapolis, the latter city stands pre-eminent in wheat. During the last decade, there has been a marked increase in the amount of wheat received at Kansas City and St. Louis; the amount at Minneapolis and Chicago has not varied; and the amount at Duluth has declined. Buffalo is a great point of interior concentration for the purpose of forwarding to Atlantic seaports.

Transportation from Primary Market to Seaboard.—In every country the extensive growing and shipping of wheat is closely dependent upon the existence of adequate transportation facilities. To the lack of these the comparative insignificance of the grain traffic of the United States in the eighteenth century was mainly due. This was before the railroad era; canals were developing but slowly; and highway transportation was too expensive to be practicable for any great distance. During

TRANSPORTATION OF WHEAT ON WATER

ABOVE—A RIVER WHEAT TUG

BELOW—A "WHALEBACK" STEAMER

the early decades of the nineteenth century, the main transportation of grain was by way of the Ohio and Mississippi rivers to the Gulf. Buffalo handled less flour than New Orleans as late as 1840. New Orleans received 221,000 barrels of flour in 1832, and this rose to over a million annually in the sixth decade. The Erie canal, opened in 1825, turned the cereal movement eastward to New York, and soon that city became the chief commercial center of the western hemisphere. Already before the Civil war, the grain traffic of the Mississippi river began to decrease in comparison with that of the Great Lakes. In 1836 the first shipment of grain from Lake Michigan took place, and two years later Chicago made its first consignment. The opening of the eastern route immediately shifted the wheat center westward and gave a great impetus to the development of the north central states. An all-rail route was established between Chicago and the Atlantic ocean in 1852. In 1859 the four leading wheat states were Illinois, Indiana, Wisconsin and Ohio, and they transported their surplus to the seaboard chiefly by water. When the Civil war closed the Mississippi river, freight rose so high "that it cost more than five times as much to transport a bushel of wheat from Iowa to New York as the farmer received for it." [1]

Shipments by rail began in 1856. By the seventh decade, the railroads had developed sufficiently to compete with the water route to the Atlantic coast. By the end of this decade the railroads were in the ascendancy in the struggle, having secured the bulk of the flour, and about two-thirds of all grains. On an average, however, only about one-third of the wheat has been carried by rail. On account of the favorable location of Chicago, the roads from this city have been most successful in the competition. As early as 1876, 83 per cent of all the grain shipped to the Atlantic seaboard was by rail. Much grain was and is shipped by a part water and part rail route, for the Erie canal has fallen into comparative disuse. A close parallel to this competition is found in the competition between the Canadian railways and the Welland canal.

The participation of railroads in the eastern grain traffic in the United States and Canada, and also of the Welland canal in Canada, besides extending the grain area and severing it from

[1] 8th U. S. Census, Agriculture, p. xli.

its dependence on the lake region, had the important effect of creating direct routes from the west to seaboard cities other than New York and New Orleans. This resulted in a competition between the Atlantic seaboard cities for the grain trade, and a considerable loss of traffic from New York to such cities as Montreal, Boston, Philadelphia and Baltimore. Of the competing roads to the Atlantic, the New York Central had the greatest natural advantages. By means of reckless competition, however, other roads wrested differential concessions from the Central. The trunk lines endeavored to equalize opportunities for securing eastward traffic by agreeing that the less favored roads should maintain rates that were lower in proportion to their disadvantages. This differential arrangement began in 1869, and in different forms it has been maintained since that date. It has been claimed that New York was not as progressive as other Atlantic ports in methods of handling grain. The net resultant of the differential and of other causes was a decline in the proportion of the grain trade done by New York, for grain could move more economically from the primary markets to Europe by way of ports north and south of New York; Chicago grain reached Europe more largely through Canadian facilities.

The southern movement of the grain traffic is the next phase to be considered. This is characterized by a competition first between the southern railroads and the Mississippi, and subsequently between the southern and eastern railroads. It resulted in southern railroads securing the bulk of the grain traffic from the Mississippi, and they are diverting a continually increasing quantity of grain from the Atlantic coast. In 1873 New Orleans participated to the extent of less than 0.5 per cent in the wheat export. But little of the south-bound grain was then intended for export, while about 20 per cent of the east-bound grain was exported. In the early seventies about 75 per cent of the south-bound grain was shipped by water and 25 per cent by rail. Before the close of the nineteenth century this ratio was reversed, less than 25 per cent of the grain being shipped by water. Thus the railroads, both on the eastern and southern routes, demonstrated their capacity to compete successfully with water transportation. For 50 years or more, competition among the railroads, and between the railroads and the eastern water route, has centered in Chicago. The most recent phase of the competition for the great bulk of the wheat grown in the north central portion of the United States is that of the competition between the eastern and southern railroads. Of Atlantic ports, New York alone is falling behind in commerce. New York once held 75 per cent of the nation's commerce, but now holds less than 50 per cent. This tendency towards a division of commerce among different cities is eminently a healthful one. The construction of the New York state barge canal has been advocated as a means of enabling New York to regain and retain the grain trade. In view of the successful competition of the railroads with water routes, however, as well as the competition of the Canadian canal and the St. Lawrence, it is not probable that the proposed canal would attain the object aimed at. The question is, however, still considered an open one.

While differentials have exerted a great influence tending to distribute export grain among the different seaboard cities, the securing of through-railroad connections has also been a prime factor in diverting traffic. Within the past 15 years, New Orleans and Galveston secured through connections, which enabled them to receive grain shipped from the primary markets of the southwest at a rate below that which was prevailing to the Atlantic seaboard. Consequently the importance of the Gulf cities as grain ports, and especially as wheat ports, has greatly increased. The percentage of wheat exported from the Gulf ports has risen steadily from 2 per cent in 1884 to 55 per cent in 1904, and the percentage for the Atlantic ports decreased from 59 per cent to 20 per cent during the same period. The corresponding variations in the percentages for both wheat and flour were from 2 to 28 per cent for the Gulf ports, and from 69 to 48 per cent for the Atlantic ports.[1] Within recent years through railroad connections have also greatly aided Newport News as an exporting city of wheat from the Atlantic coast.

Distance from Seaport to Primary Market is another factor in determining the direction taken by export grain. Some of the principle distances in miles by the best routes are as follows: From Duluth to Portland, Maine, 1330, to Boston 1400, and to Baltimore via Chicago, 1280; Chicago to Baltimore 802,

[1] U. S. Dept. Agr., Bu. of Statistics, Bul. 38 (1905), pp. 10-28.

and to New Orleans 914; St. Louis to Mobile 644, and to Baltimore 930; and Kansas City to Galveston 873. The railroads which cross the Allegheny mountains are not as level as those which follow the shores of the great Lakes, or as those which extend down the Mississippi valley, and as a consequence it costs them more to carry grain. The Gulf ports have a disadvantage on account of the tropical character of their climate, for flour and wheat, especially if northern grown, are more apt to deteriorate there than in a cooler climate.

Wheat Grown on the Pacific Coast passes into the export channels through the Pacific ports. This trade is very distinct from the rest of the wheat trade of the United States. It formed about 33 per cent of the total export trade in the ninth decade, but the amount fell to less than 25 per cent by 1900. Under abnormal conditions in 1905, however, the Pacific coast exports were 92 per cent of the wheat, and 41 per cent of the wheat and flour. It is probable that the development of Oriental commerce and western transportation facilities will increase the wheat exports from the Pacific coast.

Lake Shipments.—In 1867 the iron steamship was rapidly replacing sailing vessels on the Great Lakes. In 1900 the largest class of lake vessels, known as ''whalebacks,'' carried 250,000 bushels of wheat in a single load. This amount rose to 380,000 bushels in 1906. At 12.5 bushel per acre, one shipload represents the wheat harvested from 30,400 acres of land. In point of tonnage, Duluth, on Lake Superior, was the second port in the United States at the close of the nineteenth century, having been exceeded by New York only. The Sault Sainte Marie canal carried 2.5 times as much tonnage in eight months as the Suez canal in a whole year. During October, 1902, 14,971,318 bushels of east-bound wheat passed through this canal, and also 1,298,751 barrels of flour.

Statistics Pertaining to Rail Shipments.—An empty car weighs on an average about one-third of the gross weight of a loaded car. Twenty years ago 1,020 tons net weight was the best load of grain that could be hauled in one train in the United States. The maximum weight (dead and paying load) hauled by the New York Central in ordinary grain practice is at present from 3,300 to 3,500 tons in a train containing over 60 cars of 30 tons of paying load each; 80 cars having a gross weight of 4,500 tons, have been hauled over the line in a single train; 85 loaded cars in one train is the outside limit, and not many more empty cars can be hauled on the return trip. Even 70 cars in one train is too wearing on the engine to be profitable. The prime cost of moving a sixty-car grain train, that is, the cost at which a few extra trains could be run over the line without loss where ordinary traffic pays for maintenance and fixed charges, is estimated to be considerably below one dollar per mile, very probably under 65 cents.[1] Single wheat cars weighing 139,000 pounds and containing over 2,000 bushels of wheat have been shipped. About 25 years ago, 450 to 500 bushels were considered a large car load and a 1,000-bushel car was unheard of.

When properly shipped, all cars are sealed in transit. There is, however, considerable carelessness in shipping. For example, out of a total of 202,352 cars arriving at the five terminal points in Minnesota during 1905, 9,112 arrived in "bad order." Of these, 3,981 were not sealed; 647 had the seals broken; 1,019 had open end and side doors; 1,330 had poorly fastened doors; 878 had leaky grain doors; 259 had leaky ends and sides; 970 had doors unfastened; and 28 were without doors. Grain is lost from such cars by theft and leakage.

When shipments of grain are heavy there is often a shortage of grain cars. During such times of car shortage it is a common railroad practice to utilize the cars so as to secure the greatest possible amount of the grain traffic, taking into consideration that competing roads will also secure as much traffic as possible. In such cases buyers located at non-competitive points and having their elevators full to overflowing may lose several hundred dollars per day because they cannot secure grain cars.

The railroads often pursue a generous and far-sighted policy for the benefit of all concerned. Such cases are where transportation is furnished, often free of charge, for experts investigating questions connected with agriculture, and for farmers' meetings which are held in the interests of agriculture. The thousands of harvest laborers which the roads annually transport at greatly reduced rates, and even without any direct remuneration, is another case in point. The railroads recognize

[1] Interview with competent observers.

that their prosperity depends upon the prosperity of the farmer, and that they cannot transport the crops which he fails to harvest.

Transportation from Seaboard to Foreign Market.—The increased exportation of American cereals, especially of wheat, dates from the middle of the nineteenth century, and was coincident with an increased demand for grain abroad. Europe's extremity in grain has always been America's opportunity. The chief features in the development of wheat exportation from the United States were a decrease in cereal production in western Europe; an increase in the demand for grain, mainly in the United Kingdom, Germany, Belgium, Holland and Switzerland; the laying of the Atlantic cable; the commercial grading of cereals; and the economies effected by modern elevator and transportation methods. Prior to 1850, not more than from 1 to 9 per cent of the production of agricultural nations, or of the consumption of manufacturing nations, was a factor in international trade.[1]

The first direct shipment of grain from the Great Lakes to Europe was a cargo of wheat in 1856. Out of 125 cargoes thus going in the next eight years, only three or four carried grain. The first wheat shipped from the Pacific coast around Cape Horn was sent to New York. The grain was of such a novel character that the New York millers did not know how to manage it, and the venture was not a success. In 1860 California made its first shipment of wheat to England. The English millers sent back for instructions how to mill this grain, but it found a ready market there. By 1901 about half of the wheat flour shipped from San Francisco went to China, Japan and the East Indies, but the United Kingdom still received the greater portion of wheat. At that date, 12 ships per month were leaving San Francisco loaded with flour for the Orient, whereas only a few years before there were but two or three.

In 1900 a load of 82,000 bushels of wheat was shipped from Portland, Oregon, to Yokohama. This was the first cargo made up of wheat alone that ever crossed the Pacific to Japan. In the same year at the same port a cargo of wheat was loaded for Europe to go the route by way of Japanese, Chinese, Philippine, and Indian ports through the Suez canal to the Mediter-

[1] Emery, Speculation in U. S., p. 106.

ranean, and thence to England.   Whether the cargo of wheat
went all the way to England was not clearly stated, but it reversed the usual route to Great Britain, and full return cargoes
from Europe and the east were promised.   The following year
at least one steamship took the same route to Europe, carrying
about 3,000 tons of wheat.   In 1901 one steamer took a cargo
of 51,931 barrels of flour, besides 1,000 tons of miscellaneous
freight, from Portland, Oregon.   This was the largest cargo of
flour ever floated anywhere previous to that date, with the
exception of one of 55,000 barrels taken from Newport News.
A single shipment of about 40,800 barrels of flour was made
from San Francisco in 1903.

There is a widely prevalent opinion that the Oriental trade
of the Asiatic millions is the greatest commercial prize of the
age, and that it will absorb the entire wheat surplus of the
Pacific coast.   Great American vessels have been built especially
for this trade, but their owners have not found the traffic as
lucrative as they had hoped.   The mercantile marine system of
the United States is not the most encouraging for American
shipping.   It is claimed that the Japanese vessels carry flour
and wheat across the Pacific at over a dollar a ton cheaper than
the American vessels.   It is also claimed that the Japanese are
carrying out an ambitious plan for colonizing Manchuria on an
extensive scale in order to raise sufficient wheat to supply the
needs of Asia, and thus close its markets to American grain.

During the last decade of the nineteenth century about 333
vessels were engaged in the grain trade on the Pacific coast.
They belonged to 12 different nationalities.   Over 65 per cent
of them were English, and less than 3 per cent were American.
Regular lines of steamers carried nearly all the flour shipped
from the Pacific coast ports, but sailing vessels carried the great
bulk of wheat exported.   In California ships are often loaded
directly from the car, but in Oregon and Washington the wheat
is more generally re-cleaned and then re-sacked, before it is
loaded.   Practically all export wheat from the Pacific coast
is sacked.   From the Atlantic and Gulf ports, wheat is generally shipped in loose condition.   A very large portion of it is
carried in English bottoms.

Transportation Charges.—Early freight rates on wheat were
prohibitory.   For example, the charge for transporting the first

wheat sent from the Red river valley to Duluth was 30 cents per bushel. A general reduction in railroad, river and ocean freight rates is the condition which, more than anything else, has made possible the shipping, and consequently the growing, of immense quantities of wheat. This was the major factor involved in opening a market for the wheat grower, not only in the great centers of consumption in our own country, but in those of the world. Estimates at the close of the nineteenth century still placed the cost of carrying wheat from the northwest to the Atlantic seaboard as one-half as great as the original cost of production. By 1897, the cost of concentrating the wheat surplus at Chicago was reduced to one-fourth or one-third of the cost in 1880. In 1884 the cost of getting wheat from the farm to the consumer was 22 per cent of its Chicago value. This had fallen to 6 per cent in 1897. The Chicago price of wheat was practically the same at both dates.

From 1867 to the end of the century, the freight rate per bushel of wheat from Chicago to New York by rail decreased from 33½ cents to 12 cents. As we have seen, however, the competition between these two points was the most severe possible, and it must not be assumed that freight rates in general decreased to this extent. The rate per bushel for shipping wheat from Chicago to New York by lake and canal route was 8.8 cents in 1871, and 4.8 cents in 1905; by lake and rail it was 12.1 cents in 1875 and 6.4 in 1905; and by an all-rail route it was 20.9 cents in 1875 and 9.9 cents in 1905. From Chicago to New York, the rate by lake and rail route fell from 19.2 cents in 1870 to 5.6 cents in 1902. At the close of the century the average rate between these two points was less than one cent greater by rail and lake than by lake and canal, and railroad rates had reached the lowest notch, for the roads preferred to lose the grain trade rather than to reduce rates further.

In 1880 railroads carrying wheat to Chicago charged from 1.08 to 1.75 cents per ton per mile. In 1897 the rates were 0.78 of a cent to 1 cent, a reduction of from 0.25 to 0.74 of a cent per ton per mile in 17 years. This reduction was less than that made by the cotton and coal roads during the same period of time.[1] The average rate on all freight per ton per mile was about the same in 1890 as was that on wheat in 1897.

[1] Industrial Commission, 6:59-60.

In 1898 the farmers' organizations secured a compromise from J. J. Hill of the Great Northern according to which that road reduced freight rates on grain 14 per cent, and competition forced the other roads to meet the reduction. The reduction did not amount to quite 2 cents per bushel, but, contrary to expectations, it made no difference whatever in the price paid for wheat to the farmer. It is believed that the reduction benefited the consumer and shipper only.[1] From St. Louis to New York, the rate was 32 cents per 100 pounds in 1882 and 20.5 cents in 1905. In 1890 it cost 17.4 cents per bushel to haul wheat by rail from St. Louis to Chicago. This rate had fallen to 11.6 cents in 1901. The rate from St. Louis to New Orleans by river fell from 8.1 cents per bushel in 1877 to 4.2 cents in 1902. The cheapest transportation in the world is on the Great Lakes, 0.75 of a mill per ton per mile.

The cost of transporting a bushel of wheat to Europe from the Atlantic ports of the United States was 5.5 cents in 1902, from New Orleans 8 cents, and from San Francisco 16 to 20 cents. The rate per bushel from St. Louis to Liverpool by way of New Orleans was 22.7 cents in 1882, and 10 cents in 1903. By way of New York it was 23.7 cents in 1882, and 15.6 cents in 1905. The rate per 100 pounds from Chicago to Liverpool was 33.5 cents in 1896, and 19.2 cents in 1905. It is claimed that competition of the Gulf ports has forced the railroads carrying grain to Atlantic ports to charge a lower rate when grain is destined for export than when it is destined for domestic consumption, the only alternative being to cease exporting.

The railroad rate from Chicago to New York is only a part of the through rate from Chicago to Liverpool, and in support of the view that the rail and ocean rates are complementary, it has been cited that in 1876 there was a railroad rate war, and the ocean rates at once met the railroad rates. From April 6 to June 1 the rate from Chicago to New York fell from 24 cents to 12 cents, while the ocean rate rose from 10 cents to 21 cents during the same period. The ship owner gained and the railroad lost, while the total cost to the shipper was approximately the same.[2]

[1] Industrial Commission, 10:ccciv.
[2] Railroad Gazette, 35:722.

It was asserted in the eighties that the competition of over-constructed railroads in the United States and ships in England had caused the hauling of wheat below cost. In 1901 the surplus-cereal states still had at least as many railroads as could be profitably operated.[1] The transportation facilities proved inadequate for moving the wheat crop of 1906.

By making freight discriminations, transportation companies can exert a powerful influence upon the volume and direction of grain traffic. Discriminations are effected in various ways, and may be against certain forms of grain, against certain persons, and against certain places. It is claimed that the export flour trade is greatly injured by the fact that railroad and ocean carriers discriminate against flour in favor of wheat, thus giving the foreign miller an advantage in competing with the American miller.

The interstate commerce commission found that discriminations during the year 1898 were probably worse than at any previous time. ''It is claimed by some that direct rebates and secret rates are still frequently granted; commissions are paid for securing freight; goods are billed at less than the actual weight; traffic within a state not subject to the interstate-commerce act is carried at lower rates; allowances and advantages are made in handling and storing, etc.''[2] The large shippers generally receive the greatest favors. Laws have been enacted to remedy the evil, but their effective enforcement is not an easy task. On the whole, however, it must be said that the transportation service for wheat has improved vastly during the last 25 years, while its cost has been enormously reduced during the same period of time. Such evils as exist will doubtless be corrected in at least some measure as a result of the present wave of popular agitation against all corporate abuses.

[1] Industrial Commission, 6:48.
[2] Industrial Commission, 4:5-6.

CHAPTER XII.

THE STORAGE OF WHEAT

The storage of wheat has four aspects which correspond to the four stages of transportation, namely: Storage at the farm; at the local market; at the primary market; and at the seaboard. Under the subject of storage is included the vertical and horizontal transportation involved in getting wheat to and from wagons, cars, ships and warehouses.

Storage of Wheat at the Farm.—The granary upon the farm should have an exposed location, and should be so constructed as to make the handling of grain as easy as possible. The principal things to be guarded against are dampness, insects and vermin. Cold does not injure wheat, and it lessens the activity of injurious insects. The loss from insects decreases with increased bulk and decreased exposure of the surface of grain. Bins should be constructed with smooth, oiled, or painted walls to prevent lodgment of insects, and without air spaces where vermin can hide. If the granary is fully exposed, a single thickness of inch boards will keep out all rats and mice. Where injurious insects are likely to be abundant, the windows should be screened, the doors made close fitting, and all crevices and other means of ingress closed. If the granary is properly constructed, there is practically no loss of weight through storage. On the largest wheat farms, such as exist in the Red river valley, the grain is stored in elevators. Alongside of the railroad track which runs through the great field, two elevators of about 50,000 bushels capacity each are located on opposite corners of the farm. On the Pacific coast, where there is no danger of rain, the sacked wheat is left lying in the open field until it is shipped. East of the Mississippi river, mixed farming is generally practiced, and as a rule there is sufficient granary room on the farm to store the wheat held over, which is quite a large portion. In the Northwest, where the main feature of farming is growing grain for the market, it is estimated that 75 per cent of the grain is put upon the market before the close of the year.

TYPICAL SMALL STORAGE ELEVATORS AT A LOCAL MARKET IN NORTH DAKOTA

Storage of Wheat at the Local Market.—The unit of accumulation at the local market is the wagon load. The unit of railway shipments is the car load. The shortest time that the wheat can be stored at the local market, then, is until enough of one grade has accumulated to fill a car, which may be only a fraction of a day. For various reasons, grain may be stored at the local markets for longer periods. When the buyer lacks better facilities, the wheat is often transferred directly from the wagon to the car by means of manual labor. By far the most usual method, however, is by means of the elevator.

The modern elevator is a very essential factor in our wheat industry. Its chief functions are storage; cleaning, drying and gathering wheat; and the vertical and horizontal transportation incident to these processes and to the processes of loading and unloading from wagons, cars and ships. Steam or electric power operating the machinery of the elevator accomplishes all this work without any aid from manual labor, work that would require the manual labor of a vast army of men to accomplish it. If it were thus performed, the operations would be so slow and expensive that they would raise the cost of producing wheat to such a height as to prohibit much of the production now carried on. With one single exception, the entire process of producing wheat flour, including the raising, harvesting, threshing, shipping and milling of the wheat, may be accomplished by machinery. It remains for some genius to remove this exception by inventing a machine that can handle the sack of wheat on the Pacific coast, and one that can handle the sheaf of wheat in the Red river valley. It would seem that neither task should be beyond the inventor's power. Of the machinery used on a large wheat farm, the plow stands at one end, and the elevator at the other. Human labor has been minimized throughout all of the operations. All agricultural implements are guided by levers; threshermen are only assistants to a machine which delivers the grain into a sack or grain tank; those who unload the wheat from the wagons simply loose a bolt, and the grain is dumped; those who heave wheat into bins merely press buttons; and those who load it into cars or ships need but pull a lever. The elevator at the local market often has its machinery so constructed that it can empty 1,000 bushels an hour from wagons, and sometimes 10,000 bushels a day are received by a single elevator. These elevators are generally constructed of wood, and have a capacity varying from 10,000 to 40,000 bushels.

From the point of view of ownership and management, there are three types of elevators found at the local markets: (1) Those provided and owned by the farmers themselves; (2) those owned by the local grain dealers; and (3) those controlled by the grain buyers located at the primary markets. Hundreds of elevators situated along the railroads which extend into the grain territory are controlled from the primary markets by what are called line elevator companies. The Northern Pacific Railway with its elevators may be taken as a typical case. On this road during 1901, there were 430 line elevators, 286 local dealers' elevators, and 22 farmers' elevators. In the same year in Brown county, South Dakota, a county which is 36 miles wide by 48 miles long, and which is considered as typical of the Dakotas and Minnesota, there were 45 elevators with a capacity of from 12,000 to 15,000 bushels each. There were also 12 flat houses with a capacity of from 3,000 to 5,000 bushels each, and 3 large elevators belonging to flouring mills. Twelve line companies were operating in the county, and they owned 30 of the warehouses. 20 of them were owned and operated by independent parties.[1] When local market conditions are unsatisfactory, the farmers establish more elevators. During 1904-5, the farmers' elevators in Minnesota increased approximately 90 per cent in 18 months.[2] For the year ending September 1, 1901, 1,549 licenses were issued for country elevators and warehouses in the state of Minnesota.

The successful working of elevators as now constructed and all the principles of their machinery are entirely dependent on the flowing quality of wheat. Since the advantages of this quality for labor-saving machinery had been completely established prior to the extensive development of the wheat industry on the Pacific coast, it is a peculiar and noteworthy fact that in the subsequent development of the wheat industry, the Pacific coast differed from other parts of the country in this, as in nearly all other things, by not taking advantage of the flowing quality of wheat. The grain is handled in sacks, and

[1] Industrial Commission, 10:cccxviii.
[2] Rept. R. R. and Warehouse Commission of Minn., 1905, p. 59.

it is even resacked after it has been cleaned by the elevators. Undoubtedly one of the main reasons for this is found in the climate. During the summer season of the year, there is no rain, and the sacked wheat needs no protection from the elements. If it is not shipped at once, it is piled up in huge piles at the shipping points. This avoids the use and expense of elevators, although it is sometimes piled in warehouses. The platforms and warehouses are owned by the grain-buying firms who collect the wheat for ultimate shipment.

Storage of Wheat at the Primary Market.—The capacity of terminal elevators to handle and store grain is enormous. Chicago was perhaps the first city to develop great facilities in this line, and it is partly to this that the city owed its early pre-eminence as a grain center. Its first elevators were built in the fifties. As early as 1867 Flint wrote that "7,000 to 8,000 bushels per hour of grain may be taken from a train of loaded cars by a large elevating warehouse, and the same grain at the other end may be running into vessels, and be on its way to Buffalo, Montreal or Liverpool within six hours of time. The Illinois Central Railroad grain warehouse can discharge 12 cars loaded with grain, and at the same time load two vessels with it, at the rate of 24,000 bushels per hour. . . . It is capable of storing 700,000 bushels of grain. It can receive and ship 65,000 bushels in a single day, or it can ship alone 225,000 bushels in a day." All the warehouses of Chicago could store an aggregate of 3,395,000 bushels, and it is further said: "They can receive and ship 430,000 bushels in 10 hours, or they can ship alone 1,340,000 bushels in 10 hours, and follow it up the year around. In busy seasons these figures are often doubled by running nights."[1] By the end of the nineteenth century, however, there were single elevators in Chicago with a storage capacity greater than that of the entire city at the above writing. Some reached the high figure of four million bushels. The public warehouse capacity of Chicago in 1900 was 28,600,000 bushels, and the private warehouse capacity was 28,645,000 bushels. At that date, five cars of wheat could be unloaded in eight minutes. In 1905 one of the Chicago elevators, together with its annexes, had a capacity for storing 5,000,000 bushels.

[1] Eighty Yrs. Prog. of U. S., pp 75-76.

From 1871 to 1887, the Chicago elevators were managed by persons whose sole business was the warehousing of grain. Competition was active, and Chicago was the best market to which grain could be sent from the West. By 1892 a change took place. The elevators had passed into the control of persons who immediately embarked in the grain-buying business. Nearly every railroad terminating in Chicago favored some elevator system with concessions that gave control of the grain business of the road. As early as 1894, there was an association of all the elevator people in Chicago, and all of the great terminal elevators were owned by a comparatively few men or firms. The owners of public elevators bought a large proportion of the grain that was received, and they also controlled great private elevators.

Minneapolis had a grain storage capacity of 27,485,000 bushels in 1898, and the largest elevator had a capacity of 2,300,-000 bushels. Some 23 elevators, having two-thirds of the city's storage capacity, were operated under the Chamber of Commerce rules, 4 were operated under the state warehouse law, and the remaining 6 were private elevators. Minneapolis is perhaps the most notable city as a center for powerful houses which control elevator lines. At the close of the century it had 36 elevator companies, which controlled 1,862 country elevators with a combined capacity of about 50,000,000 bushels of wheat. St. Louis has 8 public elevators with a total storage capacity of 6,900,000 bushels, and 25 private elevators with a capacity of 2,475,000 bushels. The largest elevator has a capacity of 1,500,000 bushels. It can receive and deliver 30,000 bushels per hour. The total capacity of all public elevators for receiving and delivering grain per hour is 181,000 bushels. Kansas City, Missouri, has 24 elevators having a total storage capacity of 9,280,000 bushels. The largest elevator has a storage capacity of 1,000,000 bushels, and a capacity of receiving and of delivering 15,000 bushels per hour. A total of 215,000 bushels can be received and delivered by all elevators.

Duluth and Buffalo are the two other great inland elevator centers. Some of the elevators of Buffalo have a storage capacity of 2,800,000 bushels, are ''built of steel, operated by electricity from Niagara Falls, protected from fire by pneumatic water systems, and have complete machinery for cleaning, drying and scouring the wheat, when it is necessary.'' The 28 elevators of Buffalo have a capacity of about 22,000,000 bushels, and the estimated cost of their construction is $13,000,000. Long spouts containing movable buckets can be lowered from the elevators into the hold of a grain laden vessel. Great steam shovels draw the grain to the end of these spouts, where it is seized by the buckets and carried to the elevator. The 28 elevators have facilities for receiving from lake vessels and railroads and transporting to cars and canal boats an aggregate of 5,500,000 bushels daily. Wheat is unloaded from vessels at the rate of 100,000 bushels per hour, while spouts on the other side of the elevator reload it into cars, 5 to 10 at a time. A 1,000-bushel car is filled in 3 minutes, and the largest canal boat in less than an hour. About December 31, 1905, 6,151,693 bushels of wheat were afloat in the harbor of Buffalo.

There is often a community of interest in the management of railroads and elevators, as is shown by their methods of operation and by the fact that the same men have heavy investments in both railroads and elevators. Where the railroads owned their own storehouses they generally found it impracticable to trade in grain themselves. They made operating agreements or sales in such a manner that companies or individuals would do this work for them. These companies became the medium through which practically all the cereals tributary to the respective lines of road on which they operated must go to market. Where laws prohibited a public warehouseman from trading in grain, other companies were organized, working in conjunction with warehousemen, to handle the business.

Financially, the elevator consolidations have brought money from the great public money market of the world. On this account the rate of interest has fallen, which has been a disadvantage to the local capitalist with small capital. Without the present system of elevators a farming community would be much worse off than under existing conditions, but from the farmer's point of view there is ample room for improvement in the present system. If the competitive system is to give way to organization, the farmer must receive his proper share of the benefits arising from the co-operation of all the interests involved, for the foundation of the whole system rests on the prosperity of the wheat-grower.

Storage of Wheat at the Seaboard.—The elevators at the seaboard are not as large as those at the primary markets. The largest storage capacity of an elevator on the Atlantic coast at present is 1,800,000 bushels. Such an elevator can unload grain cars at the rate of 560,000 bushels per day and simultaneously it delivers grain to vessels at the rate of 1,000,000 bushels per day. An ocean steamship pier is usually about 250 feet wide and about 800 feet long. The railroad tracks are in the middle of the pier, and ocean vessels are moored on either side. The capacity for handling cars depends upon the size of the terminal, and varies from 65 to 1,000 cars per day. Grain in bulk is easily loaded on a vessel by transferring it through spouts running from the elevator to the hold of the ship. There are also two different arrangements for loading grain on a vessel while it is alongside a pier taking on board other freight. One arrangement consists of a series of belt conveyors which carry the grain along a gallery above the pier. The grain is transferred to the hold through spouts lowered from the sides of the gallery to the hatches of the vessel. The other method of loading is by means of a floating elevator, and it is used when the grain is loaded from boats. The latter are towed alongside the vessel, and the floating elevator transfers the grain from them through the hatches of the ship.

New York and New Orleans are the only seaports where the docks and wharves are largely under the ownership and control of city government. The stationary grain elevators of New York have a total storage capacity of about 17,000,000 bushels, and they are able to transfer over 375,000 bushels of grain per hour. From 5,000 to 14,000 bushels per hour can be transferred by each of the floating elevators, which have a combined capacity of 178,000 bushels per hour. It has been estimated that the New York elevators, working 10 hours per day, could transfer in 30 days the 157,280,351 bushels of wheat exported from the United States in the fiscal year of 1892.

Philadelphia has five stationary elevators and three floating elevators. The total storage capacity in 1904 was over 4,000,000 bushels. One thousand carloads of grain, or 800,000 bushels, could be received in a 10-hour day, and at the same time

1,380,000 bushels of grain could be delivered. The largest elevator at present has a capacity of 10,000 bushels per hour.

Baltimore has 6 grain elevators. The total storage capacity is 5,350,000 bushels. One of the elevators can store 1,800,000 bushels, and it has a daily delivering capacity of 1,000,000 bushels. The four elevators at tidewater in Boston can store 3,000,000 bushels, and they can handle, in and out, approximately 100,000 bushels per hour. Galveston, Texas, has 4 elevators, with a combined storage capacity of 4,000,000 bushels.

There are no grain elevators on the Pacific coast. Large grain warehouses supplied with cleaning and grading plants are found at the ports, however. The sacks of wheat are often simply piled on the banks of the river. When the deck of the vessel to be loaded is at a lower elevation than the grain, the sacked wheat is placed on an inclined chute over which it descends by gravity into the hold of the vessel. When the deck is at a higher elevation than the grain, the sacks are first elevated by a conveyor, consisting of a chute and an endless belt, and then descend. It requires 3 or 4 days by these methods to load a ship carrying from 3,000 to 3,500 tons of wheat. At Portland, Oregon, there are 14 wheat docks (meaning warehouses), and 350 cars of wheat can easily be put in storage in one day. One is inclined to question the economy of the whole system of handling wheat in sacks.

Legislation Pertaining to Public Elevators and warehouses was passed first in Illinois (1870). The usual subjects of legislative enactment affecting the storage of wheat are: The classification and definition of public and private warehouses; the licensing of public warehouses; the requiring of bonds with approved security from warehousemen; discriminations; warehouse receipts; grain inspection; prompt delivery; statements of grain in store; accidental losses of grain in storage; the mixing and selecting of grain by the warehousemen; combinations of warehousemen; and the negotiability of warehouse receipts.

Storage Charges.—Concentration of the wheat trade and through shipments have eliminated many of the charges incident to the storage and handling of wheat. In Minnesota and the Dakotas in 1900, storage was usually free for the first 15 days, and after that the rate was 2 cents a bushel for the first 30 days, and half a cent a bushel for each additional 30 days.

STORAGE IN OPEN ON A FARM ON THE PACIFIC COAST

Wheat was stored the entire year in elevators, and sometimes for 2 or 3 years. The expense per bushel of wheat in operating a line elevator was given as 2.25 cents if 50,000 bushels were handled annually, and 1.75 cents if 100,000 bushels were handled. Very few houses handled 100,000 bushels of wheat in a year. Three-fourths cent per bushel was charged for transferring grain from a car to an elevator and into another car. It has been estimated that the cost was only one-eighth cent.

In 1885 the country elevator charge was from 3 to 5 cents per bushel. In 1900 it was from 0.5 cent to 2 cents. Elevator transfer charges were 1.25 cents in Chicago in 1885, and 0.75 cent at the close of the century. The usual commission for selling on consignment at the terminal markets in 1900 was one cent a bushel. Inspection and weighing charges amounted to 0.01 cent per bushel. About 80 per cent of the charges involved in concentrating wheat in Chicago were railroad charges and 20 per cent were commercial charges. Charges made per car were usually those of inspection, 25 to 30 cents, and weighing, 15 to 30 cents. Storage charges at the terminal elevators were about 1 cent per bushel for the first 10 days or any part thereof, and about one-fourth cent for each additional 10 days or any part thereof. Charges for recleaning grain were from 1 to 2 cents per bushel.

In New York the charges on grain in store are, for receiving, weighing and discharging sound grain, including storage for 10 days or a part thereof, five-eighths cent per bushel, and for every succeeding 10 days or a part thereof, one-fourth cent a bushel. There is extra storage of half a cent per bushel on grain delivered to ocean vessels. Screening and blowing on receipt or delivery costs one-eighth cent per bushel. This may also include mixing. Inspection charges are 25 cents per 1,000 bushels. This, and verification of track weights, involves a charge of 50 cents per car load. One cent per bushel is the charge of weighing and discharging track wheat. Grain loaded from elevator to car is charged one-half cent per bushel, and that transferred while in store one-fourth cent per bushel. At Buffalo the cost for elevating is 0.5 cent per bushel, but this includes free storage for 10 days. If the grain is left in storage longer than 10 days, the charge is 0.25 cents for each day.

Commercial grain charges on the Pacific coast have been an argument in favor of elevator methods, especially when they were compared with the charges at terminal points in the Mississippi valley, or with those in New York city. At the very outset, the sacks add a cost of 4 cents per bushel of wheat, an expense which, according to the above statistics of charges, is probably equal on an average to the entire commercial charge involved in getting a bushel of wheat from the Red river valley through country, terminal and Atlantic seaboard elevators and transferring it on board ship at the Atlantic port. While there was no storage charged at the local warehouse on the Pacific coast for the first 6 months, the handling charge paid the local warehouseman during this time was 1.5 cents per bushel. Each month after this time involved a charge of 0.3 cent per bushel. Since there was little capital invested in local warehouses, the charge for handling the wheat must have been about 1.5 cents per bushel, even if it was sold immediately. At Portland, 60 days' storage, including the discharging of cars and truckage across the dock to ship, involved a charge of 1.2 cents per bushel. This charge became 1.8 cents when grain was also loaded on the vessel, which made the charge for merely transferring to vessel 0.6 cent. After 60 days, storage charges were one-eighth cent per bushel for 10 days. Storage charges at San Francisco were 1.5 cents a bushel per year. The charge for loading wheat on vessels was 0.75 cent per bushel, and that for weighing was nearly 0.25 cent per bushel.

The greatest portion of the expense on the Pacific coast would seem to be for handling the wheat, while on the Atlantic coast it is for storage. Receiving grain, storing it 60 days, and discharging it involves a cost of 1.8 cents per bushel on the Pacific coast and 1.875 cents on the Atlantic coast. When delivered to vessels, there is an additional storage charge of 0.5 cent per bushel, making 2.375 cents on the Atlantic coast. Storage for 60 days costs two-thirds cent on the Pacific coast, and handling the grain costs 1.13 cents. Storage for 60 days at New York costs 1.25 cents, or, if it is to be loaded on vessels, 1.75 cents, which leaves a cost of 0.125 cent for handling. It costs, therefore, about 1 cent a bushel more at the seaboard port to handle wheat in sacks than to handle it by elevator methods. At the country elevator, however, there is a gain of perhaps half a cent when the grain is sacked, but the Pacific coast warehouse is often merely a platform, which in a different climate would afford no protection to the grain. If elevators had to be built, it would materially raise the cost at the country elevator, making it higher than 2 cents per bushel. As it is, on the whole, the cost is about 4 cents per bushel greater when wheat is handled in sacks than when it is handled by elevators.

In some states warehouse rates are regulated by law. Discriminations are often practiced by elevator people, especially in eliminating independent competitors. In the interior of the country, the total cost of distributing wheat varies from 10 to 30 per cent of the price paid by the consumer. The average cost has been given as 9 cents per bushel. The cost of getting wheat to the seaboard has been given as 10 1-3 cents on the Pacific coast, and 14 or 15 cents on the Gulf and Atlantic coasts. The total cost from the United States to England was about 20 cents by way of Atlantic ports, 22 to 23 cents by way of Gulf ports, and 30 cents by way of Pacific ports.

CHAPTER XIII.

THE MARKETING OF WHEAT

The Rise and Progress of the Grain Trade of the United States is one of the greatest marvels of an age noted for its commercialism. Its entire history would form almost a complete record of the development of the American continent, for it was the major factor in the opening up of three-fourths of our settled domain. The pioneer husbandman formed the vanguard in the march of civilization. The first succeeding ranks were formed by the merchant. Then came in quick sequence the panoramic array of our ocean, lake and river fleets, of our canals, of our wonderful storage and transportation systems, and of our commercial institutions.

The cereal crop has been the distinctive feature of rural industry in the United States. Here, as in every agricultural community, the three concentric circles of distribution which arose were centered in the local market, in the city market, and in the foreign market. In the modern wheat industry, wheat farming is mainly for a commercial surplus. A minor portion of the wheat grown is consumed or retained on the farm, while the great bulk of wheat is poured into the streams of local, interstate and international commerce. The major factor in that part of the cereal crop which figures in the internal trade and foreign commerce of our nation is composed of wheat. Much more corn than wheat is produced in the United States, but only a minor portion of this corn becomes a factor in its raw form in the domestic trade of the country, while a comparatively insignificant portion is exported. Less than 3 per cent of the corn grown in the United States in 1906 was exported, and only 25 per cent of that grown in 1905 found its way into the channels of domestic trade. For the last decade of the nineteenth century, the exports of wheat from the United States were over one-third of the amount grown, while those of corn were only one-fifteenth. For the year 1902, ten-seventeenths of the wheat grown, but only two-ninths of the corn, was shipped outside of the county where it was raised.

Thus the per cent of wheat exported from the United States is five times as large as the per cent of corn, while the per cent of wheat shipped outside of the county where grown is nearly three times as large as that of corn. Wheat is the keystone in the arch of commodities which is buttressed on consumption and production, and which supports the great commercial superstructure that, with its many ramifications, unites into a threefold nexus of interests—rural producer and urban consumer, manufacturer and agriculturist, and the producers engaged in diversified extractive industries. The grain movement has a function in the national economy second in importance to that of no other factor in our agricultural life. Directly and indirectly, it is the chief feature in our commercial relations.

The Three Methods of Marketing Wheat utilized by the American farmer depend upon the amount of wheat that is grown. The largest farmers make a wholesale disposal of the bulk of their wheat, watching a good opportunity to sell, or employing agents to watch at the chambers of commerce or boards of trade of such primary markets as Duluth, Minneapolis and Chicago. A large class of less extensive growers obtain a price remarkably close to city quotations by forming close business relations with commission men at the large terminal points. By shipping their grain directly, they avoid the middleman charge of the local dealer. The great mass of smaller farmers sell to the local elevators. The profits of the local buyers, however, have quite frequently been scaled to the lowest notch by competition.

The Buyer of Wheat is always located within hauling distance of the producer's home. There are two classes of buyers, the local grain dealers and the dealers who represent the terminal grain buyers. The general policy of the railroads has been to rely upon these two classes of buyers to provide the country elevator facilities needed for receiving and shipping grain, and to enable them to do this promptly by furnishing them with adequate transportation facilities. The terminal grain buyers, controlling lines of hundreds of elevators, have been the most important factor in the producer's grain market. The local buyer is usually a dealer engaged exclusively in the grain business, but frequently, especially in Minnesota and the

WHEAT AWAITING SHIPMENT BY RIVER BOAT ON PACIFIC COAST

Dakotas, farmers' associations provide themselves with storage and elevating facilities, do their own shipping, and sell through commission merchants.

The Local Grain Dealers' Associations are one of the main features in the local elevator management. The two great purposes that they have served were the improvement of the distributive system and the securing of justice for the country shipper at the primary markets. While the absence of these associations would have been a public misfortune to producer and consumer alike, there is much evidence that they have exceeded the limits of economic usefulness in some directions. It is for the courts to determine whether they have exceeded the legal limits of rightful association.

The Independent Grain Dealer has frequently served in a most useful capacity. He has generally had the sympathy and support of the grain producer. As a rule, his capital is small, his facilities for handling grain are not elaborate, and he is subjected to the fiercest of cut-throat competition. In spite of all of his disadvantages, however, he sometimes succeeds in maintaining himself for years, to the dismay of his competitors and to the profit of the wheat growing community in which he is located. Situated at some little railway station, and perhaps not even possessing an elevator, he keeps the price at the highest notch. This draws the grain from miles of the surrounding country, and it is even hauled past the elevators of the larger towns to the little railway station. The competitors of the independent dealer buy most of the grain, but the latter secures enough for a profitable business. If he happens to be hard pressed by competition, many of the farmers will sell him their grain, even though his competitors are offering a cent or two more per bushel. The farmer can well afford to deal with such consideration, for he may secure several cents more per bushel on account of the competition which results from the independent dealer's operations. The combination, however, is often, perhaps usually, of such strength that it can stifle all competition. The larger interests endeavor to crush out the smaller ones, and the usual methods are employed. The rules of the association do not permit the farmer to consign his grain to a grain dealing firm in the primary market. Complaints have been lodged against the associations of the various states because they attempted to compel the farmer to sell to their members. The association rules permit dealing in grain only by those who do a "regular and steady business of buying and selling grain." A farmer who does not use the local warehouses or elevators, but shovels his wheat from wagon to car, is guilty of being "irregular," and he is known by the association as a "scalper." Persons who take advantage of a good market by buying and shipping grain independently of the local elevator people are also "irregular." In both of these cases, the firms to whom the grain is shipped are irregular. In the main, irregularity seems to consist in not using the local shipping facilities, or in having dealings with people who do not use them. Cars irregularly loaded are systematically traced to their destination, and the names of the offending shippers and receivers are posted. Such persons are then made the subject of a systematic boycott by the entire membership of the elevator association. Even independent dealers who are fully established in the grain trade may become "irregular." An independent dealer at Malcolm, Nebraska, shipped two car loads of corn to an Illinois farmer for feeding. For this, he was posted as a scalper, although he had $15,000 invested in the grain business, and had been a dealer for seven years. A Malcolm member of the association had traced the two cars to Illinois, and offered corn to the farmer cheaper than it could be bought in Nebraska.[1] At Lakota, South Dakota, an independent elevator was built which incurred the displeasure of the line elevators. The merchants of Lakota decided to support the new elevator in order to help the farmers, but when the old line elevators opened general stores in Lakota and sold at cost all the goods that the merchants handled, the farmers failed to support the merchants. Consequently, the independent elevator was compelled to give up business.

Railway Discriminations.—There are three ways in which the railroads can aid the elevator combination: By promptly supplying cars in the busy season; by refusing to grant sites for independent elevators along their lines; and by rebates. All of these methods have unquestionably been employed. The recent investigation by the Interstate Commerce Commission is said

[1] Industrial Commission, 6:62.

to show that combinations between elevators and railroads have practically eliminated competition. Many of the railroad officials are stockholders in elevator companies.

The History of the Primary Market has been the history of the terminal elevator systems. In the control and operation of these systems lies the key to a proper comprehension of the functions of the primary market in grain distribution. The terminal elevators, operated on such a stupendous scale, receive, store and transfer all grain that flows from the local markets to the primary market. They contain by far the larger portion of the country's visible grain supply. The importance of a primary grain market like Chicago, the historic storm center of the competitive conflict for control over the Mississippi valley grain movement, is augmented by the fact that its supremacy as a distributing center for manufactures depends largely upon its capacity to command the agricultural products which are exchanged for the manufactured products. Herein lies the ultimate explanation of the consolidation of the distributive agencies engaged in the grain trade.

The Pacific coast warehouses, located on the railroads and rivers, are generally operated in the interest of milling, exporting or speculative dealers. In recent years, many farmers have shipped directly to the coast cities and placed their grain in storage there. Each farmer makes certain of securing the return of his own wheat by marking his sacks and piling them together, for the wheat, coming from the dry interior, usually gains enough in weight to pay for storage. The large wheat dealers of Portland and San Francisco have local buyers to represent them at the railroad stations and steamboat landings.

The Inspection of Wheat.—The rigid system of grain inspection and grading maintained by various states and trade organizations not only simplifies and facilitates the movement of wheat to a surprising extent, but it also tends to minimize fraud in the grain trade. In wheat inspection the greatest care and accuracy are always maintained. The procedure at Minneapolis is approximately as follows: One man, passing along the cars, records their numbers and initials, and takes note whether their seals have been tampered with; another man breaks the seals and opens the doors; the third man is the wheat expert who is the official deputy inspector of the state. Quick and keen from long experience, the inspector looks for foreign matter mixed with the wheat, examines the quality of the grain, and smells for smut. Sometimes the cars are loaded fraudulently by placing inferior wheat in portions of the car where the cunning shipper imagines it will escape detection. Such cars are said to be "plugged." The inspector thrusts a brass plunger deep into the wheat in different portions of the car and brings up samples for the purpose of discovering improper loading. The elevator and commission houses have a sampling bureau, representatives from which accompany the official inspector. The samples which they secure are marked with the number and initials of the car from which they were taken. At the opening of the chamber of commerce, these samples are set out in pans, and form the basis of the day's trading. The state secures complete records and samples of all cars inspected. These are kept until the grain has passed out of the market, so that any dispute as to the quality of the grain could be easily settled. After the inspector has finished his work, the cars are resealed with the state seal. The wheat is rarely delayed more than a day in the cars in which it arrives.

Should an inspector make a slight error in judgment, it might make a difference of a grade in wheat, and a gain or loss of $25 per car. In comparison with this, the cost of inspection is nominal. If there is dissatisfaction with the inspector's decision, appeal may be made to a state board which is especially appointed to hear such complaints. Unless the grade of the wheat is changed, the expense of the second inspection must be borne by the objector. In 1889, 30 to 40 cars were inspected in Minneapolis in an hour. A decade earlier 60 to 90 cars could be inspected in an hour, because the wheat was cleaner. The exporting of wheat from the interior of the United States involves from three to six inspections of any given lot of grain.

At the six terminal points of Minnesota, Minneapolis, Duluth, St. Paul, St. Cloud, New Prague and Sleepy Eye, 125,564 cars of wheat were inspected "on arrival" during 1905, and for the same year there were inspected "out of store" 59,963 cars, and 19,692,490 bushels shipped in vessels. Out of 11,009 appeals coming before the Board of Grain Appeals on all grain, in 7,859 the decisions of the chief deputies were confirmed.

The Weighing of Wheat.—A few years ago the average weight of the loads weighed at Minneapolis was 20 tons. Now monstrous weighing machines weigh 50 tons at a time. Some states have a state weighing department. That of Minnesota, located at Minneapolis, has given service which steadily grows in public confidence and favor. In 1902, it employed 68 persons, and supervised weighing at 42 elevators and 17 flour mills, besides 4 feed mills, 5 oil mills, and 3 railroad yards. It weighed 233,127 car loads and 5,564 wagon loads, which included 152,-810,383 bushels of wheat. The revenue was nearly $60,000, and the disbursements were about $4,000 more. It is the intention of the law that the service shall be self-sustaining. The department has also removed from the field a notorious class of men known as grain thieves. Only 81 errors were made in weighing 259,996 cars of grain, and 6,000,000 bushels of grain have been weighed with an average shortage of only 40 pounds per car.

The Commercial Grading of Wheat.—The value of wheat varies with its quality, and with the purpose for which it is to be used. In the school of competition, manufacturers of cereal products and large consumers of raw cereals learned that it is essential to know the relative values of different lots of grain. The experience of these men, aided by science, determined the kinds of wheat that are best adapted for various purposes, and the methods of distinguishing them. This was the origin of the commercial grades of wheat. The grading of wheat consists in examining the various lots or cargoes to determine their quality and uniformity, and in assigning them to the proper grades. The principal characteristics which aid in fixing the grade are weight per bushel; plumpness; soundness; color; and freedom from smut, foreign seeds and other matter, and from mixture with a different type of wheat. These characteristics vary so in degree and combination that they are not reasonably distinct, and consequently they are difficult of measurement and definition. Gradations are continuous, and if lines are drawn to mark the limits of the grades, it is difficult to determine the grades in cases close to the lines. Consequently, grade requirements have been couched in obscure and indefinite terms and phrases, and the responsibility for their interpretation has been left largely with the grain inspectors.

Formerly wheat was sold by sample, and grading was in effect merely the determination of the value of the grain. In storage, particular lots of grain, even if of the same grade, had to be kept separate, and when called for, they had to be delivered to the proper owner. The receipts or warrants issued for the grain by the storehouse became the equivalent of the grain in the market. In the early fifties, the movement of vast crops from scattered sources became very unwieldy and difficult under the old methods of selling by sample. It was necessary to store in bulk enormous quantities of grain. The difficulties of delivering on demand particular lots of wheat to individual owners became very great. As a result, the grain trade made the most important advance of its history. Storage in bulk of all grain of the same grade was made without preserving the identity of particular lots, and general receipts were issued for the specified amount of grain of a certain grade. These receipts could be delivered in fulfillment of contracts, and when grain was withdrawn from storage, a specified amount instead of a specified lot of a particular grade was delivered by the warehouseman. Most of the wheat in Chicago was thus graded by 1860, but general receipts were not adopted in New York until 1874. In some markets, the inspection and grading of grain have reached such a degree of honesty and efficiency that samples are dispensed with entirely.

Contract Grades.—Trade organizations whose members deal in grain exist in nearly all of the larger cities of the United States. These organizations have an important function in the grain trade, for they afford means for easy communication between producer and consumer, and they aid in avoiding acute conditions of supply and demand. They have adopted rules of trade which aim at a maximum of business with a minimum of expense and friction. The established grades of grain form a part of these regulations. The trade organization of each market establishes a ''contract grade'' for its own market. The contract grades are understood in all contracts not specifying otherwise. There may be several contract grades on the same market, and there may be a difference of several cents in the actual milling value of a contract grade designated by the same name in different markets. This variation arises from a difference in the rules which regulate the respective inspecting bureaus, and is the cause of some confusion. Where there is no state inspection, the trade organizations manage their own inspection departments.

The Need of Uniform Grades.—Great as has been the value of inspection and grading to the grain trade, the service is not without its shortcomings. The greatest difficulty is lack of uniformity in grades. The different states and trade organizations establish their grades quite independently of each other, and this does not tend to give the uniform grades which the inter·market, interstate and international grain trade demands. The inspector begins with indefinite standards. He is buffeted about by opposing interests which are vitally concerned in his decisions. He must work rapidly. Sometimes the weather and light place him at a great disadvantage. Frequently he lacks apparatus for deciding doubtful cases. If reinspection is called for, he rarely knows when a change of grade is made, and why. In many cases, not only do the inspectors grade with their unaided judgment, but they also have little opportunity for correcting this judgment. The demands of the domestic and foreign wheat trade for more uniform grades are imperative.

Interest in the exact and uniform grading of wheat and other grains has come mainly from two sources, the grain dealers and the United States department of agriculture. The general concensus of opinion has been that existing difficulties can best be removed without governmental control, which, however, has some advocates. The grain-inspection work of the department of agriculture has had for its principal objects the study of methods used in the determination of different varieties of wheat, and the study of commercial grades of cereals. The work of the grain dealers has found expression in the national organization of the chief inspectors. This organization established grades of wheat which it recommended to the grain trade for uniform use.

Commercial Classes and Grades of Wheat officially recognized and adopted at Chicago and New York are given below. Wheat may be of such poor quality or condition as to be graded "rejected," or "no grade." Wheat that is wet, in a heating condition, burned, or badly smutted generally falls into the lowest grades.

CHICAGO.

White winter wheat, Nos. 1, 2, 3 and 4.
Long red winter wheat, Nos. 1 and 2.
Red winter wheat, Nos. 1, 2, 3 and 4.
Hard winter wheat, Nos. 1, 2, 3 and 4.
Colorado wheat, Nos. 1, 2 and 3.
Northern spring wheat, Nos. 1 and 2.
Spring wheat, Nos. 1, 2, 3 and 4.
White spring wheat, Nos. 1, 2, 3 and 4.

NEW YORK.

Winter wheat, Nos. 1, 2, 3 and 4.
Red winter wheat, Nos. 1, 2, 3 and 4.
Mixed winter wheat, Nos. 1, 2, 3 and 4.
Hard winter wheat, Nos. 1, 2, 3 and 4.
Western wheat, Nos. 1 and 2.
Spring wheat, Nos. 1, 2, 3, 4 and No. 1 Northern.
Macaroni wheat, Nos. 1, 2 and 3.

The rules for grading red winter wheat in New York are as follows:

No. 1. Red winter wheat shall be sound, plump, dry, well cleaned, and weigh not less than 60 lbs. Winchester standard.

No. 2. Red winter wheat shall be sound, dry and reasonably clean, contain not more than 10 per cent of white winter wheat, and weigh not less than 58 lbs. Winchester standard.

No. 3. Red winter wheat shall be sound, dry and reasonably clean, contain not more than 10 per cent of white winter wheat, and weigh not less than 56½ lbs. Winchester standard.

No. 4 Red winter wheat shall include all red winter wheat not fit for a higher grade in consequence of being of poor quality, damp, musty, dirty and weigh not less than 52 lbs. Winchester standard.

The first wheat that was raised in the Red river valley grew on a rich virgin soil that was free from weeds, and consequently the grain was of high quality and quite free from foreign matter. As the soil became impoverished and weeds became more prevalent, wheat deteriorated in quality and extraneous matter increased. In the eighties, "No. 1 hard" was the contract grade in the terminal markets, and for several years over one-half of the wheat received at Duluth was of this grade. Later the contract grade was reduced to No. 1 northern. Not 15 per cent of the crop of 1898 which came to Minneapolis was good enough for even this grade. Of 125,564 cars of wheat received at the six terminal points of Minnesota during 1905, 109,160 contained northern spring wheat, 11,118 winter wheat, 3,391 western white wheat, and 1,557 western red wheat. Of 143,375 cars received during 1902, 139,857 contained northern spring, 2,909 winter, 516 red winter, 53 northern white, 21 white winter, and 19 western white and red. The net average dockage was

21.5 ounces per bushel in 1904, and 18.6 ounces in 1905. Eighteen grades of wheat were recognized in Minnesota in 1902. Wheat contracts well illustrate Gresham's law and the action of a double standard, inasmuch as that grade which is most abundant and cheapest in any one year becomes the contract grade for that year, and other grades are delivered only at a premium.[1] The grade is always either understood or specified in contracts on the produce exchanges, and a contract cannot be settled except by a delivery of that grade, or of some higher grade. It is only in comparatively recent times that a contract can be settled by a higher grade, for this is now allowed in order to avoid "corners."

The Mixing of Wheat.—After grades became fixed, houses for cleaning grains and bringing them up to the standard were established. These branched out to include a system of mixing higher and lower grades of wheat to "bring the whole up so it would pass muster, according to the rules of the respective inspection departments for which the mixture might have been made." Grades were thus manufactured. In New York, for example, there were two classifications of grades, one for delivery on the New York produce exchange, and the other for export, both under the same name. The mixing houses were private enterprises, and under no inspection. The practice increased the profit of the mixing house, but it lowered the grades of wheat. The mixer often makes a greater profit per bushel than the producer, and the business is so important that practically all terminal elevators in Chicago have their mixing houses. In running wheat of a high quality through the cleaning house, some of a lower grade is mixed with it in such proportion that the mixture barely passes the contract grade. Two cars of No. 2 wheat mixed with three cars of No. 3 may make five cars of No. 2 wheat. The difference in price between the grades may range as high as 15 cents per bushel. The mixing of wheat tends to fix its price to the disadvantage of the producer. In order to obtain a special quality of wheat, a premium must be paid for it. Export grain sold by sample commands a premium of from 1 to 4 cents per bushel over the speculative grades held in store in American grain centers. The benefit of this premium goes to the mixer and seller of the wheat, and not to the farmer.

[1] Emery, Speculation, p. 137.

Wheat taken out of storage is not always of the same quality as that stored. The buyer who purchases in a territory where a low grade of wheat predominates is at a great disadvantage in competing with a buyer who purchases in a territory where the grade varies. Most of the mixing of wheat is done at the primary markets.[1]

The Advantages of Mixing Wheat are great, and perhaps more than counterbalance the evils resulting from the practice. Without the mixing of wheat, the farmer would be at a great disadvantage because the demand for off grades would cease. Legislative efforts have been made to stop the mixing of grain, but supervision by duly authorized inspectors is a more probable solution of the difficulty. Some elevators make an exclusive business of handling wheat that is shrunken, damp or injured, and work it up a grade by drying, cleaning and mixing. Damp wheat is turned over to them by the regular companies, who do not care to put it in their elevators.

Insurance.—There was insurance on goods in trust at least as early as 1704. On granary risks of stored grain the rate under the London mercantile tariff in 1877 was 0.76 per cent. After the fire at the King and Queen Wharf of that year, the rate was raised to 1.08 per cent. In recent years in the United States, a few companies are writing insurance on wheat in the stack or granary, upon which they charge a rate of 1 per cent per annum. Wheat in elevators at the local market is insured by most of the large fire insurance companies at a rate depending upon the construction and hazard of the elevator, and varying from 1.5 per cent to 3 per cent per annum. Grain in transit is insured under railroad schedule policies written by a syndicate of companies in New York. The rate upon this class of risks is from .60 per cent to 1.5 per cent, for it varies from year to year. The rate of insurance for wheat stored in elevators at the primary or seaboard markets varies from .50 per cent in modern elevators of steel and concrete construction to 3.15 per cent in elevators of other construction, according to type of construction and surroundings. In Canada, the law compels warehousemen to insure stored grain, and the average rate on grain in elevators is nearly 2 per cent.

[1] Industrial Commission, 10:cccxxi; cccxxix.

Marine Insurance of grain cargoes transported on lakes, rivers or oceans is obtainable. Losses on such transportation in American ships were so great during the winters of 1878 and 1879 that "many underwriters on either side of the Atlantic ceased to write them at any premium." Thirty-five years ago the insurance rate on grain-carrying sailing vessels to Liverpool was 1.25 per cent from New York and 2.25 per cent from Montreal, and on steam vessels 1 per cent from New York and 1.25 per cent from Montreal.

Financiering the Movement of Wheat.—The large money centers are not as great a factor in the moving of the wheat crop as they were at an earlier date, for to a large extent the rural sections now do their own banking. The banking power has grown much faster than the increasing money requirements for moving crops. In 1890 the banking power of the chief grain states was to the money power required to move the grain crop as 4 to 6, and a decade later the ratio was as 7 to 6. The grain growing region now has sufficient capital to move the cereals from first hands and to start them well on their way through the commercial channels. A dealer furnishing money for about 175 country elevators in Minnesota and the Dakotas sends out $500 to $1,000 to each elevator, making from $100,000 to $150,000 sent out the first day. Cars are not obtained on this day, and perhaps 50,000 to 100,000 bushels are purchased. A sort of paymaster is located in the elevator towns, and these keep the principal informed as to the amount of wheat purchased daily, and as to the amount of cash that will be required the next day. Much of this cash must usually be borrowed, but warehouse receipts for grain already in elevators are good security on which an amount of money close to the cash price of the wheat can be borrowed from the country banker. There must be a car load of the same grade of grain before shipment can be made. When the grain does begin to move it takes several days for it to get to market, and five or six days' receipts are often on hand before cash is realized. As soon as a car is loaded, the elevator man draws a sight draft on the commission house at the primary market for the amount that he borrowed from the country banker, attaches the bill of lading, and deposits the draft in the country bank as a cash item. Cables are frequently sent at night to every market of the world in order to sell wheat. What cannot be sold must be held, and future sales upon the speculative markets can be made as an insurance against loss from price fluctuations.

The country banker sends the draft to his correspondent at the market, where collection is made. As soon as the wheat reaches the terminal warehouse, it is again available for a loan close to its market value. If the terminal factor is an exporter, he also attaches bills of lading to a draft drawn against the shipment, and his banker accepts this draft as cash at current exchange rates, which include interest on the money until the draft is paid. Outside of the money used by the railroads, it requires about $500,000,000 to move the grain crops.[1] If the farmers do not wish to sell their wheat at once, they can place it in the elevator and receive a receipt on which they can borrow 90 per cent of its value from the banks.

The Marketing of Wheat in Foreign Countries.—The characteristic feature of the wheat movement in the United States consists in concentrating the surplus for export. Only a few of the larger exporting countries resemble the United States in this respect. In the non-exporting and in the importing countries, the main problem is the distribution of the wheat among the population. In this case the entire machinery of marketing and transportation must be modified and adapted to the conditions peculiar to each country. In exporting countries, the wheat is bought by buyers who are either established at the local centers, or who travel through the country purchasing grain from farm to farm. It is only the larger grain centers of Europe which employ the economical American system of elevators in handling grain.

RUSSIA.—In Russia, as is usual in foreign wheat producing countries, the machinery for buying, handling and transporting wheat is very imperfect. Where transportation facilities are adequate their use is expensive. On long distances railroad rates have been higher than in the United States. They have been estimated at 3 per cent of the cost of production. Russia is well supplied with rivers, and a decade ago the larger proportion of export grain was still moved by river and canal. The railways have now become a more important means of transportation than the rivers and other water routes, and they will

[1] Industrial Commission, 6:135-137; 11:10-11.

doubtless be the great factor in the future development of the
country. The construction of the trans-Siberian railroad has
been considered as the initial step in the opening of extensive
grain fields. This railway is about 6,600 miles long in its direct
line. Earth was broken for its construction in 1891. The road
has been completed, but ''what this country can do in the way
of wheat production is yet to be demonstrated.'' On account
of high freights, wheat cannot be shipped to the frontier by
rail, and the surplus of Western Siberia does not get beyond
the rural districts. Much of the grain from the western wheat
lands of Siberia is carried by boat down the Irtish and up the
Tura to Tiumen, from which place it is forwarded by rail to
Russia. Some is also shipped east and west on the trans-
Siberian railroad.

In Russia, grain was formerly handled in sacks. There were
no elevators at the country stations and the grain was much
damaged from exposure to the elements. The same state of
affairs existed at the seaports, where the grain was further
damaged. Here an attempt was made to classify the grain
according to its quality, but there was no machinery for cleaning it. Screenings were bought from the farmers and again
mixed with the wheat. Various other extraneous matters were
also introduced, such as manure, sand, and a species of grass,
Kukal. The latter was in such demand at times as to bring a
higher price at Odessa than rye.

In 1888 the first warehouse with elevators was erected in
Russia, and it did not pay expenses. Subsequently the Russian
government assisted in erecting grain elevators on the American plan. These were mainly along the lines of the southern
railway, and at Odessa and other southern ports. In 1895 there
were 55 warehouses with elevators, having a capacity of about
8,905,000 bushels, and 221 warehouses without elevators, having
a capacity of about 9,082,000 bushels. In 1898 over 50 per cent
of the Russian wheat contained 2 per cent of foreign matter,
and some of it contained as high as 12 per cent. No attempt
at grading and inspecting the wheat has thus far been successful. It is mostly sold on sample in Great Britain, and there
are frequent complaints of fraud. Some fruitless efforts have
been made to get Russian wheat sold on a 5 per cent extraneous

matter basis, a plan recently adopted in Roumania. Experience in other countries has shown that if such efforts were successful, the most important result would be the transportation of that much more rubbish from Russia to England.

INDIA.—An immense stimulus was given to wheat cultivation in India by the development of transportation facilities. The first of these was the completion of the Suez canal in 1869. This, however, reached its greatest importance only after some railroads were built into the wheat districts. In the eighties the movement of wheat was still greatly hampered, not only by high railroad rates, but by the entire lack of railroads in many of the best wheat districts. The situation had not greatly improved yet in 1898, when there were few branches to the railroads, the country roads were poor and freights were high. The traveler still saw the long lines of camels that were silently and majestically treading their way through the night across the plains to the seaports, in successful competition with the railroads as grain carriers. After threshing, the grain is left lying on the ground, or it is buried in pits. In the latter case, it suffers less from destructive insects than if placed in granaries. Cartmen haul it to market. Ninety per cent of them do not haul their own grain, but engage in a speculative business of buying and selling. In the eighties, there was much fraud practiced by these cartmen in handling wheat. Dirt was mixed freely with the grain. The ingenuity and resourcefulness of the cartmen seems almost incredible. In 1889 McDougall wrote: ''There are 10 or 11 villages in which the lower classes make it a trade to supply different colored earths to suit the color and size of the different kinds of grain. The earth is worked into small grains to look like grain, and the traders say it is impossible to winnow out this description of dirt. . . . Water, again, is put in to increase the weight. All these practices are resorted to by the conveying traders in self-protection against the tricks of traders, who rob them in various ways.'' [1] A poor quality of wheat was also mixed with a good one, and then the whole was given a uniform color by mixing with clay. Firms engaged openly in selling this clay. As a result of all these manipulations, the wheat did not arrive at the foreign

[1] Jour. of Soc Art., 37:644.

market in as prime condition as might have been wished. It could not be shipped to Germany, and the English buyer deducted 5 per cent "refraction." The Indian exporter soon learned to exercise care lest any wheat containing less than 5 per cent dirt should be shipped to England. He was sometimes forced to mix 2 to 3 per cent of foreign matter with the wheat in order not to sustain a loss. This caused an economic loss, not only in annually transporting 15,000 to 20,000 tons of trash to England, but the English miller was obliged to devise machinery to clean this wheat. These evils were partially remedied in the nineties. In 1898, 15 grades of wheat were shipped to England from India. In good years, the storage capacity of Bombay is exhausted by the wheat brought from the central provinces of India. The wheat of the Punjab is collected at Multan and shipped from Karachi. Considerable wheat flour is ground and exported at Bombay and other centers.

ARGENTINA.—The Argentine wheat grower has no granaries on his farm, and consequently his entire crop is marketed as soon after harvest as possible. Lack of improved facilities and methods are a source of great loss. The grain is handled in bags, which are very expensive and which are of such poor quality that there is quite a loss from leakage. The country roads are very poor. The wheat is hauled in immense two or four-wheeled wagons having wheels 8 feet in diameter. The two-wheelers are hauled by 12 to 15 horses or mules, or by 8 to 16 bullocks. One animal is fastened between the huge thills, and the others are hooked on by means of ropes tied to any portion of the cart to which a rope can be fastened. The yoke of the oxen is fastened to their horns, and the driver's seat is on the yoke between the heads of two oxen. The four-wheelers carry from 4 to 6 tons, and require more animals to draw them. The hauling is not generally done by the producers of wheat, but by men who make a business of hauling. The grain is hauled from 15 to 60 miles. Corrugated iron warehouses have been built at some of the principal wheat stations, but they are used only by the large producers and dealers. As a rule, warehouses are not available for the small farmer, nor would he store his grain if they were. He is so ignorant that he prefers to pile his wheat outdoors exposed to the weather.

Such grain is often damaged by rains, and these conditions prevail at the farm, at the railway station and at the seaport. Sometimes the piles of sacks are covered, and this greatly reduces the damage.

Transportation to the seaports is almost exclusively by rail. Of the 26 Argentine railways in operation in 1903, 22 were built mainly in order to transport wheat. The Parana is navigable to Rosaria, the only large inland city. From this point at least 5 railroads branch out into the wheat regions. The car facilities are inadequate to ship the wheat, and the bags often lie in the yards 2 months awaiting shipment. The grain is frequently shipped in open flat cars covered with canvas, but it sometimes gets wet before it is unloaded. The railways are all English, and consequently most of the cars are of the old English type. They have a capacity of from 10 to 18 tons, but the many new cars being built have a capacity of from 30 to 40 tons. The freight rates vary from 5 to 15 cents per bushel. They fell about 3·cents per bushel from 1895 to 1902. There are portions of Argentina where wheat cannot be raised for export merely because transportation facilities are lacking.

Although shipping facilities at the seaports are growing rapidly they are still entirely inadequate. Ships wait for days before they can be loaded. Then, berthed three deep in the port, it takes several days more to load, especially when men carry the bags of wheat, one at a time. Two other methods of loading are also in use. Steam winches lift the bags, or an endless belt carries them. Tramp steamers of 2,500 to 6,000 tons register usually do the ocean transportation. The rate to Europe in 1903 was from 6 to 12 cents per bushel. The grain exporters keep branch establishments at the main points where wheat is raised. They buy through an agent. A price is telegraphed to him in the morning, and this he pays all day, as he rides from farm to farm. He often buys from the machine, for the exporter gets his wheat on board ship as soon as possible. Each buyer does his own inspecting and grading. The agent is paid 1 per cent commission on all he buys.

CANADA.—In the marketing of wheat, as in nearly all other phases of the wheat industry, the development in Canada has been similar to that in the United States, only later. More than half the arable lands of Canada cannot be utilized yet because the requisite population and means of transportation are wanting. Some of these lands are among the best wheat lands in the world. The railroads are, however, rapidly ramifying through these regions. New trans-continental lines are being planned and built. As transportation facilities improve and population increases, the development of Canada will be unprecedented. Elevator building is at present very active in the Canadian northwest, both along the new lines of road and along the old lines. As far as the wheat trade is concerned, Winnipeg is the Chicago and Montreal the New York of Canada. The most noteworthy difference between Canada and the United States in connection with the marketing of wheat is in grading. Grading is entirely under the control of the Dominion government, which appoints the grain inspectors in the different markets. Uniform grades are fixed by law for the whole country.

CHAPTER XIV.

THE PRICE OF WHEAT [1]

The Factors of Price.—The price of wheat is normally determined by the world conditions of supply and demand which pertain to bread stuffs. The control exercised over price by these conditions is immediate and transient. Direct variations in price result from variations in supply or demand. Price in turn causes supply and demand to vary by reacting upon them. Such variations are, comparatively speaking, of slower action and more permanent. Legislation may also become a determining factor of price in certain countries, as, for example, when import duties on grain are established.

Supply and Demand.—Wheat and rye are the world's principal breadstuffs. There is sufficient variation in supply and demand to cause great fluctuations in price. Many causes of variation in the supply of breadstuffs exist, among which may be mentioned: (1) The great changes in climate and in abundance of rainfall to which the natural wheat and rye regions are subject; (2) the variations in acreage which result as a reaction to variations in price; (3) the increase in acreage resulting from the settlement of new countries; (4) the decrease in acreage due to planting a larger acreage of other cereals, especially corn, when there is an unusual demand for live stock feed; (5) the degree of competition, which may affect the supply at a given time or place; (6) the establishment or abolition of trade restraints by commercial treaties; (7) the hindering of transportation by war; and (8) the continuous advance of the arts of production, communication and transportation. In Europe, the average annual production of rye is approximately as great as that of wheat, while the European production of both crops taken collectively averages about 70 per cent of that of the entire world. When wheat is relatively high in price, and rye is relatively low, consumption of the latter grain increases and the demand for the former decreases.

[1] For criticism and many valuable suggestions on this chapter the writer is indebted to Prof. H. C. Emery and Dr. J. Pease Norton.

Other causes which effect a variation in the demand for wheat are: (1) War, which causes a variation in the foreign demand; (2) the standard of living is rising, and this increases the demand, especially in rye consuming countries; (3) commerce and the introduction of a foreign civilization may increase the consumption of wheat, as in some of the Oriental countries. The condition that wheat is the staple food of man in the nations of the highest civilization and of the greatest economic strength tends to keep the demand for wheat firm, while the fact that the world supply of wheat comes from all quarters of the globe tends to prevent acute scarcity of the general supply. The demand for breadstuffs taken collectively is comparatively inelastic.

The Reactions of Price.—If the price of wheat falls so low that its production becomes unprofitable there will be a tendency for capital to become engaged in the production of other crops which yield a larger return. On the other hand, there are many substitutes for wheat which are at the command of the consumer, and which he uses when the price of wheat rises too high. This lessens the demand for wheat, lowers its price, and decreases its production. Thus the forces of supply and demand always seek equilibrium. To say that the producer must get what he can for his product is not sufficient. If he does not get what is economically just, on the whole and in the long run, then he must stop producing, and his capital will seek other channels until it again receives its due return in this. The value of wheat to the consumer must on an average be high enough to cover the cost of production and the expense of distribution. As a general rule, the consumer is comparatively more free to delay purchasing than the producer is to delay marketing, and hence the interest of the latter is the more critical one. It has been said that price is "determined normally by the net cost of producing an adequate supply."[1] It is true that the price of wheat cannot normally be below the cost of production. It is no less true, however, that it cannot be above what the consumer is willing and able to pay. Cost of production and value to consumer are respectively the minimum and maximum limits of price, and they are both just as essential in determining price as the numerator and

[1] Industrial Commission, 6:32.

STORAGE AT PRIMARY MARKET, THE ST. ANTHONY, MINNESOTA ELEVATORS

denominator are in determining the value of a fraction. Historically, demand came first, while production followed and grew to such proportions as was warranted by the demand. In modern times of enterprise, however, the chain of causation may be reversed, for by decreasing the cost of production a larger supply at a lower price can be placed upon the market, and in consequence of the lower price demand increases and more wheat is consumed.

Communication and Transportation become an important factor in the price of wheat wherever the market has developed beyond the most limited local conditions. Prices that were formerly awaited for 2 or 3 months are now flashed by electricity over the whole world during the same day on which they are made. A favorable location is no longer an advantage in determining prices, for all markets are affected simultaneously by a change in either supply or demand. All improvements in communication and transportation resulting in a decrease of charges tend to lower the cost and increase the amount of production permanently, and hence they enable the producer to compete more successfully in the world's markets. If two countries have surplus wheat for export, a few cents more or less per bushel on the whole cost of moving may determine which country can sell at a price that will secure the trade. Ideally, the only difference of price which should exist between any two markets, or between what the producer receives for his wheat and what the consumer pays, is that resulting from transportation and commercial charges, and the cost of such manufacturing processes as the wheat may be put through. These are the only variations that should occur in the world price of wheat.

Competition and Price.—Competition is a powerful factor in determining the specific price paid for wheat, especially that paid to the producer. By means of competition, all charges incidental to moving wheat are kept at a minimum, while the price paid for the grain is kept at a maximum. For example, at Milton, North Dakota, a non-competitive point, 2 cents less was paid per bushel for wheat than at competitive points only 6 miles distant. When the local elevator systems combine against the interests of the farmer, the only effective remedy is for the farmers to combine among themselves and enlist the interest of the railroads. When the elevator systems also combine with the railroads, the farmer seems to be quite at their mercy. It is claimed that this condition of affairs is shown to exist by the recent investigation of the Interstate Commerce Commission, and that the elevators control prices. The only remedy would seem to be for the farmers to start independent elevators, and to secure the aid of law, if necessary, to get their wheat shipped to the primary markets, where competition has generally kept up the level of prices. The proof of the latter statement is shown by the fact that the primary market with the highest level of prices has secured the traffic.

Exportation and Price.—There is a tendency for exportation to decrease as population increases. When a country consumes all of the wheat which it produces, then its price of wheat is fixed within the country, provided there are no restraints to trade, and that the cost of production is not greater than the cost of importing grain. As soon as a country has a surplus for export, and receives more for exported wheat than the home price, plus the expense of exporting, exporting will increase, the home price will rise, production will increase, and the price is no longer fixed within the country. The country which buys the export may thus fix the price of wheat for the country which produces it, but such a price under normal conditions must always be higher than that which the producing country could possibly fix for itself, and consequently a benefit to the latter country. It is as a consumer of the world's surplus that England has held a position of such commanding importance in fixing the price of wheat. It has been asked whether the large combinations of grain interests can or do fix grain prices. The only conditions under which they could permanently do so in a large market would be that they have an approximately complete knowledge of the conditions of supply and demand, and that they would be far-sighted enough to fix the price in accordance with what it naturally should be under the existing conditions.

The Visible Supply and Price.—The consumers of wheat always have an advantage over the producers in that demand is never so tangible a factor as supply. There is always a large "visible supply" of wheat, for the grain consumed during the entire year is produced within a few months. While the American producers are now well able to carry their wheat, it is, nevertheless, still largely concentrated at the shipping points. Some of the reasons for this are the active competition of the primary markets to make sure of securing the grain by buying at once, the ample facilities for economically storing and handling wheat in great bulk at the terminal markets, and the presence of a class of men who, having capital and commercial capacity, have justified their existence by the manner in which they have handled the reserve supply. As all value is, in its last analysis, a subjective thing, the existence of this great visible supply must have the psychological effect of delaying, and perhaps lessening demand, and thus decreasing price. Since the producer has already thrown the wheat into the market, the distributer must either dispose of it at once, or, if he holds it indefinitely, run the risk of loss from depreciation as the next harvest approaches. This tends to put the consumer in a position to set the final price on wheat, a position that is further strengthened by the fact that competing wheat countries of the southern hemisphere throw their surplus into the world market about midway between North American harvests. Argentina has thus become notorious as a disturber of wheat prices, as have also Australia and India to a lesser extent. During one-fourth of the year, three-fourths of the world's wheat supply comes upon the market. This results in a congestion of supply which exerts a powerful influence in determining the price for the remaining three-fourths of the year. While this tends to give stability of price, it favors the consumer and not the producer. An increase in the local consumption of wheat by milling is decreasing the visible supply. An import duty adds the amount of the duty to the cost of production, and consequently must raise the price of the wheat imported. In the following table are given the import duties on wheat and wheat flour in 1907 in the principal importing countries having such charges.[1]

[1] Data furnished by U. S. Dept. of Commerce and Labor.

Countries	Wheat per bushel	Wheat flour per bushel
Austria-Hungary	$0.35	$0.82
Belgium	free	0.10
France	0.36	0.55
Germany	0.35	0.66
Italy	0.44	0.60
Spain	0.34	0.58

The Market.—Price is always determined in a market. The old proverb, "three women and a goose make a market," is true, assuming that the women possess some other commodities which they can exchange, for all the essential primary elements of a market are present, namely: A commodity; its owner; and one or more other persons, each of whom wishes to become the owner of the commodity by exchanging it for a quantity of some other goods. The primary origin of the "supply" and the "demand" is external to this market, which is merely the point where the forces of supply and demand meet and attain equilibrium through the exchange of commodities. A market increases in size, complexity and importance as these elements increase in number, as they are modified in form, and as consequent manipulations arise. We found a market at the center of each of the three concentric circles of distribution. The marketing or bargaining is early concentrated in the local market. The local market, so long as it is independent of larger markets, is so limited in all of its factors as to be easily known in every phase and violently affected by every local event of importance. The city market resulted from the conditions of supply and demand pertaining to a much larger territory, and is correspondingly more complex. It is not so violently affected by any single event as is the local market, and events tending to opposite results may offset one another. Prices are always a resultant of the forces or conditions of supply and demand that exist in the whole territory tributary to the market, and that have been there concentrated, directly and indirectly. Prices can no longer be predicted from a knowledge of conditions in any one locality. What is true of the city market is pre-eminently true of the foreign or international market. Under normal conditions, the wider market determines the price. After world markets had arisen, the local market became an insignificant factor in determining prices, even within its own circle of distribution. This change took place in the United States about the middle of the nineteenth century. Before this date prices of grain were determined chiefly by local conditions.

Concentration of Price-Determining Influences now occurs in the world market. Modern transportation, which enables the California wheat grower to send his product to the Liverpool market, and modern communication, which has practically eliminated the time element in sending news, are the factors which have made the whole world tributary to the great central markets, where the changes that affect supply and demand are continually recorded, and given their due weight by the keenest of experts in the modification of prices. These changes vary greatly in character, and they are reported from every wheat raising quarter of the globe. If a telegram is received saying that the monsoon in India is overdue; that the drought in Kansas has been broken; that a swarm of grasshoppers has been seen in Manitoba; that a hot wind is blowing in Argentina; that navigation on the Danube is unusually early; that bad roads in the Red river valley are preventing delivery; that ocean freights to China have risen; or that Australian grain "to arrive" is freely offered in London, prices rise or fall to a degree that corresponds to the importance attached to the news. New inventions and discoveries, legislative enactments and international agreements, political, commercial and industrial complications—all have their effect upon prices.

The Rise of the Speculative Market.—When the local market was the center of all distribution, the producer, having under his own observation all of the factors which determined price, endeavored to hold his products for sale until such a time as when he would receive the highest price. This was the first form of speculation, and it must have arisen very early in civilization. It has had a continuous existence until the present day. Practically every wheat grower who holds his crop for a rise in price, instead of selling it as soon as it is threshed, is a speculator of this class.

Dealers in Grain as a class became differentiated from the producers at an early date. They frequently bought and sold wheat, not merely for trade profits, but to make an additional profit by taking advantage of the fluctuations of price resulting from variations in supply and demand. They bought wheat outright, and held it for a higher price, and thus they belonged to the same class of speculators as did the producers who held wheat. With the great development of the arts of transportation and communication during the middle of the nineteenth century, there arose the market which covered the entire civilized world. In this world market, great, sudden and unforeseen changes in the conditions of supply and demand occurred, and the uncertainties of trade became so great that the possibility of a total loss of the capital of the dealer grew very burdensome. Before the great and varied mass of phenomena which affect the price of wheat, the producers and ordinary dealers stood quite helpless, as far as forming an adequate judgment of effect on prices was concerned, even if they could secure timely reports of changed conditions. As a result, dealers became differentiated into two classes. One of these classes, the wheat dealer proper, is in the market simply to secure those trade profits which always exist independently of speculative profits. The other class is that of professional speculators. This special class formed organizations in the large exchanges, all of which existed as commercial institutions in pre-speculative times. The organized speculative market arose in direct response to conditions which brought risks that were intolerable to the ordinary dealer and its development was hastened because it took place at a time when the risks usually incident to the wheat trade were greatly augmented by those resulting from the Civil War. While two typical classes of persons, dealers and speculators, are engaged in the grain trade, it must not be understood that these classes are mutually exclusive. There are large millers and producers, for example, who keep well enough informed on the market to engage properly and profitably in speculative dealings.

The Machinery of Speculation.—The early speculator stood ready to purchase wheat at the current price, and he assumed the risk of a fall in price in the hope that he might gain from a rise in price. "Bull" speculation, which consists of first buying, and then selling at a later date, is the term by which the operations of this speculator are designated. He always desires a rise in price, and endeavors to bull the market by buying. He operates on the "long" side of the market. This speculator also contracts in the present to purchase wheat at some future date at a price which he now fixes. Here also he assumes the risk of a fall in price in the hope that he may gain from a rise, for if the price rises above what he has agreed to pay at the fixed future date, known as the date of delivery, then he is able to sell his wheat on or before this date at a higher price than he paid. In the present contract for future purchase is involved one form of the transaction technically called the "future." This term is defined by Emery as a "contract for the future delivery of some commodity, without reference to specific lots, made under the rules of some commercial body in a set form, by which the conditions as to the unit of amount, the quality, and the time of delivery are stereotyped, and only the determination of the total amount and the price is left open to the contracting parties."[1]

The other type of speculation is "bear" speculation, which consists of first selling, and then buying at a later date. In such speculation, the operator stands ready to sell wheat at the current price for present delivery, or at a fixed price for delivery at a given future date. This speculator assumes the risk of a rise in price in the hope that he may gain from a fall in price. His operations generally consist of selling in the present for future delivery. Most frequently he owns no wheat at the date of sale, but hopes to secure the contracted grain before the date of delivery (which is called covering the sale), and at a price below that at which he sold. He always desires a fall in price, and endeavors to bear the market by selling. He operates on the "short" side of the market, and his "short-sales" are always "futures."

Grain Privileges, or "Puts and Calls."—Insurance against loss in wheat transactions may be secured by buying a "put" or a "call" from a maker of privileges. For example, if a dealer is holding wheat that is worth 80 cents per bushel, for a certain price he can buy the privilege of selling the wheat to a speculator at 79½ cents per bushel during any period of

[1] Speculation, p. 46.

time that may be agreed upon. Having done so, he cannot lose more than one-half cent per bushel, plus what he paid for the "put." If the price advances, he sells at a profit, but if it falls, he delivers or "puts" the wheat to the speculator. Similarly, a "call" is the privilege of buying wheat at a certain price within a given time, and it is most frequently used in protecting short sales. Grain exporters sometimes protect their contracts with privileges. Dealings in privileges, however, have not always been held in the highest repute, and they have even been prohibited by the rules of some commercial exchanges. A privilege has no value unless its maker can meet his engagements.

Deposits Securing Contracts for future delivery may be demanded. In this case, each party makes a money deposit large enough to secure the other from loss in case of failure to fulfill the contract. If one party thus "calls an original margin," he himself must perforce deposit an amount equal to that for which he calls. In New York, the maximum deposit that can be called for wheat is 10 cents per bushel. Additional margins, equal to the fluctuations in price, may be called for, and usually are, even if there was no original margin.

Delivery.—The rules of grain trading on the various speculative exchanges contemplate the actual delivery on maturity of contracts of all wheat sold. Each contract mentions the time for which it is to run, and its maturity is on the last day of this term, which is usually the current month. In the exporting and forwarding of wheat, the time is generally determined by special contract, but in the general speculative markets the current trading is in the deliveries for July, September, December and May. The price for immediate delivery is that current for the next succeeding delivery, less the carrying charge to the beginning of the period of the next delivery.

The operator who sells 100,000 bushels of wheat in a speculative deal has three ways in which he can settle the contract on or before the date of maturity: He must either deliver the actual wheat; buy the same quantity of wheat on the same exchange; or lay himself liable to a damage suit for non-delivery. In comparatively few instances does he deliver the actual wheat, which he might possess, or which he might purchase in another market. If he buys wheat on the same exchange, his operations are settled through the clearing house, which is the same in principle as that of a bank or stock exchange. If A buys wheat from B and sells the same quantity to C, the clearing house settles both contracts for him by having B deliver to C. The great bulk of transactions are settled in this manner, which involves only the payment of differences. The latter may arise from differences in amount or grade of wheat bought and sold, or from differences in price. Thus, if a speculator buys May wheat in April, he can avoid having the actual grain delivered to himself by selling the same quantity of wheat before the date of maturity of the contract. The man who buys wheat for May may do so in two different ways: he may buy actual wheat and store it until May, or he may buy a future. In the same way, the seller for future delivery may sell actual wheat, or he may sell short and cover the sale before the date of maturity. If the speculator has bought May wheat, and wishes to hold the grain longer than until that date, he can do so by selling his May wheat at the date of maturity, and at the same time buying July wheat. He will pay the cost of storage, and he will pay or receive the difference in price, according as July wheat is higher or lower than May wheat. Speculators as well as dealers sometimes buy actual wheat and store it in anticipation of a rise in price. Contracts calling for immediate delivery are called "cash" or "spot" sales.

General Warehouse Receipts.—It is obvious that a commodity can be the subject of extensive "future" dealings only on condition that it has the representative quality. Early lake and canal shipments of wheat were sold ahead by sample "to arrive" and "for shipment" as an insurance against fluctuating prices, but the system of grading and general receipts alone made possible the real future, which is the great modern contribution to the machinery of speculation. These general receipts are usually reliable, although at least one gigantic swindle has been perpetrated by means of fraudulent warehouse receipts. From the beginning, however, the receipts have been considered as good as the wheat which they represented. In other words, wheat had become a perfect representative commodity. Being a staple article when classified, receipts issued against graded wheat are as current and negotiable as a bank check. They have the same meaning in Liverpool or Antwerp as in Chicago or New York. This greatly facilitates dealing in wheat, for a contract can be fulfilled by delivering a receipt. Under ordinary conditions such receipts can be purchased in the open market at any time, and consequently it is possible to make a contract to deliver in the future receipts not yet owned. Short sales of receipts complete the speculative machinery.

"**Hedging Sales.**"—Two lines of compensating contracts are frequently carried by dealers and manufacturers, one in their business and one in the speculative market. For example, if a merchant buys 10,000 bushels of wheat for export, he sells the same quantity of wheat in the speculative market. Later, when he sells his wheat, he buys in the market. Hedging in this manner is widely practiced by persons who deal in wheat for trade profits only, for it eliminates all risks due to fluctuating prices.

Arbitrage Transactions consist in buying in one market and selling in another when there is difference enough between the prices of the two markets to make such operations profitable. Arbitrage continues until the relative supply and demand of the two markets is so changed that prices in both markets are practically the same. This does not mean, however, that speculators change prices at will by manipulating supply and demand in the various markets, for the speculative supply and demand in the market are, under normal conditions, entirely dependent upon the actual supply and demand existing outside of the market.

The Functions of Speculation.—Speculation is the flywheel which imparts to the modern commercial machine a motion so uniform that all of its parts operate continuously and simultaneously. As men produce and consume, as well as exchange, according to comparative prices, it also directs the production and consumption of commodities into the most advantageous channels. Professional speculators are the men best equipped for securing and interpreting news of variations in supply and demand, and they determine a price that, with slight local variations, prevails throughout the wheat industry of the entire world. Price, in turn, is a sensitive barometer which records the influence of every event which immediately or mediately affects supply or demand. Speculation anticipates price-determining events to such an extent that it relieves the producer from the risk of growing wheat that he may be obliged to sell below the cost of production on account of an unforeseen change in the market. The directive control exerted through prices "is its service to society in general." The "risk-bearing function is its service to trade as such." That the need of speculation is proportionate to the magnitude of the risk element is axiomatic. Through the speculative market flows a continuously moving stream of business which will carry the risks of merchant, producer, manufacturer and consumer alike, and at any time or place. Speculation alone makes hedging transactions possible. By its anticipations, it lessens price fluctuations. The short seller is the most potent influence in preventing wide fluctuations in price, for he "keeps prices down by short sales and then keeps them strong by his covering purchases."[1] The producer always finds a ready market, and large stocks of wheat can be carried over from a season of abundance to one of scarcity without great risk of loss.

The Speculator.—The American is unquestionably the greatest and most typical of all wheat speculators. The stupendous undertakings which he sometimes assumes are characterized with an importance, as well as with a boldness and a brilliancy, that excites world-wide interest. He is practically the manager and director of the world's wheat movement. If objection is made to the great scope of his influence, it must be remembered that experience has already taught him that he cannot continue long in his position of importance unless he solves the mighty world problem that is ever presenting itself, the problem of providing bread for the non-producers of wheat. Eminently practical and clearheaded, his future vision is as keen and penetrating as was that of the prophets of old. He is necessarily a cosmopolite, and he knows the traits and needs of many races. His facilities for acquiring information are unsurpassed. The governmental weather map shows him the rising storm which threatens Kansas wheat. The experiment station bulletin informs him that the next year's crop will be damaged 20 per cent by the Hessian fly. The state government weighs and grades his wheat. He knows the progress of harvesting in Australia and Argentina. The transportation companies give him regular quotations of freight rates to all parts of the world. Telegrams and cablegrams give him immediately the changes of

[1] Emery, Speculation, p. 121.

price in the principal markets. He has an intimate knowledge of the visible supply of wheat that is stored in the world's great terminal elevators, and of the wheat that is being transported in car and vessel. His eye is always on the fine wavering ratio line between supply and demand, and from its movements he determines the form of his price line. The markets make wheat so liquid for him that the banks will advance him money at the lowest rates on elevator certificates in larger proportion to their value than they will on the safest real estate.

The Pit—The Chicago Board of Trade is perhaps the most powerful and famous institution which furnishes an organization for dealing in wheat. When the speculators assemble in the pit, the board become a clearing house of opinion that forms a very picturesque and dramatic institution. Their methods of business are an excellent illustration of development brought about simply by utility. The very life of the institution depends upon the scrupulous honesty of all its members. It is more profitable for the operators to make certain honest gains than to destroy the institution by endeavoring to make dishonest ones, and the degree of integrity that can be attained under these circumstances is truly remarkable. Not that brokers and speculators are an unusually upright class of men, as judged by their actions when not operating on the exchange, but that it simply pays to be honest. Any quantity of wheat can be bought on the floor of the exchange by a sign, a nod or a shout, or by a scrawl on a trading card. Either party to the deal could easily claim that the sign had not been noticed or understood, and the contention could not be disproven, nor could the contract be enforced before any court in the land. Considering the great confusion and excitement of the pit, the ease and rapidity with which fortunes are often made and lost, and the many opportunities and temptations for dishonest dealings, it is certainly an exceptional record that the Chicago Board of Trade finds it necessary to expel on an average only five members a year.

The Volume of Transactions.—Perhaps 90 per cent of all transactions on the Chicago board are pure speculation, neither side expecting to receive or deliver a bushel of grain. The "spot" sales on the New York produce exchange in 1895 amounted to 43,405,076 bushels, while the "futures" amounted to 1,443,875,000 bushels. The New York market for wheat is small compared to that of Chicago. Record of the amount of trading in options is no longer kept, but in a lively market it runs into millions of bushels daily. Under very exceptional circumstances it is said that ten million bushels of wheat have been sold in the Chicago pit in less than ten minutes.

The unit on the Chicago and New York exchanges is 5,000 bushels of wheat. In Chicago 1 per cent variation is allowed on the contract, and in New York 5 per cent. When wheat is sold in "boat-load lots to arrive" 8,000 bushels is understood as the unit, and 10 per cent deficit or excess does not vitiate the delivery.

The Evils of Speculation.—The modern speculative system is of such recent origin, and its operations seem so complex to the ordinary layman who is unacquainted with produce exchanges, that it was and is little understood. Its evils were more easily recognized than its benefits. Without an understanding of speculation, it was easy to ascribe many evils to it with which it had no connection. "The modern system of 'futures' has proved itself a convenient scapegoat for all the evils of the grain trade. It is charged with being the cause of low prices and of high prices, with increasing trade risks, and with diminishing them till there is no chance for profit. A few years ago the farming class clamored for the suppression of the speculative market, while recently the Kansas farmers started a movement to contribute a cent a bushel on all their wheat to a fund for the benefit of the most daring speculator of the Chicago market."[1] As the functions of modern speculation were better understood, its advantage became more apparent, and the speculator was looked upon as something more than a mere gambler. Opposition became more rational and less intense. The evils of speculation may be divided into three general classes: (1) Corners; (2) public gambling and "bucket shops;" and (3) manipulations in general.

Corners.—To "corner" wheat is to secure such a control over the existing supply as to be able to dictate its price. Success in this is so difficult that it is very rare. Corners and the opposition to them are not of modern origin, for even in antiquity there were prohibitions to cornering grain.[2] Perhaps the first

[1] Emery, Econ. Jour. (1899) 9:45.
[2] Lexis, Handworterbuch d. Staatswissenschaften, 3:861.

great corner was that of Joseph in Egypt. He bought grain outright. This was the only type of corner that could be effected before the advent of the modern speculative market. The appearance of the world market has made impossible the perfect control of the whole supply of actual wheat. Even partial control is possible only under very unusual and favorable conditions. Neither can any great or extended control over local supply be maintained, on account of the ease and rapidity with which wheat can be transported to any market.

The speculative corner arose with the practice of selling short. This is not a corner of the world supply of wheat, but only of wheat for delivery at a particular time and place. Such a corner is always run by the bulls, who effect the overselling of the market by securing control of local supplies and by inducing short-selling. The result is that there is no wheat with which to cover short-sales. Such a corner is absolutely effective, for the short sellers must cover their contracts before the end of the month, or default. It is not without its difficulties, however. In order that the shorts can move no wheat for delivery except on the terms of the cornerer, he must buy at rising prices all that is offered. Almost invariably the amount that can be offered at the increased prices is more than was calculated. Corners in Chicago have been broken by the big Minnesota millers who, at the last moment, have found it profitable to sell their large stores by telegraph. After the corners were broken, they could buy back most of this wheat before it had left their elevators.

After the supply has been successfully cornered the hardest part of the game is still to be played. The grain accumulated in cornering the supply must be disposed of. This is what Hutchinson, the first great cornerer, called "getting rid of the corpse." High prices were paid for enormous quantities of wheat that must be sold on a continually falling market, for after the cornerer settles with the shorts the price falls at once. He must squeeze enough out of the shorts to make himself whole in selling his own accumulation at the lower price. In such a corner wheat in general does not rise in price, but only wheat for delivery at a particular time in that market where the corner was run.

As Chicago is the great center of the wheat trade, it is the most advantageous place for running a corner, and the Chicago Board of Trade has been the scene of the great corners. The first one was run by B. P. Hutchinson in 1867. He bought the million bushels of contract wheat stored in Chicago warehouses, and all of the options, or privileges, that he could induce the shorts to sell. At the maturity of their contracts, the sellers were unable to deliver the wheat which they had sold. They "walked to the captain's office," and settled their accounts at $2.85 per bushel. Within an hour after they had settled, the price of wheat fell 50 cents, and within a day it fell 90 cents. It was an attractive manipulation, and looked easy. John B. Lyon repeated the operation during the next year, and the price rose to $2.20 per bushel. In 1872, Lyon started another corner, but the Northwest now had more wheat than he could control with his limited capital. The corner broke ruinously, and within two days the price fell 50 cents. Corners were run on the Chicago board in 1880, 1881 and 1882, but they were of no great magnitude.

In 1887 a mysterious "bull clique" was buying strongly of the May option. The clique was variously accredited as being John W. Mackay and his bonanza friends; as the Standard Oil millionaires; and as E. L. Harper and some of his Cincinnati associates. The conflict between the clique and the trade resulted disastrously to the former. When the wreck was cleared away, E. L. Harper was found in the débris. Accused of looting the Fidelity National bank of Cincinnati, of which he was vice-president, he was sent to the Ohio penitentiary, but was subsequently pardoned.[1] The last chapter of the corner was written in 1906, when the United States circuit court rendered a verdict against E. L. Harper for $5,280,333 in favor of the receiver of the Cincinnati bank.[2]

Another corner put wheat to the two dollar mark in 1888. This was a corner in September wheat, and B. P. Hutchinson was again a prominent manipulator. He figured that not more than three million bushels could be delivered to him on his contracts, but this amount was exceeded by 330,000 bushels on the last day of September. It was these, and not the three

[1] Payne, Century, 65:748.
[2] Wall St. Jour., Jan. 6, 1906.

million, that seemed excessive.[1] It was the same old story of the supply at the increased price being underestimated.

The difficulties of running a corner increased with the world's production of wheat. Not only did it require more capital on account of the greater supply of wheat, but there were always many sources from which wheat unexpectedly poured into the market and broke the price. In the eighties, India wheat became a factor in defeating corners in the Chicago and San Francisco markets. As the operations of the bull clique increase in magnitude they cannot be concealed, and "the shorts become extremely wary about getting in too deep." Traders as a class had a pronounced aversion to corners, "for they broke people, unsettled values, and made the pit as dangerous as a powder mine." By 1878, many of those who were best qualified to know did not believe that it was still possible to run a successful corner.

In spite of all these opinions, however, in spite of the gigantic magnitude and numerous difficulties of the task, it remained for a young man with great command of capital, with amazing audacity, with unlimited self-confidence, with an unusual capacity for appreciating and comprehending extensive business situations, and with a prodigious recklessness, to show that even under conditions existing at the end of the nineteenth century, not only is a successful speculative corner possible, but also a corner in actual wheat.[2] Joseph Leiter appeared in 1897, and his operations extended over about a year before they closed in June, 1898. He began with the strongest position ever held in the wheat trade, for the world's wheat crop in 1897 was less than that of 1894 by over 400,000,000 bushels, and less than that of 1895 by over 300,000,000 bushels, while the production of Europe was over 200,000,000 bushels less in 1897 than in 1896. The United States was practically the only country that had a large surplus for export. Leiter's plan was to control this surplus, and make Europe pay his price for it. With this end in view, he sent an army of purchasers into the Northwest, "contracted for vast storage space, chartered miles of cars and a

[1] Hutchinson, N. Amer. Rev., 153:416-7.
[2] For verifying the correctness of this account of the Leiter corner and for furnishing important statistics used in the account, the writer is indebted to Mr. Joseph Leiter, who ran the corner.

whole fleet of vessels, secured large contracts for delivery abroad, and prepared to supply all comers at good prices."[1]

Opposed to Leiter were the elevator interests, headed by Philip Armour, as wily and dangerous opponent in a wheat deal as could well be found. Leiter was endeavoring to establish his corner by buying more wheat than Armour could deliver. Armour was endeavoring to deliver more wheat than Leiter could pay for, and thus break his price. The battle for supremacy which followed is one of the most spectacular in our commercial history. Leiter soon held not only millions of bushels of actual wheat, but also contracts for millions of bushels of the December delivery in Chicago. The latter were chiefly short sales by Armour and the elevator people, who already held enormous quantities of wheat, and who expected to deliver actual wheat for every bushel contracted. Being in the elevator business, they were thoroughly equipped for extensive buying and rapid delivery. Their agents and those of Leiter frequently were competitors in securing grain. With such competition, the price of wheat began to jump. At every upward movement of the price "grain appeared as if by magic." By December it was thought that Leiter had the Chicago market cornered, but Armour used steel prowed tugs in plowing through the ice at the head of the lakes, and made a midwinter movement by lake and rail of 6,000,000 bushels from the interior. Unprecedented quantities of wheat were poured into Chicago. With perfect equanimity, Leiter not only paid for every bushel of it, but marked the price up from 85 cents to $1.09. He is reported to have taken over nine million bushels in one month. Armour was able to deliver all that he had sold, and Leiter was able to pay for all that he had bought. A great battle had been fought, but which man out-generaled the other, and with whom was the victory? The bond was paid, but just what its nominations were will perhaps never be known.

After the deal, Leiter owned enormous quantities of wheat. He seemed in no haste to sell, however, and began buying May wheat. His ambition seemed boundless, and his confidence unparalleled. The tension was great, and his movements were watched by the trade and by the public with the intensest interest. The foreign demand remained strong, and all of the

[1] Emery, Econ. Jour., 9:56.

market factors were bullish. The Spanish-American war could not have come more opportunely if it had been contrived for the deal. Europe now desired to purchase its wheat at once, for a grave vision of Spanish men-of-war cutting off American wheat shipments arose. The French import duties of 36 cents per bushel were suspended. Other countries suspended similar duties. Anticipations of bearish crop news were not fulfilled. These conditions were most favorable for the exportation of wheat, and Leiter took every advantage of them. He seemed to have a monopoly of the wheat business. How profitable a business it was, however, is not known, for many claims were made that he was paying freight charges and granting large discounts on export wheat. That the demand was not purely speculative is shown by the fact that low grades of wheat were bought heavily. Leiter's profits were figured far into the millions by the newspapers, and the pluck and coolness with which he had carried through the great deal largely won for him the admiration of the American public, in spite of the prejudice against speculation. He continued operations by selling off his May wheat and buying about all the cash wheat that came into the market. His further purchases may have been necessary in order to maintain prices, but it was a widely prevalent opinion that he courted the inevitable by not furling sail.

It is claimed that at one time in his wheat corner Leiter had $5,000,000 profits, but in the end he lost this and millions more. Wheat bought by him as low as 64¾ cents per bushel sold at $1.85. At one period he controlled 35,000,000 bushels of cash wheat and over 140,000,000 bushels under options. He exported and sold 25,000,000 bushels during the course of his famous deal. He was carrying about 15,000,000 bushels of cash wheat in the Northwest and in the course of transportation to Europe on June 13, 1898, when the tremendous load became too heavy to carry, and his deal ended.

The details of his manipulations cannot be known. He doubtless lost a fortune, and he completely disorganized the wheat business for 10 months. It is claimed that "Leiter's gambling in human food" caused a great rise in the price of bread in England and on the Continent, and that it brought about riots and bloodshed in Italy. While the operations of Leiter undoubtedly had a marked influence on the price of wheat, an influence that was not merely that of a speculative squeeze, but such as to be felt throughout the world, it is entirely unjust to attribute to them the great rise in price and consequent hardships, for "the high prices of wheat from August to June were not mainly the work of Mr. Leiter. For the first time in many years the bears in the wheat market were destined to learn the lesson that the production of wheat might run far short of the required needs, and, whatever direction the efforts at manipulation had taken, the price of wheat was bound to make remarkable advances in the season 1897-98. Leiter was wise enough to recognize the way things were going and to early put himself in a positon to profit from the inevitable outcome, and it was only when he tried to control the market in the face of adverse conditions that he failed." [1]

It is claimed that an international corner of the surplus wheat of the world was proposed to the United States by the Russian government in 1896. The two governments were to buy wheat at $1 a bushel, and were to sell none below the price which would cover all expense of buying it. The theory was that all of the wheat which could be produced at that price would be needed for food, and that the consumers would pay the price without either government having to buy any wheat. This visionary plan met with no support from the United States.

Public Gambling and "Bucket Shops."—The ordinary dealer or producer can do nothing more foolhardy than to risk his small capital in speculating and "playing the market," for he has no means of adequately knowing the world-wide conditions which determine price, he has not the judgment for properly interpreting such conditions even if he could know them, and those conditions often bring about results of such a magnitude as to sink a fortune completely in a very short time, if the speculator does not keep in touch and harmony with price-determining events. The character of speculation has changed somewhat with the increase in wheat supply, and fortunes are now made by men who watch the drift, and shape their way from day to day, "like prudent merchants, according to the current."

The "bucket shop" made its first appearance about a quarter of a century ago. It is always ready to take the opposite side

[1] Emery, Econ. Jour., 9:62.

of any speculative transaction which may be proposed. It deals on margins only, and as a rule its transactions are never executed either in a market or on a board of trade. It acts as a clearing house for the deals of its patrons by matching contracts, that is, purchases and sales. Those contracts that are matched cancel each other, at least as far as the bucket shop is concerned. Being simply booked, they never come into the market, and can have no effect on prices. It is only when the bucket shop has a large balance of contracts on one side of the market that it sometimes fears a loss and seeks insurance by itself making the counterbalancing transactions in the speculative market. It is only this small fraction of the bucket shop's transactions that really comes into the market and affects prices through the medium of the speculative supply and demand.

According to the law of chances, the bucket shop has an entirely safe and sound basis from its own point of view. By matching contracts it makes its patrons carry the greatest part of its insurance. The remainder of the insurance is carried by the bucket shop itself. Its advantages over its patrons are in three ways. It carries the risks of only a small fraction of the contracts involved. As a result of this, on a relatively smaller amount of capital, its chances of permanency are greatly increased. By being in continuous existence it secures the effects of advantageous changes in price as well as the disadvantageous ones. These will, at least in a measure, offset each other. It charges the same commissions as the exchanges and thus has a substantial income. The speculator, however, not only must carry all of his risks himself, but usually his capital is also very limited and he has no regular income from transactions. When his capital has been engulfed by a disadvantageous change in price, his operations must cease, and he secures no benefits from subsequent advantageous changes. The great revenue of the bucket shop consists chiefly of its commissions, but it is also continually acquiring the capital which is sunk by its patrons. How certain a process this is, is shown by the fact that the list of names of those dealing with the bucket shop usually changes completely within a few years. If the game were a profitable one to the speculator, it is quite safe to assume that his name would remain permanently on the list.

Not only is it true that "the local bucket shop is as efficacious" as the board of trade in enabling the "novice prophets" to exploit their theories and lose their money, but it is proving itself to be more so. Varying estimates assume that there are from 10,000 to 25,000 bucket shops scattered throughout the whole country. This is ample evidence of the profits of the business. Competent observers have estimated that these establishments have deflected from the regular speculative exchanges from 50 to 75 per cent of the business that would otherwise go to them. The New York stock exchange requires the speculator to deposit a 10 per cent margin on his transactions, and the smallest unit traded is 100 shares. The consolidated exchange in Chicago requires a 5 per cent margin, with 10 shares as a minimum unit traded. These minima are large enough to keep out a large class of persons with small capital whose operations are pure gambling. The bucket shop, however, usually requires a margin of only 1 per cent, and the minimum unit traded varies from ten to two shares. The bucket shop is prepared to do business with a half point margin on two shares. What is true of stock is also true of wheat as to the relative size of margins required and units traded on regular exchanges and on bucket shops. Consequently the bucket shop affords greater opportunity than the exchange to "play the market" and it is more frequently sought by those who have little capital and less knowledge on which to "speculate."

The professional speculator has a true economic function in our system of distribution, but gambling by outsiders is pernicious, and should be done away with as soon as this is practicable. Such gambling, however, is not more pernicious when done on the floor of a bucket shop than when done on the floor of the Chicago Board of Trade, a fact recognized by the United States Circuit Court of Appeals in more than one instance. In view of the business which the bucket shop is diverting from the regular exchanges, it is but natural that the latter should oppose the bucket shop with all their powers. Greatly as the boards of trade pride themselves over the high and lofty plane of honor upon which they transact business, they do not hesitate to designate[1] uncompromisingly as "pernicious gambling"

[1] Hill, N. Y. Commercial, Sept., 1903. Indus. Com., Vol. 6.

the operations of bucket shops—the very operations, in a large measure, that were formerly carried on upon the boards of trade. This opposition is merely an effort to regain that part of their business which has attached to it the bulk of the evils of all their business. Brokers "bucket" orders on their own account, a practice which the exchanges endeavor to stop.

Manipulations in General.—Speculation depends upon price fluctuations, but price fluctuations are decreasing. There is then the tendency to resort to all possible "manipulations" in order to cause abnormal fluctuations in price. Fraudulent and immoral means are often utilized in such efforts. In the produce market price can be influenced by the operator in only two ways. "He must either buy or sell the commodity himself, or he must persuade others to buy or sell." By such operations, however, a speculator with sufficient capital may bring about a rise or fall in price, but it seems to require unexpected crop conditions favorable to the manipulation in order to bring success.

Legal Restraints.—The feeling against speculation in the United States has been strongest when prices were depressed. In the early nineties it resulted in the introduction of several "anti-option" bills in congress. None of them became a law, but two of them passed one branch of congress, and there was much public sympathy in support of the measures. Much earlier some of the state legislatures made an effort to stop trading in futures. Some very stringent laws were passed by various states during the eighties. They generally considered short selling and trading on margins as gambling. The tendency to legislate in this direction seems to have lessened, and some of the statutes which were passed were soon repealed.

Speculation in Foreign Countries.—Sales of grain before it was threshed were forbidden in England in 1357, and in the Hanse cities in 1417. Time dealings in grain were forbidden in Antwerp in 1698. During the first half of the eighteenth century, practices similar to those of the modern speculative market were common in the European grain trade. The system did not become widely developed, however, until the nineteenth century. In 1883 the Liverpool clearing house was established in connection with the maize and wheat trade. It began after the American fashion, but was not extensively used before the close of the decade. The London produce clearing house began business in 1888, and wheat was one of the products dealt in. Business in futures, however, seems to have appeared in London as early as 1887. This first effort was a failure on account of the prejudice against speculation, and because several standards of wheat were dealt in. The London clearing house made another effort in 1897. Only Northern Spring No. 1 wheat was dealt in, and to make certain of the grade, a Duluth elevator certificate was always demanded. The grade in futures grew slowly at first, but seems to have become permanently established now.

In Germany, a drastic law forbidding grain futures was passed in 1896. On the whole its effect does not seem to have been advantageous. As a result, Berlin, formerly one of the most influential grain markets of Europe, has fallen to the rank of a provincial market. Berlin merchants transferred some of their speculations to Liverpool, New York and Chicago. Commission merchants have disappeared, price fluctuations have been greater, and the small dealer has been at a special disadvantage.[1]

Legislation against speculation seems to have been very general by the time of the last decade of the nineteenth century. One exception seems to have been France, where options and futures were practically sanctioned by law. In Hungary, Sweden and Norway there was no legislation on futures. The courts of Belguim treated futures under the general betting law. In Switzerland the law prohibited dealings in which there was no intention to deliver the goods. The deal was not legal in Austria unless the amount concerned had actually been paid or deposited. In Greece dealing by payment of differences was held to be null and void under the old Roman law, which was still in force. Sales entailing payment of differences only were illegal in Argentina.

General Results of Speculation.—The speculative grower who held his wheat until it seemed an opportune time to sell was the far-sighted, conservative man of the first part of the nineteenth century. Conditions have so changed that, unless there is lack of transportation facilities, or lack of competition among buyers, he is the greatest and most reckless of all speculators, if the

[1] U. S. Consular Reports, 64:438-444.

degree of ignorance under which he is operating is the test of recklessness. Even many of the large milling and elevator companies insure themselves regularly by hedging.

Statistics show that since the advent of speculation, fluctuations in the price of grain have been of smaller extent, comparing year by year. Such fluctuations as do remain are changes of a more gradual nature, and the gradations are much finer. For example, wheat was formerly quoted in fourths of a cent a bushel, while now it is quoted in sixteenths of a cent a bushel. A half century ago, traders required a margin of 10 cents a bushel for carrying wheat. Now the margin is 2 cents. With the minimizing of risks, profits for carrying them fell. In part, these changes are doubtless the result of other concomitant developments, but there is no question of their being chiefly due to the development of speculation.[1] As to the agreement of present prices of futures with future cash prices, little, if any, increased accuracy of prediction is shown. While there have been improvements in the methods of speculation, there has also been an increase in the size and complexity of the world wheat market, a factor which would tend to decrease the accuracy of prediction.

The increasing uniformity of price tends to decrease the amount of business done upon the exchanges, for this business is dependent upon price variations. This tendency of prices to remain more steady will be increased with the further concentration of commercial wheat interests. A similar development has taken place in the case of other commodities, such as oil and pork. In these commodities prices are practically fixed by a small group of men who know, and, in a measure, control supply and demand, and there are few price fluctuations left to serve as a basis for speculative dealings. Consequently operations no longer have their former magnitude. Local consumption of wheat will increase with the growth of population, and less actual wheat will be bought and sold at the terminal and export markets. The force of these influences is certain to be reflected on the speculative exchanges. The opportunity for men to indulge their gambling proclivities by means of the bucket shop; the growing prosperity of the country; the increasing steadiness of the price of wheat on account of the

[1] Emery, Speculation, pp. 124-165.

growing perfection of the speculative machinery and of the knowledge of the conditions of supply and demand; and the increasingly great combinations of commercial wheat interests, already foreshadowed by the large combinations of transportation, elevator and milling interests, will continually reduce the importance of wheat as a commodity in the speculative markets of the world, perhaps to the extent of finally eliminating it entirely.

CHAPTER XV.

THE MILLING OF WHEAT

Methods of Milling.—''The first miller plucked the berry from the stalk, and using his teeth for millstones, ground grist for a customer who would not be denied—his stomach.'' All millers who have succeeded this first pioneer have made use of various forms of apparatus to make the grinding process easier and more effective. There have been three distinct types of mills from which all others are only variations, and each one of which effects the reduction of the grain by a method peculiar to itself: (1) The mortar and pestle type, in which the work is done by grinding and rubbing; (2) some form of the machine having two roughened surfaces, between which the grain is crushed or cut through the motion of one, and sometimes of both, of the surfaces; (3) the roller system of milling, involving a gradual reduction or granulation process in which the grain of wheat is separated into particles and reduced to successive degrees of subdivision by being passed between rolls, first corrugated and then smooth, each successive series of which has an increased approximation of surfaces.

The Mortar and Pestle Type.—The second miller was always a woman. This initial stage in the development of milling was marked by several types of grinding devices.

Handstones.—Our knowledge of handstones, or ''corn'' stones, goes back to the paleolithic period. Such stones were

doubtless first used for pounding nuts and acorns. The same type was used the world over, and there is an abundance of specimens. The grain was placed upon a second stone with a flat surface. Pounding with the globular stone caused a cup to be hollowed out of the lower stone. Within a few

MEXICAN HANDSTONE

feet of each other, 26 such hollows have been found in the rock near an Indian settlement at El Paso, Texas. They are found in many parts of the world.

THE MORTAR AND PESTLE.—In time, the globular crusher became oval in form, which was of great advantage when the cups became deep. Eventually, it elongated into the pestle. Nomadic tribes found it advantageous to utilize a portable rock

AMERICAN INDIAN CORN MORTAR AND PESTLE

for the under stone. Shaped outside as well as inside, this became the grain mortar. Wooden mortars and pestles were now also made in imitation of those made of stone. The wooden mortars were sometimes 2 feet in diameter, and the pestles 4 feet in length. The first development in the direction of grinding instead of pounding was when the pestle was ridged at the bottom, and the grain was partly pounded and partly grated by giving a rotary motion to the handle of the pestle or pounder.

THE "SADDLE" STONE is another type of primitive milling devices. The upper surface of this was made concave, and in the hollow thus formed the grain was rubbed or ground by another stone, the muller, which was not rolled, but worked backward and forward. This was the first real grinding. Experience proved that the upper stone should be ridged. From the saddle stone evolved all later forms of milling stones.

These early forms of the mill have been used throughout the world. Babylon, Nineveh, Assyria and Egypt used them, and they are found in the prehistoric Swiss lake dwellings. The Romans of Virgil's time ground their grain by hand between two marble slabs. Many of the early forms of mills have been used in the United States. The settlers of Plymouth, Massachusetts, used the mortar for a decade or two. In the "hominy block" of early Pennsylvania, the bowl was a big block of wood burned or dug out. Sometimes it was found inside of the cabin, and also served as an article of furniture. At other times it was merely a convenient stump in front of the cabin door. In the latter case a nearby sapling was often bent over and attached to the pestle, which it helped raise. Such mills were replaced by power mills as soon as population had increased sufficiently to make the latter profitable. The Greeks of the mortar period made the first recorded milling revolution by using male operatives. These were called ''pounders.''

THE QUERN was the first complete grinding machine in which the parts were mechanically combined, and it also introduced the circular motion. It was apparently unknown before 200 B. C., but it was widely used at the dawn of the Christian era, and it is often mentioned in the Bible. The first form of the lower stone was conical, but the later type was flattened. The upper stone conformed to the pattern of its mate. Hollowed out in the centre until there was a hole at its base, the upper stone also served as a grain hopper. The stone was turned by a

THE QUERN, AN EARLY FORM OF STONE MILL

handle inserted in its side. An important improvement was made at an early date when the grinding faces of the stone were grooved, for the edges facilitated grinding and the grooves served as channels through which the meal was forced to the rim of the stones. This was a rude foreshadowing of the principles of methodical furrowing, a process which was not fully developed until the era of water mills. The quern was the original British flour mill, and it is claimed that it was still used in the mountainous parts of Scotland a decade ago. When the quern was large the upper stone had two handles. Women did the grinding. In Homer's time, the millers were also women, and it required the labor of from one-tenth to one-sixteenth of the community to prepare flour. Ever since the making of flour became a distinct trade, milling has been esteemed as an honorable occupation. A sturdy and independent character was always ascribed to the miller, and he and his mill have been a favorite theme with the writers of all ages.

Slave and Cattle Mills.—For many centuries the greatest changes in the milling industry were made in the motor power rather than in the grinding process itself. The advent of the quern and its improvements brought the professional miller, who marked the beginning of manorial or village milling. As the quern increased in size it ceased to be a hand mill, and power was applied. At first slaves, and even criminals, supplied the power. A circular piece of wood was placed around their necks, so that they were unable to put their hands to the mouth and eat of the meal. There were also cattle mills which were similar to the slave mills, and for many years in Rome, "the human animals and their brute companions performed the flour-making of the Eternal City." Cattle mills increased in number after the abolition of slavery in the fourth century. As early as 1537, treadmills were worked by convicts in Europe. They are still found in some countries, and are the sole survivors of the old Roman slave mills. The slave and cattle mills were supposed to have preceded the water mills, but the latter have existed in northern and western Europe prior to all historic records. They were also found in Greece, and later in Rome. Besides the hand querns, the ancient Egyptians had a larger quern that was worked by oxen.

Wind and Water Mills.—In many cases the wind mill appeared before the water mill. In early England wind was utilized to a greater extent than water, and wind mills were in existence at least as early as 1191. With the development of the mill stone, the grist mill appeared. The miller now ground for a larger district, and exacted toll, called "millcorn," from the farmers. The mills were generally owned by the lords of manors, who farmed them and their appurtenant privileges to the millers. The water mill was introduced into England at the time of Julius Cæsar. In France, Italy and elsewhere mention of it became common in the fifth century. It was exactly like the hand mill, except that water was used for power. Tidal mills were worked as early as 1526. The water was impounded at high tide, and the mills worked during the ebb. The wind mill seems to have come into use in England about 1200. The first milling by steam was in England in 1784.

The earliest mills in the United States were operated by horse power, and the toll was higher than at those where water

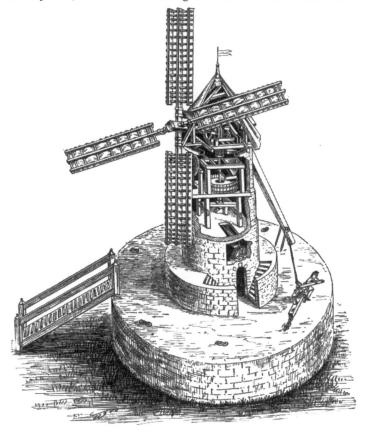

DETAILS OF AN OLD DUTCH WIND MILL

or wind power was used. The first mills of the Red river valley were operated by oxen, or by wind power. In 1870 there were 22 flour mills in South Carolina that were operated by horse power. In Texas there were 50, and 17 more were driven by oxen, while wind furnished power for five. Many of the primitive forms of mills can still be found in operation in various parts of the world.

Modern Improvements and Processes.—In the first milling, the entire wheat went into the flour. There was no "bolting" or classification of the product by separating it into several grades. Usually not even the bran was separated. The first distinctively modern improvements were in the line of bolting the flour. The primary sieve was an extended bag which was shaken by machinery. Its first introduction was in the power mills at the beginning of the sixteenth century. A German miller seems to have the credit for bringing forth this reel as a flour-dressing device. It was the predecessor of all subsequent bolting apparatus and of all appliances for purifying and separating the various grades of flour.

The old Roman system of cylinder milling, which is similar in principle to an ordinary coffee mill, was developed in Hungary. Elsewhere the system known as "low milling" was more common. In this the grain was ground in one process between two crushers placed as near together as possible.

In the United States the flour making industry was early developed in Pennsylvania, and in connection with this was given the first patent to a citizen of the new world for an invention (1715). A Philadelphia woman invented the device, which was in its essential portion a series of mortars driven by mechanical power. In few industries has there been so much litigation and controversy as in the manufacture of machinery for milling. Many patents for machines with the same object in view were taken out almost simultaneously. In the invention of all kinds of milling machines, competition has been so brisk that it is difficult to determine questions of priority and relative efficiency. New York city and Philadelphia had good bolting facilities even before 1698, but such facilities did not become general until the beginning of the nineteenth century. Oliver Evans (Philadelphia, 1756-1819) invented the elevator, conveyor, drill, descender and hopper-bag, from which "dates the long period of so-called 'American' milling, which produced flour as economically and of as good a grade as that of foreign millers." There was little progress from the days of Evans until the introduction of the "new" process about 1870. During the time of Evans, wheat was cleaned with rolling screens and blast fans. About the middle of the nineteenth century smutters were introduced, and a little later, separators, by means of which a more thorough system of wheat cleaning became established.

"Low" Milling.—Before 1850, the millstones in the United States were run at a comparatively low speed, and the grinding was slow. By this date the milling industry had assumed such commercial importance that it was necessary to increase the speed of the stones in order to get the work done. From 1850 to 1875, hard, low grinding was the rule, and the prime object was to make the largest possible percentage of flour at the first grinding. The change in process, due to greater speed, increased the output and improved its quality, "the outcome being a white, soft flour that met with favor in all he leading markets of the world where American winter wheat flours were handled." By this process, however, it was impossible to get the flour entirely free from contamination, and some of the bran always remained. There were two parts to this old process, reducing the wheat to flour by passing it through a run of stones, and bolting the resulting material in order to separate the flour from the bran and other undesirable parts of the kernel. The percentage of flour obtained by this single grinding depended on four things: (1) The dress of the millstone; (2) the face of grinding surface; (3) the balancing of upper or runner stones; and (4) the speed of the runner. As there was but one grinding, the making of middlings was avoided as much as possible. By this method of milling, some of the bran was pulverized so that it could not be separated from the flour. This gave the flour a darker color, and caused it to gather more moisture, which injured its keeping qualities, especially in moist or hot climates.

"High" Milling was the next step in advance. In this the speed of the stones was again decreased, and they were set far apart. This advance was made possible by the middlings purifier, which was not invented in the United States much prior to 1870, although its principle had long been known and applied in Europe. It was a machine for separating the dust, fluffy material, particles of bran, and the flour, from the middlings. It was now possible to make an excellent and pure flour from winter wheat, for the middlings thus purified were reground to superfine flour, which brought more per barrel than the best flour formerly in the market. As an indication of its superior grade it was called "patent" flour. The "new" process consisted of four parts, for purifying and regrinding the middlings were added to the "old" process. In the first operation the wheat was "granulated," not ground. . These particles, technically known as "middlings," were run through the middlings purifier and then reground. Being of great advantage, the process was further developed by introducing more stages. The grain was now ground very coarsely and the endeavor was to make as little flour as possible at the first grind, and the largest possible amount of middlings.

Ever since the sixteenth century, when good flour began to be manufactured at the mills, and bolting had been introduced, winter wheat at all times and places had commanded a larger price than spring wheat. Spring wheat flour was usually of a dark and inferior grade, valued considerably below winter wheat flour. With the opening of the wheat regions of north central United States, however, spring wheat was produced in enormous quantities, even at the lower price which was paid for it, and there was great need of an improved process of milling which would produce a high grade of flour from spring wheat. Spring wheat proved to be better suited to grinding by the continually improving process of high milling than winter wheat, for being harder, it yielded a greater percentage of middlings. This had been its great disadvantage under the old processes of milling, where the purpose was to get flour at the first grinding, and not middlings. All unpurified middlings are foul, and when reground they produced a low grade of flour. When the purifier remedied this difficulty, the best grade of flour was that made from the middlings, and almost at a single bound spring wheat took front rank as a flour producer. The winter wheat flour now became second grade instead of being the best.

Roller Milling.—The hard quality of spring wheat and the increasing number of "breaks," or stages in the milling process, necessitated new improvements. Rolls made of porcelain or of chilled iron were now devised to take the place of the time-honored millstone. The "new" process of high milling was first developed in the stone mills of Austria (1820-30). With an extension of principles, it became the Hungarian or gradual reduction process.[1] Experiments with roller mills date from 1820 in Switzerland, and rolls were used in Hungary in 1874, although minor experiments date vaguely back to 1861. These Hungarian rolls were 7 inches long, $4\frac{1}{2}$ inches in diameter, and made from 180 to 200 revolutions per minute. The first complete roller mill was erected at Budapest, and for years the mills of this city produced the leading flour in the world's markets.

In the United States, the principles of the gradual reduction process were taken from Hungarian millwrights, and rolls were first used in 1878. A complete outfit of roller mill machinery was brought to Minneapolis from Hungary, and Americanized. By 1880 rolls were rapidly coming into use, but it necessitated a change of machinery, and the change was stubbornly fought by the conservative old burr millers of this country. The spring wheat interests were large, however, and it seemed a useless fight. The thousands of small country millers held out longest, for the expense of the change bore most heavily upon them. The larger millers very successfully adopted the new process with all its intricate mechanical details. "Patent" flour had been fully recognized and established in commercial circles some time before 1876. Spring wheat brought 6 cents a bushel more in the market by 1882 than any other sort. Winter wheat formerly sold at from 5 to 30 cents a bushel more than spring wheat.

The Process of Milling wheat by the gradual reduction methods in the early eighties was quite complex. The grain was first passed through separators until it was perfectly free from foreign matter. It was then conveyed to ending stones, made of sandstone slightly harder than that used for buildings, and having the shape and size of ordinary millstones. These removed the "whisker" and fuzz from the wheat, after which it went to the brush machine—always by machinery. Here the clinging dust was removed, and then it passed through a series of five break rollers, each successive pair being set a little nearer together than the last. The flour and middlings were

[1] Smith, Hist. of Milling, Northwestern Miller, March 20, 1907.

removed between each breaking. The flour which was thus removed came from the center of the wheat grain, which is softer and first reduced in milling. This flour was so dirty as to be fit for only a low grade. The middlings were purified from bran, and then passed to rollers which reduced them. By a bolting operation, the coarse particles were now removed, and the unreduced portion of the middlings were again purified, then reground and rebolted. They passed through eight such operations. The residuum of the last process passed to the bran duster, and the refuse from the bran duster was sold as "shorts." The flour from the middlings was the "patent" flour. It required several hours for wheat to pass through the different processes. The richest part of the endosperm, the outside, was to a certain extent lost, being closely attached to the tough bran coats, or so contaminated with small pieces of bran as to injure the color of the flour, throwing it into the "baker's grade."

Revolutions seem to be continually taking place in the milling industry. After the process of milling had become long and complicated, an effort was made to shorten it again, and with considerable success. "It was the triumph of the 'short system' over the long system, and resulted in affording every small mill owner in the country an opportunity to adopt the roller system at an expense that was within his reach." The reform extended to Great Britain and the Continent, even affecting Hungarian methods and systems. It granted the small country miller a new lease of life.

The Present Processes of Milling.—The milling of wheat has become a very scientific and exact business, especially in the largest mills.[1] Prior to the milling of flour comes the selection of the wheat to be ground. The grain should be bright-colored and plump. Grain which is dark-colored from exposure to rains, or from heating in stack or bin, is of an inferior grade, for the rising quality of the gluten has been impaired. In selecting wheat, the miller does not rely upon external appearances, however, and all wheat is selected by chemical and baking tests, which are made before the wheat goes to the mill.

[1] For the major portion of the data bearing on this phase of the subject the writer is indebted to Messrs. James F. Bell and Frank W. Emmons of the Washburn-Crosby Company of Minneapolis.

The daily output of the Minneapolis mills is so enormous that every effort is made to maintain uniformity of character and quality in the flour. A special expert with several assistants is employed for this purpose. He grinds the samples of wheat by little mills designed for the work. In the selection of wheat, the quality of flour desired must be borne in mind.

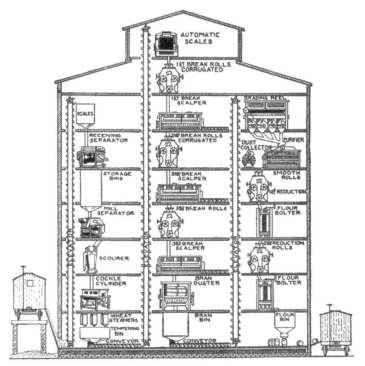

SECTION OF A LARGE MODERN FLOUR MILL

The mixing of the wheat is another preliminary process. This is the most recent and scientific method of keeping the different grades of flour uniform from year to year. The practice seems to have been adopted in the eighties, and it became well established in the United States and Europe during the early nineties. The mixing is done by means of the elevator machinery. Each mill generally has its own elevators for storing the grain. Different bins receive the different grades of wheat from the cars. The mixing and other manipulation of grain may take place in the elevators, or by a separate system of machinery in the mill proper. The quality of wheat varies so much with climate and season that it is practically impossible to prevent corresponding variations in flour without mixing. In our country much winter wheat is grown, and the best grade of flour can be made from it only by mixing it with the hard spring wheat. The spring wheats bear a higher and more uniform percentage of gluten, and herein lies their great value for mixing purposes. It is also claimed that spring wheat flour is more regular in the time required to mature in the bread dough. These are the reasons why spring wheat brings a higher price in the markets.

Three Fundamental Processes are passed through by the grain in the milling: (1) Cleaning; (2) tempering; and (3) grinding or milling proper.

CLEANING.—In this three objects are held in view: The removal of foreign seeds from the grain; the securing of clean wheat berries that are free from dust and other adherent foreign matter; and the removal of small particles of bran which would drop off afterwards and find their way into the flour. A special machine has been designed for the removal of each kind of foreign seed, such as that of other grain and of weeds. In the main, two different methods are used in the removing from the wheat berry all undesirable matter adhering or attached to it. In each method, machines adapted to the purpose are utilized. One method is known as dry cleaning, in which the wheat is passed through scourers. In the other method, the wheat is washed with water and subsequently dried. Each method has its advantages for different conditions of the grain, but some millers wash all wheat.

TEMPERING consists of putting wheat in the best of condition for milling. The coats of the berry must be so tough that the bran flakes out in one large piece in the grinding, and the interior of the grain must be in such condition as to give the largest yield of flour. There are nearly as many methods of tempering as there are mills, for each miller uses a process that will yield the results which he desires in the final milling. Heating the wheat to a certain temperature is a part of the tempering process, and moisture in some form is always applied. This may be accomplished by one or more applications of water, of steam, or of both water and steam.

MILLING PROPER.—The wheat is passed between corrugated steel rolls, each of which moves at a different speed from its mate. The berry is not crushed, but ruptured and flattened out, so that its interior can be separated from the bran coats in the largest pieces possible. As much of the interior as is thus separated from the bran coats is sifted out, and the residue is again passed through steel rolls so that more of the interior may be separated. This is what is meant by gradual reduction.

The interior of the berry which has been separated from the branny portion is known as "middlings." This material is now passed through the middlings purifier, which removes any particles of bran that may be present, the cellulose structural material of the interior of the berry, and the germ of the grain. The latter would give the flour a yellow appearance, and impair its keeping qualities. After the middlings have been purified, they are reground, and again purified. These processes are repeated until the material is of such fineness that it will pass through the finest silk bolting cloth. The material is tested at every stage of the process, and finally the finished flour is again tested before it is shipped to the trade. An expert can determine its quality largely by feel and color. Only the largest mills have facilities for making chemical tests. The smaller millers frequently have their products tested at chemical laboratories.

The Bleaching of Flour is a process of recent origin, and there has been considerable controversy as to its merits. The most common method employed is to pass air through an electric discharge of high voltage and low amperage. This results in the formation of oxide of nitrogen. The treated air is piped to an agitator or spraying machine, through which the flour is passing in a thin running stream. The latter operation requires from 7 to 10 seconds, and during this time the flour is aged and whitened. It is claimed that the only effect upon the flour is the decolorization of its oil. Bleaching gives whiteness only, and it does not enable the miller to increase the amount of flour,

nor to change lower grades into higher ones.[1] Of the many
processes for bleaching, the only ones having any industrial
value seem to be "based on the use of peroxide of nitrogen, prepared either by chemical action or by the action of a flaming
arc upon atmospheric air."[2] Flour naturally grows whiter as it
grows older.

The Dust Collector.—The milling of wheat always produces
flour dust. The ignition of these particles suspended in the air
caused disastrous explosions in the Minneapolis mills during
1877-78. This led to the development of the dust collector, the
first form of which was a filtering diaphragm. Its essential
principle is now the vortical or rotary air current, which masses
and precipitates smaller particles than the finest filter could
arrest.

The Grades of Flour most usually made are four in number:
(1) Patent; (2) first class; (3) second class; and (4) red dog.
The feeds comprise the remainder of the milled product. The
basis of flour grading is mainly its purity, that is, its freedom
from the bran and germ portions of the wheat kernel. The
best flour comes from the center of the grain. The strongest
gluten is nearest to the outside of the kernel, but the outside
can never be perfectly separated from the bran. The degree
of purity varies with the different processes of milling. The
equipment of the miller, his special process of milling, and the
market to which his products go determine the number of
grades of flour that he makes. He may omit some of these four
grades, or he may further separate them and make a larger
number of grades. Patent flour, for example, may be separated
into first and second patent. The requirements of some buyers
are for a flour that is sharper or more granular, while those of
other buyers are for a flour with fine soft granulations and very
white color. After all of the flours have been collected from
the various machines into their respective grades, they are conveyed to the packing bins, from the bottom of which they are
drawn by automatic power packers into packages varying in
weight from 2 to 280 pounds.

[1] Letters, Frank W. Emmons, Washburn-Crosby Co., and John
E. Mitchell, Alsop Process Co.
[2] Sci. Am., S. 61:25263 (1906).

The Large Typical Mill of the Winter Wheat Belt.[1]—A milling plant of five buildings and six grain storage tanks or silos, having a storage capacity of 300,000 bushels and a daily milling capacity of 1,500 barrels, involves a capital investment of about $200,000. The buildings include: The mill, 88 by 42 feet, and five stories high; the warehouse, 98 by 42 feet, and three stories high; the power house, 68 by 72 feet, and 25 feet high, with a tile smoke stack 125 feet high; and the grain elevator, 48 by 42 feet, and 118 feet high. The steel tanks or silos for storing the grain are each 30 by 60 feet. All of these tanks are connected at the top with a gallery for delivering the grain, and at the bottom with a conveyor belt for discharging the grain. Four thousand bushels of grain can be received and discharged per hour. It is moved from car to elevator and conveyors by means of steam shovels. The number of men required when the mill is running day and night consists of 3 officials, 5 office employees, and 35 other employees.

When the grain is received in the mill, it is given one cleaning over ordinary separators, and then stored. Before grinding, it is given such additional separations as may be required. It takes about one hour to complete the milling. The finished product is either loaded into cars or stored in the warehouse, and it is disposed of to the local and foreign trade. The average amount of wheat carried is about 200,000 bushels, and the probable average amount of flour on hand is about 10,000 barrels. Mills located at interior country points depend largely for their supply of wheat upon the deliveries of farmers and of country grain merchants in contiguous territory. About 5 per cent of the capital invested is devoted to products other than those of wheat. About $25,000 of cash must be ready for immediate requirements. Five to 10 per cent of the business is done on credit. The average expenditure for salaries and wages amounts to about $45,000 per year. Approximately 5 per cent of the original cost is annually charged to the depreciation of plant and equipment.

One of the largest mills in the world is the ''A'' mill of the Pillsbury-Washburn company of Minneapolis. The ordinary capacity of this mill is 15,000 barrels of flour per day, but

[1] These data are furnished by the Kansas Milling Company, Wichita, Kansas.

15,500 barrels have been ground. The company is now expending $500,000 in enlarging the plant so that it will have a capacity of 17,000 barrels. The maximum milling capacity of the Minneapolis mills aggregates a total of 82,765 barrels daily, and this is to be increased to over 90,000 barrels before the close of 1907. The great progress of the industry is best understood when it is remembered that the first crude mill of the ancients could not produce over 3 bushels of partly ground meal in one day. Later, the Greeks ground from 5 to 10 bushels of meal per day.

The Flour Yield of Wheat.—Soft wheat weighing 64 pounds per bushel has been found to yield about 80 per cent of flour, while that weighing 54 pounds yields about 65 per cent. The heaviest hard wheat yields about 74 per cent of flour, while the lightest yields 67 per cent. McDougall found that India wheat yielded from 77 to 81 per cent of flour, English wheat 65 per cent and American wheat 72 per cent. The flour yield will, of course, vary from season to season, for it is dependent upon the quality of the wheat. In 1905, it required 4 5-6 bushels of wheat to make a barrel of flour in Minneapolis, while in previous years it required only 4 1-3 bushels. According to a miller in Kent, England, 4.2 bushels of wheat made a barrel of flour in 1876. Two centuries ago in New England, it required between 6 and 7 bushels. There has been a continual increase in the amount of the highest grade of flour obtained.

Toll.—The first toll dish was the hand of the miller. In England in 1300, the toll was one-twentieth of the wheat ground. During the middle of the seventeenth century, the miller's toll in New England was one-sixteenth of the wheat ground. Of the thirteen original colonies, all but New York and Pennsylvania had laws for regulating tolls, which varied from one-fourth to one-sixteenth. The amount of labor required to grind a barrel of flour at the close of the seventeenth century, if expended at the close of the nineteenth century, had a market value approximately equal to the cost of grinding a barrel of flour in the latter period. In 1891, the legal toll in Minnesota was one-eighth. Measured in wheat, this is twice the toll which the miller received in New England 200 years ago. In the cities and large towns, however, where wheat was exchanged for flour on a cash basis, the cost of a barrel of flour would

THE NEW BUFFALO WASHBURN-CROSBY FLOUR MILL, DAILY CAPACITY 7500 BARRELS

purchase enough of No. 1 hard wheat to make 1.5 barrels of flour. In some places in Minnesota only one-eighth, the legal toll, was taken, and the amount taken varied from this to one-third at Minneapolis. It was stated that 700 years ago the English miller with his small toll made several times the profit that the Minneapolis mills made in 1891.

Geographical Location and Extent of Milling Industry.—The first development of the milling industry in the United States was in New York and in Pennsylvania. These states exported flour to the other colonies and to the West Indies. They long held first rank, and still mill large quantities of wheat, having held fifth and seventh places respectively among the flour producing states of our country in 1900. In the number of establishments and the amount of capital invested, they have always held first rank, at least until after the early nineties. Virginia, Ohio, Illinois, Missouri, Indiana and Michigan later assumed importance as milling states. Thirty years ago, the flours most sought after in the home and foreign markets were those of St. Louis and the south. St. Louis was then much the largest flour-making center of the United States. The improvements in milling processes changed the whole situation, for the best flour was now made from spring wheat. In the great rush to obtain Minnesota flours, St. Louis and southern flours were for the time forgotten. The first Minnesota mill was erected in 1823. The development of the milling industry in Minneapolis was most remarkable and rapid, chiefly by reason of the cheap water power obtainable from the falls of the Mississippi river. Other factors were the nearness of the wheat fields and the subsequent improvements in the art of milling. Before 1860, the annual output of the Minneapolis mills was about 60,000 barrels. This increased to 98,000 barrels in 1865, 193,000 in 1870, 585,000 in 1873, and over 1,000,000 in 1876. A conflagration then impeded the industry by destroying many mills, and it was not until 1879 that the output again exceeded a million barrels. By the end of the century, the average annual output was approximately 15,000,000 barrels.

As the milling industry developed, it moved toward the wheat fields. From 1877 to 1888, the receipts of flour at Buffalo were 22 per cent of the receipts of both wheat and flour, while from 1889 to 1898 they were 42 per cent. As the freight rates per hundred pounds were about the same for grain and flour, it was comparatively less expensive to ship flour than wheat, for an equal weight of flour had the greater value. This was true in both the domestic and foreign trade. On the other hand, it has been maintained that transportation companies can ship and handle wheat more easily and cheaply than flour, and that consequently there is a tendency for foreign countries to buy our wheat and manufacture it into flour themselves. Chicago annually grinds between four and five million bushels of wheat, which is about one-seventh of its total receipts.

The rank as to production of flour in 1900 of the twelve chief flour-producing states of the United States was, in decreasing order of importance: Minnesota, Ohio, Illinois, Indiana, New York, Missouri, Pennsylvania, Wisconsin, Kansas, Michigan, Tennessee and Kentucky. In portions of the south, it is thought that wheat growing would become more profitable and would increase, if local flour mills were established. There are some roller mills in northwestern Georgia, and even in central Georgia, but an increase in milling capacity would increase the demand for wheat.

In the table below is shown the flour milling industry in the United States as given by the last census. The most rapid increase in the number of establishments was from 1860 to 1870. From 1880 to 1890 there was a decrease in the number of establishments on account of combinations. From 1890 to 1900 there was again a remarkable increase. The annual milling capacity of the United States is over one billion barrels.

(All figures are in round millions, except the number of establishments,)

	1850	1860	1870	1880	1890	1900
No. of establishments.	11 891	13 868	22 573	24 338	18,870	25,258
Capital invested	54	85	152	177	208	219
Salaries paid	....	....	....	....	9	5
Total wages	6	9	15	17	18	18
Cost of materials used	113	208	367	442	434	476
Value of products	136	249	445	505	513	561

Milling in Foreign Countries.—Excepting for the United States, Hungary leads the world in the manufacture of

flour. Budapest was the star milling city of the world until
about 1890, when it was eclipsed by Minneapolis. The mills of
Hungary have the best equipment obtainable, and the wheat is
carefully graded for milling. Hungarian flour of the first quality commands a higher price in the English market than the
best Minneapolis flour. Sometimes it sells as much as a dollar
per barrel higher than any other flour. The reason for this lies
not so much in a superior process of manufacture, as in the
fact that this flour is the product of the very best wheat obtained by the close system of grading. American millers find it
more profitable to make more flour of a slightly lower grade. It
may be that the difference in price is also partly accounted
for by English prejudice.

The first roller mills of Great Britain, dating from
1878, were said to be unsuccessful. It was not until the
middle eighties that a respectable body of roller millers had
sprung up. It is estimated that they numbered 400 to 500 in
1891. Two years later there were 664 complete roller plants,
while at the present time 900 is the estimate. These mills have
a daily capacity of about 247,000 barrels of flour, and a yearly
capacity of 61,715,000 barrels. This is over ten million barrels
more than is annually consumed in the Kingdom, and takes no
account of the millstone flour production. In 1878 there were
10,000 millstone flour mills in Great Britain. Perhaps 6,000 or
7,000 of these still exist, but few of them grind wheat. There
has been active competition between American and British milling interests for the milling of the Kingdom, with the advantage
slightly on the British side on account of the freight discriminations between wheat and flour. The flour mills are now being
built at the quayside instead of inland, as formerly. Liverpool
is one of the largest milling centers of the world.

The modern roller system has been in operation in Russia
nearly 30 years. The work of the mills along the Volga and in
south Russia compares very favorably with that of mills in the
United States or Hungary. Dampening the wheat is an important part of the milling process, for most of the grain is extremely dry, and their softer red wheats are fully as hard as
our hard spring wheat from the Red river vally. The flour is of
a golden color and highly nutritious. Their product seldom

reaches the world's markets. After they become accustomed to it, most persons prefer it to any other.

Argentine wheat growing began to develop in 1880, and before 1895 over 300 mills had been built, an increase of nearly 100 per cent. The milling industry was so overdone that many mills went to ruin. In 1901, the annual producing capacity of the Argentine mills was stated at over 13,000,000 barrels, but the exportation and internal consumption did not equal half of this amount. It is especially the large mills of the interior that have had little to do. High taxes were a great disadvantage. New mills were, however, erected in the ports in 1903. These mills were equipped with the most modern machinery, and turn out an excellent product. The flour yield averages about 66 per cent. There is little home demand for by-products, and they are disposed of chiefly by exportation. It requires great economy to make milling profitable, and the industry will very probably be confined to the chief river and ocean ports, and to the small and comparatively unimportant local gristmills. On the whole, milling in Argentina is progressing slowly, and in other South American countries it is only local.

American competition crippled the Dutch mills in Holland, but they are regaining their trade on account of freight discriminations. In 1902, The Netherlands ranked second in importance as a market for American flour, Great Britain being first. Tariffs drove American flour out of Belgium, but Belgium millers suffer from ruinous competition among themselves. In Canada, mill-building is active, and both foreign and domestic trade is carried on. During 1903 flour-milling in New Zealand and Australia was temporarily at a standstill on account of crop failures, but it is usually an important industry. Progress in New Zealand seems to have been slow in this industry during the last few years, apparently on account of over-capitalization and over-production. The Chinese and Japanese have erected some flour mills, and they are ambitious to do their own milling, but success in this is not yet assured.

CHAPTER XVI.

THE CONSUMPTION OF WHEAT.

The Whole Wheat was used by the ancients for food. Pliny describes "amylum," a food prepared from unground wheat, which was first soaked, and then hardened into cakes in the sun. At an early date in England whole wheat, known as "frumity," was used as food. Here the grain was also soaked, and then boiled with milk and sweetened. Ordinarily wheat is no longer used as human food without first being ground or crushed. Where mills are wanting, as is sometimes the case in frontier and in savage life, the grain is often simply parched or boiled. The Arabs, for example, have a dish known as "kouskous," which is made by boiling fermented wheat.

The Uses of Different Flours.—When wheat is ground by the modern processes many different grades of flour result, not only from different kinds and grades of wheat, but also from the same grade or variety. Over 50 direct milling products may result from grinding one grade of wheat. These products differ so in quality that many of them are each most suitable for a certain purpose of consumption. What is true of one grade or variety of wheat in this respect is true also of different grades and kinds of wheat, and the products differ more widely yet.

Hard-Wheat Flour.—Hard wheat, of which the spring wheat of the Red river valley and the Turkey red wheat of Kansas are excellent examples, produces the flour that stands for the world's white-loaf bread, or "light bread." This flour is rich in gluten, which readily absorbs a considerable quantity of water. As gluten becomes wet, it swells to several times its dry bulk, and it grows elastic and tenacious. Gluten is the nitrogenous or tissue-building part of the wheat, and it supplies the same important food elements as are furnished by lean meat and the casein of milk.

Soft-Wheat Flour.—The flour made from soft wheat is the best flour for crackers (English "biscuits"), cake, pastry, and the hot 'soda biscuits'' so common in the southern portion of the United States. The respective uses of hard and soft-wheat flour are well defined and clearly recognized by bakers, millers, and wholesale dealers. Soft-wheat flour has more starch and less gluten than hard-wheat flour. It makes a whiter, and, in a certain popular estimation, a more attractive loaf, but it is less nutritious, and has a poorer flavor. Tenacity of gluten, so essential for good bread, becomes undesirable "toughness" in pastry and cake. In pastry, porosity is rendered unnecessary by "shortening," and in cake it is obtained with greater delicacy by adding the beaten albumen of eggs. Soft-wheat flour, having less gluten, is most suitable for these products. The thinly rolled and thoroughly baked cracker has the best color, texture and crispness when made from soft-wheat flour. Pastry and cake in some of their many varied forms are so universally a part of the daily diet of America and Europe that soft-wheat flour is sometimes designated in the markets as "pastry" flour.

Durum-Wheat Flour.—The flour from durum wheats has hitherto been used chiefly in the manufacture of macaroni and similar products. Its special fitness for this is its high gluten content. Bread made from this flour has a fine flavor, but a dark color. Because of the latter fact, and because of the fact that durum wheat requires special milling processes, there has been a prejudice against it as a bread wheat. With the great increase in the production of durum wheats in the United States, these difficulties are being removed, and it is very probable that its use for bread-making will greatly increase. It has long been used as a bread wheat in parts of Russia and France.

Graham[1] Flour contains the whole grain, and is made by cleaning the wheat and grinding it to a moderate degree of fineness. Soft wheat is the most suitable for making this flour, which, however, is used chiefly for bread.

Entire-Wheat Flour is prepared by a process similar to that used in milling graham flour, only that between the cleaning and grinding it is run through a machine which removes the three outer layers of the berry. This leaves the cerealin in the flour, but removes the bran. This also is a bread flour.

[1] So called from Graham, a temperance reformer of a century ago, who advocated bread made from unbolted meal as an aid in curing alcoholism.

Self-Raising Flour is produced by mixing leavening agents with flour, such as form the essential constituents of ordinary baking powder. The addition of water liberates the carbon dioxide, and a spongy dough results. Self-raising flour has had little commercial importance.

The Comparative Value of Different Flours.—The nourishment that can be obtained from flour depends upon its chemical composition and digestibility. Of the different flours that can be made from the same lot of wheat, graham flour contains the greatest proportion of protein and phosphates. Experiments have shown, however, that patent flour has the greatest amount of available or digestible protein and other food elements. More phosphates are present in white bread than are needed or absorbed by the body. The lower digestibility of graham flour is due to the bran, both because of its resistance to digestion, and because of its physiological action. The lower grades of flour, although not of such a fine white color, are yet highly nutritious, and yield a bread that is quite thoroughly digested. Since nitrogenous foods are proportionately more expensive than starchy foods, and since wheat is cheaper than lean meat, all wheat products are economical food, and those containing a high percentage of gluten are especially so.

Commercial Brands or Grades of Flour.—To a greater or less extent each miller manufactures a flour that, on account of the closeness of grinding, the proportions of the different kinds of wheat, or for other causes, is peculiar to his mill. His flour is branded, and a trade arises for his particular brand. As he has a monopoly of this brand, his business is largely non-competitive. While the brands of flour reduce competition for the wholesaler, they increase competition for the retailer, who must meet in the brands that he handles the prices of all other brands. The wholesale baking trade generally demands a sharp granular flour with a great capacity for absorbing water, whereas the household trade requires a finer granulation and a whiter color. The foreign trade prefers a strong granular flour with little regard to color, for the flour bleaches during the time consumed in transportation. In some of the larger markets, authorized flour inspectors stamp the packages with a brand which indicates the date of the inspection, the weight of the package, and the condition and quality of the flour. In St. Louis the standard grades are, in descending order of quality and whiteness, Patent, Extra Fancy, Fancy, Choice and Family. Besides improving in color, flour also yields a larger loaf as it grows older. When properly stored, the only loss is in the power of absorbing water. Flour readily absorbs undesirable odors, such as those of pine wood, kerosene, and smoked meats.

Human Foods Made from Wheat.—Not only does wheat have great superiority in sustaining life, but a large variety of healthful, palatable and attractive foods are made from it, either wholly or in part. Breads, pastries, crackers, breakfast foods and macaroni, of almost endless variety in composition, form and appearance are now found on table and market. Many of these have a comparatively recent origin, while others of a more remote origin have come into general use only in recent times. Wheat foods alone do not furnish proper nutrition for the body, for an amount sufficient to supply the requisite protein would furnish more than the requisite carbohydrates.

Bread is the oldest and most important product made from wheat. It supports life better than any other single food except milk, and it is the most staple food of modern civilization. The baking of bread is older than history. The prehistoric Swiss Lake Dwellers baked bread as early as the Stone Age. From the burnt specimens that have been disinterred, it was found that they did not use meal, but that the grains were more or less crushed. The ancient Egyptians carried the art of baking to a high perfection. Lippert maintains that the baking of leavened bread was practiced longest by the Egyptian and Semitic peoples. The Jews, however, still hold one feast in memory of the old form of unleavened bread. The bread of the Homeric Greeks is supposed to have been a kind of unleavened cake baked in ashes. The ancient Greeks had at least 62 varieties of bread. An oven containing 81 loaves of bread similar to the bread of modern times was found in Pompeii.

"Strong" and "Weak" Bread Flour.—The higher the gluten content of flour, the more water it will absorb in the dough; consequently it will yield more bread, and is known as "stronger" flour. Baker's bread is sold according to its weight in the dough, and a barrel of hard-wheat flour will make several pound loaves more than a barrel of soft-wheat flour. The weight of the dough and the size of the baked loaf are largely determined by the quantity and quality of the gluten. One hundred pounds of flour will make about 160 pounds of dough and about 140 pounds of bread. The flavor of the bread depends to a great extent upon the gluten and oil of the flour. These two compounds give the desirable "nutty" character so prominent in hard-wheat bread.

At the present day, first-class bakers generally use but one grade of standard flour for making bread. Every barrel of such flour is numbered at the mills where it is made, and if the quality should happen to be inferior, a report is made to the mill, and from the number of the barrel the mill determines the date when the flour was milled, its composition, and whether other similar complaints have been made concerning the same flour. The difficulty is thus located and remedied. Flour of the first-class standard grades costs from 10 to 25 cents per barrel more than other flours which are often just as good, and which are frequently used, although less reliable.

Yeast.—The making of leavened bread requires the use of yeast, a fungous plant. Three forms of yeast have been used in making bread: Brewer's yeast, which is that used by brewers in malting; German yeast, also called dried or compressed yeast, which consists of sporules only, and contains little moisture and no gas; and patent yeast, which is a thin watery liquid prepared from an infusion of malt and hops.

Mechanical Processes.—The most primitive method of making bread consisted merely in soaking the whole grain in water, subjecting it to pressure, and then drying it by natural or artificial heat. Perhaps the simplest form of bread and the rudest baking of modern times are found in the Australian "damper." Dough composed of flour, salt and water is made into cakes, which are baked in the dying embers of a wood fire. There have been no great modern improvements in machinery for making bread. A quarter of a century ago it was still made and baked much as it was in ancient Greece. The sponge was mixed and the dough kneaded by machinery, but as yet there had been failure to make loaves by machinery.

Except in the formation of loaves, perhaps, there seem to have been no marked improvements during the last 25 years.

The Modern Bakeshop.—The statutes generally require bakeshops to be inspected and kept in healthful condition. Each baker contracts by the year with a specialist to keep insects out of his establishment. The specialist visits the place at least once every three months whether insects appear or not. He receives a notification if but a single bug appears. His work is performed so thoroughly that it is exceptional if a bug is seen at all.

The following description is of a representative, moderately large-sized bakeshop which uses from 25 to 60 barrels of flour per day, and daily bakes from 7,500 to 20,000 loaves of bread. Each day the flour for the next day's baking is sifted. The sifter consists of a rotary brush running over a sieve, and it sifts the flour as fast as an attendant empties the barrels, about one each minute. All machinery is operated by electricity. The bakeshop is three stories high, and the sifting is done on the third floor. The sifted flour descends to a bin under the ceiling of the second floor. Under this bin, and on the second floor, is located the mixer. It has a capacity of four barrels of flour. The water, milk, lard, sugar, yeast, malt extract and salt are first placed into the mixer, and then the flour is added. Two parts of moisture are used to one of flour. Compressed yeast is used, and more is required in winter than in summer. The arms of the mixer revolve at a comparatively slow rate, about once in every two seconds, throwing the dough from side to side. The mixing operation requires 30 minutes. A large spout extends through the floor to the room below. As soon as the dough is in proper condition, the mixer is turned over, and the bread descends through the spout to the floor below, into the large bread trough which has been rolled under the spout. In this trough it rises about three hours. Thus far no hand has touched the bread, but some handwork now becomes necessary. Enough dough is weighed for 12 loaves, which are then cut out at one operation with an air pressure machine. After the loaves are cut they are molded by being run through the molding machine, of which the capacity is theoretically 60 loaves a minute, but in practice only about 40. The loaf is molded or rolled by an endless apron underneath, which carries it along in a rolling motion, and by a fixed top-piece, which is lined with sheep's wool in order to prevent sticking. The loaf travels about three feet. This is the Corby patent. After molding, the bread is placed in pans to rise. The best temperature for this is from 70 to 75° F.

The bread is baked in a continuous oven fired by coal. The temperature for baking should be from 450 to 550° F., so that the interior of the loaf will be at the boiling point, 212° F. When baked, the loaves are tipped out of the pans upon racks to cool, after which they are ready for sale. It is by varying the proportion of ingredients, the quality of the flour, the size of the loaf, and the time of rising and baking, that each baker produces bread of a quality in accord with his own ideas. The amount of bread produced from the same flour also depends to a great extent upon such variations. Rolls are cut by a special machine, 36 at a time. They are placed to rise, after which they are shaped by hand. They rise again, and then are baked. There are also special machines for mixing and cutting cake.

Kinds of Bread.—Common or leavened bread needs no description. Unfermented or unleavened bread is of two kinds: That in which substitutes producing carbonic gas are used in place of yeast, and that in which nothing but flour and water, and perhaps salt, are used. The former, also known as a vesiculated bread, is made in three different ways: (1) Carbonic acid is developed within the dough through fermentation of the flour; (2) the dough is mixed with water that has been previously mixed with carbonic acid; or (3) carbonic acid is disengaged from chemicals introduced into the dough. Maryland, or beaten biscuit, is an interesting variety of unleavened bread. Air is introduced into the dough by means of folding or pounding. These small portions of air expand in the baking, making a porous bread.

The original graham bread was made from graham flour without yeast or any of its substitutes. The dough was left standing several hours before baking. It was heavier than ordinary yeast bread, but somewhat porous, probably owing to fermentation started by bacteria accidentally present in the flour and acquired from the air. It was sweet, and by no means unpalatable. It is now baked like common bread.

Gluten bread is made from strong flour and water. The dough is pressed and strained under a stream of water until the starch has been worked out, when it is kneaded again and baked. It gives a light and elastic loaf which is often prescribed for diabetic patients. Aerated bread, which has had considerable popularity in London, is made by a method invented in 1856. The water used is charged with carbon dioxide gas. Another form of bread that has been made is the salt-rising bread. Hot water and cornmeal are mixed into a stiff batter, which is left at blood heat until it is fermented. The ferments originally present or acquired from the air produce fermentation, which leavens the batter. A thick sponge is then made from wheat flour and warm milk in which a little salt and sugar have been dissolved. This sponge and the fermented batter are thoroughly kneaded together and set in a warm place for several hours.

Chemical Changes and Losses in Baking.—Below is given in per cents the average composition of white bread and of the flour from which it was made.

	Water	Protein	Fat	Carbohydrates	Ash
Bread	35.3	9.2	1.3	53.1	1.1
Flour	12.0	11.4	1.0	75.1	0.5

In mixing the bread, the water was added, the fat was added as butter or lard, and the ash was added as salt. The protein and carbohydrates which were lost went to nourish the yeast plant. This feeds mainly on the sugar in the dough, and in its growth gives off alcohol and carbon dioxide gas. The gas and the generated steam expand with heat, force their way through the dough, and thus lighten it. Yeast also acts as an agency to turn starch into sugar. It is the tenacious quality of gluten (wanting in other than wheat flours, however nutritive), which retains the gas in its tendency to escape. Being elastic, the gluten expands, and the bread becomes porous.

"In bread making the action of the yeast and heat results in: (1) The fermentation of the carbohydrates and the production of carbon dioxide and alcohol; (2) the production of soluble carbohydrates, as dextrin, from insoluble forms, as starch; (3) the production of lactic and other acids; (4) the formation of other volatile carbon compounds; (5) a change in the solubility of the proteid compounds; (6) the formation of amine and ammonium compounds from soluble proteids and (7) the partial oxidation of the fat. In addition to these changes there are undoubtedly many others which take place. Inasmuch as many of the compounds formed during the fermentation process are either gases or are volatile at the temperature of baking, appreciable losses of dry matter must necessarily take place in bread making. These losses are usually considered as amounting to about 2 per cent of the flour used. In exceptional cases, as in prolonged fermentation, under favorable conditions, the losses may amount to 8 per cent or more."[1] It is claimed that the losses need not exceed 2 per cent and that they may be reduced to 1.1 per cent. Liebig calculated that in Germany the yeast plant consumed as much food daily as would supply 400,000 persons with bread.

On account of this consumption of nutritive elements by yeasts, and on account of the uncertainty of their working, chemical substitutes were sought 50 years ago in the United States and Germany. The substitutes for yeast are easily adulterated, they must be prepared with great care in order that they may not be inefficient or harmful, and even when successful the bread is usually rather tasteless. As a consequence, they have not met with success. Another loss occurs when the bread is baked. The carbon dioxide is largely retained in the dough, but the alcohol passes off. In 1858 it was estimated that 300,000 gallons of spirits were lost annually in London from baking bread, a loss of over a million dollars. Over $95,000 were spent in an effort to devise means to save these fumes. It was given up, not on account of failure to secure the alcohol, but because the bread baked in the process was dry, unpalatable and unsalable.

In baking, the starch is rendered soluble by the heat, the fermenting growth is killed, and the gluten is solidified, so that the cavities formed by the carbonic gas retain their figure.

[1] U. S. Dept. Agr., Off. Exp. Sta., Bul. 67, p. 11.

A FIELD OF DURUM OR MACARONI WHEAT IN NORTH DAKOTA

The crust and the crumb of the bread differ physically and chemically. This is due to the sudden and intense heat to which the crust is subjected. In the crust the starch is rapidly decomposed into dextrine and maltose, which are caramelized by the heat, making the crust darker and sweeter than the crumb. When bread grows stale, the moisture passes from the damper crumb to the drier crust, and it is supposed that the starch undergoes a chemical change. ''The whole question of staleness is one about which little has been absolutely proved.''

Cost of Baking.—A barrel of flour will make nearly 300 loaves of bread as ordinarily baked. A 10-cent loaf weighs about 1¼ pounds. The consumer thus pays 8 cents a pound for bread. A pound of bread can be made from about three-quarters pound of flour. At 2 cents per pound for flour, it is estimated that the cost of a pound of bread, exclusive of fuel and labor, is about 2 cents, which allows a half cent for shortening and yeast. While the fuel and labor add materially to the cost, these figures verify the statement that all the combined operations of raising wheat in Dakota, transporting it to Minneapolis, grinding it, and shipping the flour to Boston or New York cost less than to bake the flour into bread and carry it from the bakery to the home.

Macaroni in its numerous forms is a palatable and nutritious food. It is comparatively inexpensive, and is largely replacing meat dishes, which are continually becoming more costly. In food value and in use in the dietary, macaroni is very similar to bread. As a rule, the harder the wheat, that is, the more gluten it contains, the better it is suited to the manufacture of macaroni. Many wheats are used, however, which are not real macaroni wheats. The true varieties are quite widely grown, and have long figured in commerce. Algerian durum wheats are exported for this purpose, and form a standard type. Not a little macaroni wheat is grown and used in South Argentina. The wild goose wheat of Canada, rejected as a bread wheat, now finds use as a macaroni wheat, especially in France. The Japanese use home-grown wheat. The metadiné wheat of France is a half-hard wheat that is being largely used, but with a mixture of durum wheat. Indian and Turkish wheats are often mixed with such wheat as the

Algerian. **Russia** grows some of the finest macaroni wheats, chiefly known in France as Taganrog, because Taganrog is the principal point of export. Some of the wheats of Italy, the native land of macaroni, are second to none. One of the best varieties is Saragolla wheat. Even the common bread wheats have been quite extensively used, especially in the United States. Such wheat, however, does not produce a high grade macaroni, and this is one of the reasons why the quality of American macaroni has generally been below that of the imported product. Austria has also manufactured a low grade product from bread wheats.

The Macaroni Industry had its birth in Naples, and before 1875 the Italian product had not yet been equaled in any other country. The Neapolitan manufacturers gained their fame on account of the excellent quality of the native wheat. The cultivation of this has long been neglected. In the main, the spread of the macaroni industry seems to have taken place during the last quarter of a century. It developed a great wheat growing industry in Algeria and Tunis. ''Semolina'' or ''Semoule,'' the coarse flour from which macaroni is manufactured, has become an article of commerce beyond mere local trade. Not only has the macaroni industry developed greatly in France and Italy during recent years, but also in the Levant and in many other foreign lands. In 1903 France produced about 330,000 pounds of pastes per day, one-third of which was exported, chiefly to the United States, but also to Austria, Germany and Belgium. Italian exports go principally to the South American nations, and to a limited extent to England and the United States. In Japan, macaroni is extensively manufactured and consumed.

In the United States, the macaroni industry began with the use of bread wheats. During 1900, it became established on a durum wheat basis in North Dakota. From 1896 to 1901, about 15 to 20 million pounds of macaroni, vermicelli and similar preparations were annually imported by the United States. These imports amounted to nearly 30,000,000 pounds and were valued at over $1,000,000 during the fiscal year of 1902-3. The very finest quality of Italian macaroni is rarely exported to America, because it retains its quality only a few months, ''while the commonly exported article remains good for a year." The high quality of the best Italian macaroni is doubtless largely due to the wheat from which it is made, but it may also be due to the action of bacteria.

Process of Manufacture of Macaroni—In the manufacture of semolina, the wheat is first cleaned, which includes washing by water. Sometimes it is then dried, and moistened a second time. The water is considered essential to the cleaning, and it also aids in decortication. Special machinery has been devised for cleaning and dampening the wheat. It is milled in much the same way that soft wheat is ground into flour. In the best quality of semolina the resultant product is from 60 to 65 per cent of semolina, from 12 to 15 per cent of flour, and from 18 to 20 per cent of bran. Some of the lower grade wheats yield only from 30 to 40 per cent of semolina of an inferior quality. The miller's object is to get as much semolina and as little flour as possible. A special machine known as a "Sausseur" is used in grading the products. Semolina is not flour, but a much coarser product. As a rule, the manufacturers of macaroni do not grind their own wheat, but obtain their semolina from millers of that product. The semolina must be mixed in order to maintain a certain standard, the same as wheat is mixed in order to obtain a uniform flour. The product which goes into the macaroni should have from 45 to 50 per cent of gluten.

Before mechanical methods came in vogue, macaroni was kneaded by means of a wooden pole, or by piling up the dough and treading it out with the feet, after which it was rolled with a heavy rolling pin. By having a fire under the vessel, it was partially baked while being reduced to tubes and strips. "Modern mechanical methods are simply enlargements of the old family process by which the housewife mixed flour and water, kneaded the batch, rolled it into sheets, cut it into strips and hung it out to dry. In the modern factory the semolina is measured into a steel pan about 8 feet in diameter, within which travels a stone wheel. Water is added, the machine is put in motion, the wheel moves slowly around the pan, thus kneading the batch until it attains proper consistency. Just ahead of the wheel is set a small steel plow, to gather and turn over the mass, so that it falls under the rim of the approaching wheel, thus guaranteeing an even kneading of the whole amount of semolina measured out.''[1]

There are also other mechanical methods of mixing the dough. A small quantity of saffron is added to give a yellow color. After mixing, the dough is placed in a cylinder with a perforated bottom, through which the product is forced by means of a piston. The strings of paste are cut to the proper length as they issue, and are then thrown over reed poles to dry. In two hours they will dry sufficiently in sunlight, but if the weather is unfavorable longer time is required in sheltered terraces. When slightly dry, they are cellared in damp underground vaults for at least 12 hours. By this time the dough is moist and pliable again, and the poles are carried to storehouses which are open on all sides, but shaded from above. Here the strings hang from 8 to 20 days, according to the dryness of the weather. This gives them a horn-like toughness which prevents breaking from rough handling. In winter, the drying rooms are kept at a temperature of about 70° F. Thousands of reed poles bending under the weight of the yellow strings of macaroni cover the housetops, the courtyards, the narrow streets, and the hillsides of the little suburban towns about Naples. Mats spread upon the ground are covered with many kinds of short-shaped ''pastas.'' If the holes in the iron plate through which the dough is forced are very small, *vermicelli* is formed. A still smaller and finer sort is called *fedelini*. When the holes are larger and have a conical blade inserted, tube macaroni is formed. Paste rolled thin and cut in various shapes is called *Italian paste*.

In producing the various kinds of pastes, there is a very slight difference in the amount of water needed. Vermicelli requires a little less than any other form. To meet competition and changes in public taste, eggs are kneaded into the paste, rice flour, corn flour and potato flour are introduced, and the juices of carrots, turnips, cauliflower and cabbage are mixed with the paste. So much is mixed with the semolina that the macaroni consists of wheat to the extent of only

[1] U. S. Dept. Agr., Bu. Plant Indus., Bul. 20, p. 25.

60 per cent. Our own homely dish of "noodles" can be traced back to a macaroni ancestry

Crackers, Often Called Biscuits (*bis cuit*, twice baked), are a variety of unleavened bread. They find their way to almost every table in the land. They are usually made from soft wheat flour. "Milk, butter, lard, spices, dried fruits—anything or everything desired to give them particular consistency, color or flavor—is mixed with the flour and water." The manufacture of crackers is a trade by itself, different from ordinary baking, and requiring machinery and processes peculiar to itself. As early as 1875, crackers were made by a rapid and continuous process. Machines mixed the flour and water, pressed the dough into a sheet, cut it and even fed the biscuits into an oven. A traveling stage carried them through the oven. The patent traveling ovens were 30 to 44 feet long, and fitted with endless webs of plates or chains. The chains were used for small fancy biscuits, and the plates for large and plain water biscuits. The rates at which biscuits of different sizes and degrees of richness traversed the length of the oven in order to bake varied from 5 to 40 minutes, and the temperature of the oven was modified to suit various qualities. Both the heat and rate of motion were "under easy and adequate" control.[1] Crackers are rarely made in the home. Formerly they were placed upon the market in the bulk, but the package form of the trade has increased so greatly that some companies are rapidly doing away with the less profitable business of selling crackers in the bulk. About 50 different package biscuits are placed upon the market by one company. Perhaps the most interesting form of unleavened breads is the Passover bread, which has been used during Passover week by orthodox Jews from the time of Moses until now. It is not unlike the plain water cracker.

Ready-to-Eat Wheat Foods.—These foods are also known as breakfast foods. Their manufacture dates from 1895, and seems to be confined to the United States. The pioneer in this business was Henry D. Perky, who patented "Shredded

[1] In spite of many efforts, the writer was unable to secure any considerable data on any phase of the modern cracker industry. The business is largely monopolized by a few men not affected by the recent wave of publicity. It is rumored that the profits of the business are too great to make publicity advisable.

Wheat Biscuit'' in 1895. This product contains every portion
of the wheat kernel. The whole wheat is cooked without
being flavored, and then mechanically ground into filaments.
It is formed into miniature loaves and baked. The distinctive-
ness of this food has always been retained and has never
been successfully imitated. It stands in a class by itself and
is in great favor with American consumers.

The great development of the breakfast food industry has
centered at Battle Creek, Michigan. John H. Kellogg patented
''Cranose Flakes'' in 1895. It consisted of the whole wheat,
which was cooked, slightly flavored with salt, rolled into thin
flakes, and baked. It was the first flaked wheat food that
met with considerable sale. Charles W. Post began the manufacture of ''Grape Nuts'' in 1896. This product is made from
wheat and barley ground together into flour, baked into bread,
toasted, and finally crushed to granular form. The food is
distinguished by its hardness, its amber color, and its large
percentage of dextrine. The products ''Malta Vita'' and
''Ready Bits'' were the result of experiments conducted at
Battle Creek in 1898. The former consisted of cleansed whole
wheat seasoned with salt, and treated with malt extract for
the predigestion of starches before it was finally baked.
''Force,'' brought out a few months later, was manufactured
in a similar manner. ''Ready Bits'' was not perfected until
1903. ''Its form is distinctive, consisting of readhering particles of disintegrated cooked wheat, from which the excess
starch has been removed by the use of an enzyme.'' All of
these three foods attained national distribution. By 1903 at
least 50 undistinctive brands of ready-to-serve wheat flakes
were upon the market, and nearly all of them were made from
whole wheat cooked, salted, rolled and baked. Their merit
depended upon the quality of the material and the care and
skill used in their preparation. Their success was proportional
to the vigor and intelligence with which they were advertised.
The total annual output of ready-to-serve wheat foods was
estimated to have a value of $11,000,000.

In 1903, 18,191 families were visited in a house-to-house
canvass of the city of New Haven, Connecticut. Seventy-six
per cent were found to be users of ready-to-eat cereal foods.
The number of the families of the different nationalities who

were users and non-users of these cereal foods appears in the table below: [1]

Nationality	Families Visited	Users of Cereals	Non-Users
American	10,047	8,697	1,350
Irish	1,512	912	600
German	1,874	1,466	408
Italian	955	371	584
Swedish	174	111	63
French	20	18	2
Hebrew	3,073	1,957	1,116
Negro	491	327	164
Polish	8	1	7
English	32	24	8
Japanese	1	0	1
Chinese	3	0	3
Bohemian	1	0	1
Total	18,191	13,884	4,307

The analysis [2] of some of the leading ready-to-serve wheat foods indicates the following average percentages of constituent substances:

	Water	Protein	Fat	Carbohydrates (including crude fiber)	Crude fiber	Ash
Cracked and crushed (average of 11 analyses)	10.1	11.1	1.7	75.5	(1.7)	1.6
Flaked (average of 7 analyses)	8.7	13.4	1.4	74.3	(1.8)	2.2
Germ preparations (average of 10 analyses)	10.4	10.5	2.0	76.0	(.9)	1.1
Gluten preparations (average of 3 analyses)	8.9	13.6	1.7	74.6	(1.3)	1.2
Parched and toasted (average of 6 analyses)	8.6	13.6	2.4	74.5	(.8)	.9
Shredded (average of 6 analyses)	8.1	10.5	1.4	77.9	(1.7)	2.1
Patent flour, high grade, spring wheat	12.3	11.7	1.1	74.5	(.1)	.4
Patent flour, high grade, winter wheat	13.3	11.0	.9	74.4	(.3)	.4
Flour, low grade	12.0	14.0	1.9	71.2	(.8)	.9

Cereal breakfast foods have been more extensively and ingeniously advertised than any other class of foods. Such a bewildering variety is upon the market that it is difficult to make an intelligent choice between them. They are very convenient

[1] For all of the preceding data concerning ready-to-serve wheat foods, the writer is indebted to Mr Burritt Hamilton, formerly President of the Ready Bits Company.
[2] U. S. Dept. Agr., Farm Bul. 105, p. 20.

for use and give a pleasant variety in food. It is claimed, however, that "at the usual prices the nutrients in ready-to-eat cereals are considerably dearer than those furnished by bread and crackers." Where strict economy is not essential, the special convenience and variety is often considered to be worth the additional cost.

During the first few years of the twentieth century, the most active competition prevailed between the numerous companies manufacturing ready-to-serve breakfast foods. Events in this business happened with kaleidoscopic rapidity. During the years of 1902 and 1903 there was an overproduction of cereal foods which caused a protracted glut in the market. Many of the younger companies were unable to continue in the business, and failed. The survivors are now doing a satisfactory business, and the making of cereal foods has settled down to a staple milling industry.

The Natural Food Company, the present manufacturer of shredded wheat, has a conservatory overlooking Niagara Falls. It is one of the finest food factories in the world. Power is furnished by electricity from the Falls, and the total cost of the building and equipment was $2,000,000. The united structure covers an area of 55,653 square feet. It has 5.5 acres of floor space, and a frontage of 900 feet on the upper Niagara Rapids. Educational features have been established, and there is an auditorium, seating over 1000, for entertainments, lectures, and conventions. Its food has been a great commercial success, and is one of the best selling products on the American market today. Some of its products are also exported.

Grape Nuts is an unpatented food. The manufacturing company relies on its trade marks for protection. By vigorous advertising it has created an extensive demand for its goods in the United States and in some foreign countries.

"From 100,000 to 125,000 one-pound packages are put up daily, representing a daily consumption of 1,500,000 of portions. In the manufacture of Postum Food Coffee and Grape Nuts, about 2,200 bushels of wheat are consumed daily. These two products are mostly used by the English speaking race, but are being gradually introduced in all the commercial centers of the world. Stocks of both products are carried in all the prominent cities of the United States, Canada and England. Some 625 male and female employes find employment throughout different parts of the factory. The capital employed by the Postum Cereal Company is $5,000,000. Their expenditure for advertising is one million dollars per annum." [1]

Adulterations.—There are many substances that have been used to adulterate and cheapen flour. Among the vegetable substances are rye flour, corn and rice meal, potato starch, and meals from leguminous plants, such as peas and beans. Among the mineral substances are alum, borax, chalk, carbonate of magnesia, bone, and various clays. Alum in any form is harmful and the use of the others is reprehensible, for they often make a poor bread seem good. The addition up to 20 per cent of cornstarch can be used with high glutinous flours, but it produces a much drier loaf, lacking flavor. Terra alba has been widely used for adulteration in foreign countries, but at least as late as 1894 there was no knowledge of its having been used in the United States. Mineraline, one of its forms, was, however, subsequently used. [2] The poorer classes of people sometimes adulterate the flour themselves. For example, it is said that the Scandinavian peasants at times mix half flour and half ground tree bark in their loaf. In the United States, an internal-revenue tax was levied on mixed flour by the war-revenue act of 1898. It largely stopped the mixing of cornstarch or corn flour with wheat flour, a practice that had been frequent.

Wheat Products as Animal Food.—All of the grain of wheat which is unfit for flour is generally fed to animals. Wheat that finds poor sale for any reason, as for example goose wheat and durum wheat in former times, is often fed to stock. In times of very low prices, even the bread wheats are extensively fed. During 1893 over four million bushels, or 16.5 per cent of the total wheat crop of Kansas, were fed to farm animals. Authorities, however, do not seem to be agreed as to the value of wheat for feeding. For certain feeding purposes it seems to have advantages over corn and other grains, while for other purposes it has disadvantages. It should generally be fed with other grains, and its food value is slightly increased by grinding. Wheat should not form more than half the grain ration. All classes of domestic animals are fond of wheat in any form.

[1] Letter, Postum Cereal Co., Ltd.
[2] Industrial Commission, 11:2.

Growing wheat is often pastured in the fall or spring. At times this can be done without injury. On the Pacific coast as much as 10 per cent of the wheat is sometimes cut green for the purpose of making wheat hay. This practice is often followed in Oregon. After the wheat is threshed the straw is often used as fodder in the United States, and also in other countries.

Other Uses of Wheat Straw.—In the time of Fitzherbert wheat straw was used in England to thatch houses. In the Old World some varieties of wheat are grown solely for making hats and other articles of plaited straw. It is also used for various other purposes, such as packing merchandise and making mattresses and door-mats. Another great use is in paper mills, where it is at times bought at $3 to $4 per ton. Efforts have been made in this manner to save some of the straw that is going to waste at the rate of millions of tons per year in North Dakota. The problem of using wheat straw economically is no nearer solution than it was 20 years ago. In the Northwest and on the Pacific coast it is often worse than useless, because it must be burned to get it out of the way.

The Per Capita Consumption of Wheat is not an index to the bread consumption of countries where rye bread is used. Including the amount required for seed, the estimated per capita consumption in the United States for 1902 to 1904 inclusive was 6.23 bushels. The following estimate of the per capita consumption of wheat in certain countries was presented to the British Royal Commission on Supply of Food and Raw Material in Time of War, by Mr. W. S. Patterson of the Liverpool Corn Trade Association.

PER CAPITA CONSUMPTION OF WHEAT

I. IN IMPORTING COUNTRIES

	Bushels		Bushels
United Kingdom	5.6	Spain	5.3
Germany	3.2	Portugal	2.3
Belgium	7.2	Sweden	2.0
France	7.8	Greece	3.3
Holland	3.9	Austria-Hungary	3.6
Italy	4.4	Switzerland	5.7

II. IN EXPORTING COUNTRIES

	Bushels		Bushels
United States	4.7	India	0.7
Canada	5.5	Australia	5.5
Russia	2.6	Argentina	4.0
Balkan Provinces	4.3		

CHAPTER XVII.

PRODUCTION AND MOVEMENT OF WHEAT

The United States Wheat Production.—With the development of any agricultural community, farming becomes more diversified. This tendency is already manifesting itself in the great wheat regions of the North Central States, not only in the diversification of crops on the smaller farms, but in the rotation of crops beginning to be practiced on the larger farms. There is also a tendency for even the largest farms to become divided into smaller holdings, and this will further increase the growing of diversified crops. All this diversification will tend to decrease the wheat acreage in the best wheat lands of the West. With the development of our whole country, land values are certain to rise. This is a factor of the greatest importance, for it will make certain lands too valuable for the production of wheat, while it will sufficiently raise the price of other lands now lying idle so that their cultivation will become profitable. Some wheat will be grown on many eastern and southern farms which are not cultivated at present. With the development of drought resistant varieties of wheat, the wheat acreage in the semi-arid regions of western United States will be increased.

It is probable that all of these developments will result in a reverse in the historic westward movement of the center of wheat production, and that this center may begin to retrace its course and proceed eastward, for it is probable that the decrease of western acreage by diversified farming, and the increase of eastern and southern acreage resulting from the raising of wheat on lands formerly abandoned, will more than counterbalance the increased acreage in the semi-arid regions. On the whole, it has been concluded by some students of agricultural statistics that the limit of wheat production in the United States has approximately been reached. With the future growth in population, and especially with the further development of mining and other non-agricultural industries, the home consumption of wheat in the West will be greatly increased. This will have a tendency to diminish wheat exports from western United States, and may even divert to the West some of the grain from the Central States which is now

ACREAGE, PRODUCTION, VALUE, AND DISTRIBUTION OF WHEAT OF THE
UNITED STATES IN 1905, BY STATES [1]

(In round thousands.)

State or Territory	Crop of 1905			Stock in farmers' hands March 1, 1906		Ship'd out of county where grown
	Acreage	Production	Value	Bushels	Per cent	
	Acres	Bushels	Dollars	Bushels	Per cent	Bushels
Maine	8	181	192	58	32	0
Vermont	1	27	25	10	35	0
New York	491	10,301	8,859	2,472	24	1,442
New Jersey	110	1,805	1,589	397	22	361
Pennsylvania	1,629	27,861	24,239	10,030	36	2,786
Delaware	121	1,670	1,369	417	25	785
Maryland	810	13,197	10,821	2,903	22	8,050
Virginia	738	8,417	7,408	2,273	27	3,115
North Carolina	593	3,975	4,055	1,073	27	199
South Carolina	318	1,942	2,156	369	19	19
Georgia	305	2,107	2,254	485	23	63
Alabama	108	1,041	1,051	177	17	10
Mississippi	3	28	27	0	0	0
Texas	1,249	11,118	9,784	1,668	15	3,113
Arkansas	198	1,565	1,408	344	22	63
Tennessee	882	6,349	5,777	1,206	19	1,968
West Virginia	356	4,373	3,892	1,268	29	670
Kentucky	780	8,810	7,665	1,586	18	3,083
Ohio	1,883	32,198	26,402	9,015	28	16,421
Michigan	1,027	19,003	15,013	5,131	27	7,981
Indiana	1,932	35,351	28,988	8,131	23	16,615
Illinois	1,872	29,952	24,261	5,691	19	13,778
Wisconsin	474	7,893	5,999	2,842	36	947
Minnesota	5,446	72,434	51,428	20,282	28	54,326
Iowa	964	13,683	9,715	4,242	31	3,421
Missouri	2,260	28,022	22,138	5,324	19	12,610
Kansas	5,536	77,001	54,671	13,860	18	57,751
Nebraska	2,473	48,003	31,682	12,961	27	31,202
South Dakota	3,221	44,133	29,569	11,033	25	34,865
North Dakota	5,402	75,623	52,180	15,125	20	64,280
Montana	119	2,843	2,019	995	35	768
Wyoming	29	748	539	217	29	52
Colorado	254	6,359	4,451	1,526	24	3,815
New Mexico	43	948	853	227	24	28
Arizona	15	332	388	56	17	10
Utah	178	4,710	3,156	1,837	39	1,178
Nevada	27	724	557	116	16	14
Idaho	367	10,342	6,785	1,861	18	6,722
Washington	1,322	32,517	21,326	5,203	16	25,038
Oregon	718	13,383	9,100	2,409	18	7,227
California	1,886	17,542	14,384	1,403	8	10,876
Oklahoma	1,435	11,764	8,117	1,882	16	7,058
Indian Territory	270	2,703	2,081	297	11	1,351
United States	47,854	692,979	518,373	158,403	22.9	404,092

[1] Yearbook U. S. Dept. Agr., 1906.

exported by way of Gulf and Atlantic ports. With the increase of population and local consumption, the internal and export movement of wheat will greatly decrease, and American wheat will be a factor of declining importance in the international grain trade.

VISIBLE SUPPLY OF WHEAT IN THE UNITED STATES AND CANADA, FIRST OF EACH MONTH, FOR TEN YEARS [1]

(In round thousands)

Month	EAST OF ROCKY MOUNTAINS		PACIFIC COAST	
	1896-1897	1905-1906	1896-1897	1905-1906
	Bushels	Bushels	Bushels	Bushels
July	61,354	20 476	1,927	839
August	58,414	21,314	1,917	581
September	57,588	21 705	3,512	1,130
October	63,955	28,894	5,454	3,156
November	76,716	53 745	6 883	4 486
December	76,433	62 402	6,548	5,866
January	73,270	71,634	4,189	5 511
February	68 092	73 151	3,005	5,295
March	61,664	70 530	1 857	4 898
April	55 946	66 599	1,730	4 947
May	49 684	54,856	1,614	3,917
June	37,975	40,347	1,221	3,349

QUANTITY AND PERCENTAGE OF DOMESTIC WHEAT, INCLUDING FLOUR, EXPORTED FROM LEADING PORTS FOR   EARS ENDING JUNE 30,1884-1904. [2]

Customs District	ANNUAL AVERAGE				1904	
	1884-1888		1894-1898			
	Bushels	Per cent	Bushels	Per cent	Bushels	Per cent
Atlantic	82,757,000	67 6	102,780 000	64 4	57,361 000	47 5
Gulf	2,061 000	1 7	10,843 000	6 8	33 315 000	27.6
Pacific	31,865 000	26 0	36,833,000	23 1	22 334,000	18 5
All other	5,737,000	4.7	9,118,000	5.7	7,717,000	6 4
Total exports	122,420 000	100 0	159,594,000	100.0	120,728,000	100 0

[1] Yearbook U S. Dept. Agr, 1906.
[2] U. S. Dept. Agr., Bu. of Sta., Bul. 38, 1905.

AN AMERICAN REAPER IN A RUSSIAN WHEAT FIELD

A feature of the wheat industry in the United States that merits special mention is the increased production of durum wheats. These wheats are now widely grown in the semi-arid regions where the annual rainfall does not exceed 10 or 12 inches. In the early years they were a product very difficult of profitable sale, but they are now assuming a strong commercial position. The nature of the grain was not generally understood by American millers until it had been on the market for several years. In Russia it is blended with about 25 per cent of red wheat, and the same practice has been followed with some success in the United States. Many mills are now grinding the grain. A large portion of the durum wheat grown in the United States is exported, chiefly to Marseilles and other ports of the Mediterranean sea. About 10,000,000 bushels were exported during the year ending June 30, 1906. About 2,000,000 bushels were produced in 1902, 6,000,000 in 1903, 20,000,000 in 1905, and 50,000,000 bushels in 1906.

Russian Wheat Production.—Viewed solely from the point of view of its natural resources and economic aspects, Russia is the United States of Europe. It has immense undeveloped areas that would form ideal wheat lands, lands very similar to those which constitute the wheat belt of the United States. European Russia may be divided into two regions distinct as to the nature of their soil by a line running from Bessarabia in the southwest to Ufa in the northeast. In the southeast is the chernozium (black soil) region, and in the northwest the non-chernozium region. Clay, sand and rocky soils are all found in the non-black soil region, which lacks fertility and is chiefly devoted to the production of rye. The black soil zone is an arable plain, vast in extent, very fertile in soil, arising through centuries from the decomposition of accumulated Steppe grasses and sheltered by outlying forests. This plain stretches across the empire to the Ural Mountains, extending completely over 15 provinces and partially over 12, and even reappearing in Siberia. It is one of the largest fertile sections of land on the globe. In European Russia, the 18 provinces which lie chiefly in the black soil region produce two-thirds of the wheat and only one-third of the rye. Of the 328,000,000 acres of arable land, 59 per cent, or 193,000,000 acres, is located in the black soil region. Of the 197,000,000 acres of cereal crops, 72 per cent, or 142,000,000 acres, is found in the chernozium area. The black soil is of great uniformity in type and composition, varies in depth from a few inches to about 4 feet, and owes its dark color to its high proportion of organic substances (4 to 16 per cent). The Russian Steppes have fully as great a similarity to the Great Plains of the United States in climate as in soil, although greater extremes prevail.

The similarity between Russia and the United States in the natural resources of the wheat growing regions is quite equaled by the dissimilarity in political practice, social theory and economic condition. The Slav does not possess the Anglo-Saxon's proud institutional heritage. The Russian proletariat have no "Uncle Sam" who is rich enough to provide farms for all. There is, indeed, plenty of land, and they do have the Little Father, who is supposed to exercise a paternal care over his people. Sadly lacking in the institutions that are fundamental for progress and prosperity, however, the Russian people have found the Little Father to be far less capable and generous in aiding their material advancement than is essential to its realization. Consequently they have been unable to rise above their ignorance, poverty and misery. A population of exuberant fertility residing in a land of unlimited natural resources, the Russian peasantry have had neither means nor opportunity to attain a higher plane of life. The poor system of land ownership and the antiquated methods of agriculture made Russian wheat a dear wheat in spite of cheap labor and a low standard of living. The future possibilities of Russian wheat production depend upon the social, economic and educational progress of Russia. There are symptoms of improvement in this direction. The extension of peasant land ownership is improving economic conditions. It seems that political and social conditions are at last changing and popular education is growing. In agriculture, better machinery is being introduced, and crops are being rotated. The production of wheat increased 122 per cent in European Russia from 1870 to 1904. From 1881 to 1904 the acreage in wheat gained 57.3 per cent, while that of rye gained only 1.7 per cent, and the ratio between wheat and rye changed from 45:100 to 70:100. The yield of wheat per acre decreases from west to east.

Since the construction of the Great Siberian Railway the actual and potential productive powers of Asiatic Russia, and especially of Siberia, have been an interesting subject for speculation in Europe and America. In the popular conception previous to this event, Siberia was a land of polar nights and eternal snow, the monotony of whose dreary wastes was broken only by the clanking chains of Russia's exiles—exiles who were not always criminals, but who, according to Occidental ideas, frequently represented the very flower of Russian citizenship.

AREA UNDER WHEAT IN THE RUSSIAN EMPIRE[1]

(In round thousands)

Year	Total	European Russia	Poland	Northern Caucasia	West Siberia	Middle Asia
	Acres	Acres	Acres	Acres	Acres	Acres
1895	42.233	31,894	1,170	5.808	2,594	767
1896	45,869	34.848	1,198	5.589	2,905	1.329
1897	46,733	35,606	1,210	5.589	3,021	1.206
1898	47,019	36,008	1,221	5.263	3,377	1,150
1899	49.743	38 045	1,305	5,966	3.179	1,248
1900	52 313	39,967	1.318	6,228	3,660	1 140
1901	54,306	41,921	1.247	6,416	3,537	1 185
1902	55,112	42 590	1,301	6,817	2 933	1,471
1903	57,221	43 753	1.292	7,189	3.455	1,532
1904	59,186	45,635	1.242	7,473	3.352	1,484

PRODUCTION OF WHEAT AND RYE IN THE RUSSIAN EMPIRE

(In round thousands.)

Year	Wheat						Rye
	Total	European Russia	Poland	Caucasia	Siberia	Middle Asia	
	Bushels	Bushels	Bushels	Bushels	Bushels	Bushels	Bushels
1895	413.341	292.272	17.386	67,127	29.093	7 462	801,413
1896	412.038	300.423	19,477	45,148	34.160	12,830	789,562
1897	340,171	238.557	17,808	29.883	42 835	11,087	712,319
1898	459,289	334,246	21,691	52.251	36 157	14 944	737 501
1899	454,145	314,877	21,545	57.313	45 473	14,938	911,633
1900	422,994	319.193	19,722	56,948	20,172	6 959	920 134
1901	427,781	319,992	14.409	67 232	16 504	9,645	754.927
1902	607,370	463 259	20 349	77,069	30,796	15,897	919 019
1903	621,459	454,598	19 256	77,941	48,670	20,995	911 944
1904	666,752	519,966	21 241	81,132	31 590	12,823	1 008 440
1905	636,285	568,274	20.239	96,708	42.411	....	....
1906	506,400	450,900	19,000	73,000	35,000	....	....
1907	511,000	455,000	....	....	....	....	....

[1] U. S. Dept. Agr., Bu. of Sta., Bul. 42, 1906, p. 16.

With the completion of the railway, foreign conception underwent a great change, and Siberia suddenly became the "future granary of the world." Subsequent developments have not met expectations, for the true Siberia is a mean between these conceptions. This enormous country, which is 24 times as large as the German Empire, and nearly twice as large as the United States proper, has a very rigorous climate, and perhaps only half of it is habitable, while a still smaller portion is suitable for agriculture. This still leaves an immense area, however, upon which the cultivation of wheat is not only possible, but probable. Wheat is at present the most important crop of Siberia. It is exceedingly difficult to foretell the rôle which the Russian Empire is destined to play in the world's future wheat production. The possibilities are tremendous. Since, however, they are so largely dependent upon social, economic and institutional evolution, it is very improbable that Russia will duplicate the rapid development of wheat production which took place in the United States. While the development will be gradual, it is probable that Russian production will be one of the great permanent factors in the wheat industry.[1]

India's Wheat Production.—The two factors which enabled India to become a large exporter of wheat were the completion of the Suez Canal in 1869 and the subsequent development of the railroads. The former gave an enormous stimulus to wheat cultivation. Wheat thrives best on the dry plains of the Punjab and on the plateaus of the central provinces. Agricultural conditions in different parts of India, and meteorological conditions in different parts and in different seasons, are so diverse that the annual production varies greatly and is extremely difficult to predict. India wheat as a factor in the world market is made still more uncertain by the fact that domestic consumption is unusually susceptible to variations resulting from changes in the price that may be obtained in the export markets.

In recent years the annual wheat area in British India has been approximately 28,000,000 acres. About one-fourth of this is planted in the United Provinces, and about one-fourth in the Punjab. Of the remaining wheat area, the Central Provinces

[1] Rubinow, Russia's Wheat Surplus, U. S. Dept. Agr., Bu. of Sta., Bul. 42, 1906.

have annually about 3,000,000 acres; Central India and Bombay about 2,000,000 acres each; Bengal about 1,500,000 acres; and Rajputana, Hyderabad and the Northwest Frontier Province each about 1,000,000 acres. Beror and Sind are the only other important wheat-growing provinces, and each has an annual area of about 500,000 acres. The wheat is harvested during our spring months. The wheat from the Central Provinces is shipped from Bombay. That of the Punjab is collected at Multan and shipped from Karachi. There has also been a large export of flour, which is ground at Bombay and other centers.

AREA, PRODUCTION AND EXPORT OF WHEAT IN INDIA, AND THE
GAZETTE PRICE OF BRITISH WHEAT

(In round millions)

Years Ending March 31	Area	Bushels	Exports in following year in bushels		British Gazette Price of wheat per quarter	
			Total	To United Kingdom	S	D
1890	24,773	204.100	23 866	15,451	31	11
1891	26,576	229,200	50,511	23,110	37	0
1892	24,482	184 500	24,955	16,543	30	3
1893	26,429	239,766	20,261	12,381	26	4
1894	25,778	225,700	11,483	8,480	22	10
1895	25,994	209,310	16 673	12,893	23	1
1896	23,242	183,698	3,185	2,510	26	2
1897	19,024	163,095	3 988	1,798	30	2
1898	22 954	222,891	32 533	17,790	34	0
1899	23,923	211,320	16,173	10,915	25	8
1900	17,183	162,323	83	11	26	11
1901	22,922	225 523	12,203	8,131	26	9
1902	23,477	202,116	17,153	14.823	28	1
1903	23,092	258 870	43,185	32,888	26	9
1904	27,773	312,916	83.128	....	....	....
1905	....	283,000	37,477	....	....	....
1906	....	319,600	....	....	....	....

Argentina's Wheat Production.—The wheat industry of Argentina is similar to that of Russia in some of its most important phases. While the country is of much smaller extent, its land, climate and railroad extension are available potentials for an enormously expanded wheat production. As in Russia, the wheat area cannnot be definitely determined without years of experimentation and a great increase in population. Here, too, there are vast arable plains of great fertility, a fertility of which little is known to the world on account of poor methods of farming and on account of the fact that much of the land has not been under cultivation. The cattle industry was first developed in Argentina, and for many years it completely overshadowed agriculture. Thousands and even hundreds of thousands of acres were owned by the great cattle kings who had no desire to have their land broken up, because they knew nothing of its agricultural value. Another controlling factor is the dependence of agricultural work upon immigrants and their descendants. These immigrants differ greatly in character from those found upon the new lands of the United States and Canada. The great number of illiterate peddlers, laborers, cobblers, and what-not of Italy, Spain and Russia do not become intelligent farmers. They do not endeavor to become permanent additions to the population by securing ownership of the land which they cultivate. They are chiefly Italians having a very low standard of living and little efficiency as laborers. Many of them return to Italy within a year after their coming. According to the census of 1900, not one farmer in three is a renter in the United States, but in Argentina two out of every three do not own the land which they till. Two systems of renting are in vogue in the latter country, the "medianero," or share system, and the "arrendatario," or cash system. The government encourages immigration by offering free transportation from Europe and by making easy the acquisition of land. There are Jewish, Russian, Swiss, German, Austrian, Italian, Spanish and Scandinavian settlements. The number of immigrants averages about 100,000 per annum, and the number of emigrants at least half this number. Generally speaking, the Argentine wheat farmer will submit to life conditions that would not be endured in North America, for he has been accustomed to hardships in Europe. He is slow in understanding what a republic means. Class distinctions between rich and poor are sharply drawn.

Agricultural methods and conditions are improving, however, and Argentina is certain to assume a higher rank as a producer of wheat and other cereals. Twenty-five years ago not enough wheat was produced for domestic consumption. During the last decade wheat has been the principal crop, and approximately 50,000,000 bushels have been exported annually. The total area of Argentina is over a million English square miles, an area equal to all of that portion of the United States which lies east of the Mississippi, with the Dakotas, Minnesota and Iowa added. Wheat growing began in the north and extended in that direction farther than was advantageous. It is estimated that there are at least 60,000,000 acres of land that will eventually be producing wheat. One great advantage is that

WHEAT STATISTICS OF ARGENTINA[1]
(In round thousands)

Year	Production in bushels	Average Price in cents per bushel in Argentina	Acres	Exports Wheat Bushels	Wheat Value	Flour Barrels	Flour Value
1880	....	....	....	48	$45	16	$97
1890	....	....	....	12,048	9,493	135	580
1892	39,319	77	....	17,273	14,182	212	988
1893	59,109	63	....	37,042	22,639	427	1,272
1894	81,129	48	....	59,092	26,169	458	984
1895	59,164	59	....	37,121	18,790	607	1,816
1896	46,738	73	....	19,547	12,381	582	1,881
1897	31,594	86	....	3,742	3 349	466	2,327
1898	49,423	85	7,506	23,705	21,586	359	1,537
1899	88,067	58	7,826	62,957	36,746	669	1,870
1900	99,075	63	6,793	70,903	46,926	576	1,65º
1901	74,752	75	6,045	33,227	25,322	807	2,016
1902	56,380	....	8,893	23,696	17,934	439	1,544
1903	103,759	....	....	75,000	58,000	810	....
1904	129,672	....	....	84,684	....	1,207	....
1905	150,745	....	....	105,391	....	1,628	....
1906	134,931	....	....	....	....	....	....
1907	155,993	....	....	....	....	....	....

the land can be worked at almost any time of the year, for the climate is comparatively moderate. It is probable that the development of the wheat industry in Argentina will be more rapid than in Russia.[2]

Canadian Wheat Production.—Canada has greater possibilities of an immediate and rapid increase in wheat production than any other country. It holds this position of pre-eminence

[1] U. S. Dept. Agr., Bu. of Sta., Bul. 27, 1904.
[2] Bicknell, Wheat Production and Farm Life in Argentina, U. S. Dept. Agr., Bu. of Sta., Bul. 27, 1904.

by virtue of its large area of fertile land, land so well suited to the growing of wheat that the grain produced is of a quality not generally equaled by other countries, and by virtue of the intelligent and industrious settlers who are rapidly taking up the unoccupied lands. Estimates vary greatly as to the actual wheat area available in Canada. The best lands are located in Manitoba, Assiniboia, Saskatchewan and Alberta. It is probable that there are at least 150,000,000 acres within these limits upon which wheat could be profitably grown, an area approximately three times as great as that annually sown in wheat in the United States. As yet there is not more than about 5 per cent of this land under cultivation, but over 100,-000,000 bushels of wheat are annually produced. The hard wheat of the Canadian Northwest ranks with the world's best wheat, and the Toronto papers quote it at a price about 15 cents above that of Ontario wheat. In some years over half of the crop grades No. 1 hard, and it is greatly desired by the millers for mixing with lower grade wheats for the purpose of maintaining a desirable and uniform strength of flour. The yield of wheat per acre is larger in Canada than in the United States. The average yield of spring wheat in Manitoba from 1891 to 1900 was 19 bushels. During the same period of time the yield in the Dakotas was about 11 bushels, while that for the whole of the United States was 13.3 bushels. The land of Canada seems to be more productive, the climate more favorable, and the methods of farming better. About one-fourth of the country is capable of tillage.

The settlement of Canadian lands is progressing rapidly. A large proportion of the immigrants and a great amount of capital come from the United States. From March to August of 1902, about 25,000 emigrants went from the United States to Canada. 12,000,000 acres of land have been settled in one year. In effect, the homestead laws of Canada are similar to those of the United States. Transportation facilities are being rapidly developed in order to meet the demands of the increased population, and some of the largest modern grain elevators are being constructed. It appears as if Canada is destined eventually to produce the bulk of North American export wheat. The cold climate is unfavorable to the production of corn and many other crops, and it is very likely that the growing of wheat will be one of the great permanent industries of Canada, especially as the population is so largely agricultural.[1]

Wheat in the United Kingdom.—The imports of wheat by Great Britain are far greater than those of any other country and approximate two-fifths of those of the world. It is this fact which gives the United Kingdom its position of unusual importance in the wheat industry. About the time of Christ the Normans made England so productive of "corn" (wheat) that a large amount of grain was exported, and England was known as "The Granary of the North."[2] At the close of the eighteenth century the average crop of Great Britain was over 60,000,000 bushels. In 1852 the wheat acreage was over 3,500,-000 acres. With the development of wheat production in the United States and other countries having great natural advantages over the United Kingdom, the price of wheat declined to such a degree that it became more profitable for the latter country to grow other crops and to import the bulk of its wheat. By 1868, less than 2,500,000 acres of wheat were grown in Great Britain, and the acreage continued to decline for over a quarter of a century. Less than 2,000,000 acres of wheat are now annually grown, but the yield is over 30 bushels per acre. During the decline in wheat acreage the price fell in still greater proportion. Wheat imports to England began about 1846.

Australian Wheat Production.—Wheat growing has not always been a profitable industry in Australia. It has been claimed that there is less return there for the farmer's labor than in any other civilized country. Wheat thrives best on the cooler and drier lands of the southern part of Australia. Many farmers, however, have abandoned wheat raising for the cultivation of the grape vine, which is a more profitable crop in good seasons. Victoria, New South Wales and South Australia are the chief wheat growing states. The yield per acre is never large, and short crops often result from severe droughts. For this reason Australia is not a reliable exporter. The production of wheat has been increasing, however, and

[1] Saunders, Wheat Growing in Canada, 1904.
[2] Warner, Landmarks Eng. Indus. Hist., pp. 8-11.

now averages about 75,000,000 bushels annually. Wheat is one of the chief crops of New Zealand, and is exported.

Miscellaneous Countries.—The two other South American countries besides Argentina which produce a surplus of wheat are Chile and Uruguay. Wheat is the leading agricultural product of Chile, which exported grain to California and Australia in the early years. Its export wheat now goes chiefly to Peru, Ecuador and the United Kingdom. The exports of Uruguay go to Brazil and Europe. Some wheat is grown in southern Brazil. The production of wheat in Mexico is steadily increasing, but it is insufficient for domestic needs. The per capita production of wheat in France is large, and about one-seventh of the agricultural territory is devoted to this industry. By reason of the liberal encouragement given by the government, and on account of the conservatism of the French peasantry, the area and production of wheat in France has been practically uniform for over a quarter of a century. Excepting Russia, France produces more wheat than any other European country. Austria-Hungary ranks next, and then come Italy and Germany. The latter country stands next to England in wheat imports. Roumania and the Netherlands each export over 25,-000,000 bushels of wheat annually, and Belgium exports about half of this amount.

In the time of the Pharaohs and in the time of Rome's greatness, Egypt was the most important commercial wheat center of the world. It is estimated that Egypt annually furnished 20,000,000 bushels of wheat to Rome. Ancient Mauritania and Numidia, the present Algeria and Tunis, were also long the granaries of the Eternal City. Although wheat is still exported from Northern Africa, it does not form the principal crop. Most of the wheat produced is of the durum varieties, and its chief commercial use is for the manufacture of macaroni. Wheat thrives well in parts of southern Africa, and several million bushels are annually produced.

The World Production and Movement of Wheat.—Ever since the time of Malthus there have been periodical predictions of a scarcity of food supply for mankind. Less than a decade ago Sir William Crookes, President of the British Association for the Advancement of Science, predicted that a serious shortage

WHEAT CROP OF COUNTRIES NAMED, 1901-1906[1]

(In round thousands of bushels.)

Country	1901	1902	1903	1904	1905	1906
United States............	748,460	670,063	637,822	552,400	692,979	735,261
Canada.......................	91,424	100,051	84,583	74,834	113,022	131,614
Mexico.......................	12,021	8,477	10,493	9,000	6,000	5,000
United Kingdom......	55,581	60,065	50,320	39,083	62,188	62,354
Norway......................	300	265	307	212	300	300
Sweden......................	4,193	4,757	5,538	5,417	5,419	6,227
Denmark....................	942	4,528	4,461	4,302	4,500	4,400
Netherlands...............	4,231	5,105	4,258	4,423	4,400	4,700
Belgium.....................	14,143	14,521	12,350	13,817	13,000	13,000
France.......................	310,938	327,841	364,320	298,826	338,785	324,725
Spain.........................	136,905	133,523	128,979	95,377	83,605	154,090
Portugal....................	10,000	10,400	8,000	6,500	5,000	8,000
Italy.........................	164,587	136,210	184,451	150,664	160,000	168,000
Switzerland...............	4,400	4,200	4 000	4,000	4,000	4,000
Germany....................	91,817	143,315	130,626	139,803	135,947	144,754
Austria-Hungary......	180,900	235,022	226,856	204,535	227,646	268,574
Roumania..................	72,386	76,220	73,700	53,738	100,000	113,867
Bulgaria....................	24,000	35,000	36,000	42,000	39,000	55,000
Servia.......................	8,102	11,409	10 885	11,700	12,300	13,211
Turkey in Europe.....	22,000	25,000	26,000	23,000	20,000	22,000
Greece.......................	6,400	7,000	8,000	6,000	6,000	8,000
Russia in Europe......	401,772	560,755	551,942	622,487	568,532	450,000
Russia in Asia...........	61,149	84,718	110,102	86,412	107,903	56,000
British India.............	264,825	227,380	2 '7,601	359,936	281,263	319,586
Algeria......................	32,244	33,896	34,035	25,244	20,000	28,000
Egypt........................	12,000	12,000	11,000	12,000	12,000	12,000
Australia...................	49,877	39,753	12,768	76,488	56,215	70,681
New Zealand.............	6,733	4,174	7,693	8,140	9,411	7,013
North America.........	851,905	778,591	732,898	636,234	812,001	872,771
South America.........	87,417	74,625	119,013	155,185	171,445	155,337
Europe......................	1,513,797	1,795,336	1,831,193	1,726,084	1,790,693	1,825,733
Asia..........................	395,574	382,122	467,115	518,589	456,135	444 786
Africa.......................	50,672	52,023	54 313	50,003	41,500	46,813
Australasia...............	56,610	43,927	20,461	84,628	65,626	77,694
Total...................	2,955,975	3,126,624	3,224,993	3,170,723	3,337,400	3,423,134

in the supply of wheat would exist by 1931 on account of the increasing population. Such predictions generally over-emphasize the numerical increase in population which is current, and fail to give due regard to the laws which control the production of the food supply and its ratio to population. A scarcity of wheat simply raises its price and increases its production. In the world markets a sudden and acute scarcity of the general food supply is impossible. A gradual decrease in the general food supply until a serious shortage exists is equally impossible, for, whatever the standard of living, population will limit itself long before acute conditions are reached. While several countries each possess many millions of acres of the finest lands—

[1] Yearbook U. S. Dept. Agr., 1905-6.

THE VISIBLE SUPPLY OF WHEAT

(In round thousands of bushels.)

World's visible supply of wheat, first of each month			Year March 1st	Stocks in farmers' hands in United States	Visible supply of the United States and Canada	Visible supply of the world
Month	1896-1897	1905-1906				
			1892	171,071	68,007	181,400
			1893	135,205	110,693	229,300
			1894	114,060	105,863	222,400
July	137,454	114,302	1895	75,000	110,546	212,400
August	124,292	106,838	1896	123,045	98,834	191,900
September	126,485	113,511	1897	88,149	63,521	155,500
October	151,271	138,759	1898	121,321	49,445	140,600
November	190,559	157,735	1899	198,056	56,189	151,100
December	202,329	189,323	1900	158,746	91,749	181,500
January	184,616	192,690	1901	128,098	86,272	192,700
February	173,496	188,030	1902	173,703	88,332	191,900
March	155,533	193,520	1903	164,047	79,771	163,700
April	139,049	183,687	1904	132,608	58,389	152,000
May	121,491	159,406	1905	111,055	54,580	165,400
June	106,912	139,154	1906	158,403	75,428	193,520

THE WORLD'S INTERNATIONAL TRADE IN WHEAT AND FLOUR IN BUSHELS,

(In round thousands)

Country	Exports		Imports	
	1900	1905	1900	1905
Argentina	73,495	112,718	....	....
Australia	14,517	31,730	....	....
Belgium	9,701	18,496	41,847	64,977
Brazil	....	....	10,800	14,983
Bulgaria	5,140	17,508	....	....
Canada	14,773	47,293	....	....
Chili	422	671	....	....
Denmark	....	....	3,171	5,936
Finland	....	....	3,464	3,581
France	....	....	5,810	7,347
Germany	12,464	10,513	49,246	85,137
Greece	....	....	6,873	5,864
India (British)	1,633	37,477	....	....
Italy	....	....	26,941	38,744
Japan	....	....	3,057	7,874
Netherlands	28,488	53,951	44,416	70,380
Portugal	....	....	5,029	4,673
Roumania	27,665	65,247	....	....
Russia	74,140	181,158	....	....
Servia	3,641	3,521	....	....
Spain	....	....	8,502	35,503
Switzerland	....	....	14,685	17,898
United Kingdom	....	....	182,100	212,089
United States	215,990	97,609	....	....
Other Countries	15,658	26,870	33,075	26,287
Total	497,727	708,393	439,016	608,788

lands that lack only the application of human industry to make them productive of wheat, there is no occasion for any fear of a shortage of grain. The wheat industry of the world must undergo great developments before even its approximate limits can be defined.

The Northern Hemisphere produces about 95 per cent of the wheat crop of the world. This half of the globe not only consumes its entire product, but a large part of the crop of the Southern Hemisphere as well. About 75 per cent of the total wheat crop is produced in seven countries north of the equator. Europe produces over half of the world's wheat, but her population is so great that she consumes the world's surplus in addition to her own product. It was not until after the middle of the nineteenth century that large masses of trans-oceanic wheat appeared in Europe. In the seventies of this century India wheat made its advent into the world market, and two decades later there was a sudden and enormous influx of Argentine wheat. The world's production of wheat is continually increasing, and in 1906 it approximated three and a half billion bushels.

APPENDIX

CLASSIFICATION OF WHEAT

Following the classifications of Carleton,[1] of Haeckel,[2] and of Kornicke and Werner, and perfecting them by adding new data, by extending to smaller subdivisions, by giving world distribution, and, for the sake of unity and completeness, by giving the essential characteristics of each division, there is given below a descriptive and distributive outline of the division Hordeæ given on Page 2.

1.1 Hordeæ (Sub-tribe).
2.1 Lolieæ (Rye Grass).
3.1 Leptureæ.
4.1 Elymeæ (Barley Wild Rye).
5.1 Triticeæ
  1.2 Agropyrum (Genus) (Quack. G
  2.2 Haynaldia.
  3.2 Secale (Rye).
  4.2 Triticum.
    1.3 Aegilops (section). Species *ovata* taken as type. Found in southern Europe to Turkestan in Asia. Twelve species in all are recognized.
    2.3 Sitopyrus.
      1.4 Triticum monococcum.
        1.5 Name: None in English. German *Einkorn* preferred. French *Engrain*.
        2.5 Characteristics: Spikelets three flowered but one grained; hardy; non-shattering; short, thin narrow-leaved plant seldom over 3 feet high. Very constant in fertility; does not give fertile cross with common wheat; only species in which paleæ fall in two parts at maturity; spikelets awned; spike compact.
        3.5 Distribution: Found from Achaia in Greece to Mesopotamia. Present in Swiss Lake dwellings of stone age. Cultivated to a limited extent in Spain, France, Germany Switzerland and Italy. Unknown in America except to experimenters.
        4.5 Varieties: *Einkorn*; *Engrain* double (two grains).
        5.5 Use: Rarely for bread; usually for mush and "cracked wheat," and for fodder.
      2.4 Triticum Polonicum.
        1.5 Name: Polish wheat a misnomer; Giant or Jerusalem rye. Perhaps native in Mediterranean region.
        2.5 Characteristics: Only species in which lowest flower has palea as long as its glume; outer glumes at least as long as flowering glumes; two to three seeded; tall; stems pithy within; heads and kernels extremely large; macaroni gluten; drought and rust resistant; resembles rye.
        3.5 Distribution: Spain; Italy; Abyssinia; Southern Russia and Turkestan: Brazil; Northwestern United States.
        4.5 Varieties: Only one, White Polish, is widely known.
        5.5 Use: Principally for macaroni.
      3.4 Triticum sativum dicoccum.
        1 5 Name: None in English, often erroneously called spelt; German *emmer* preferred.

[1] U. S. Dept. Agr., Div. Veg. Phys. and Path., Bul. 24, p. 6.
[2] Minn. Bul. 62, p. 392.

2.5   Characteristics: Probably derived from *Einkorn*; leaves usually velvety hairy; plants pithy or hollow; heads very compact and flat almost always bearded; threshing does not remove chaff; spikelets two-grained; non-shattering; some varieties drought and rust resistant.

3.5   Distribution: Extensively in Russia and Servia; Germany; Spain; Abyssinia; Switzerland; to some extent in France, and Italy; also perhaps in northern India Thibet, and in portions of China; in the United States; cultivated in prehistoric times.

4.5   Varieties: Red chaff; white chaff; etc.

5.5   Use: Quite extensively for human food in portions of Russia, Germany, Switzerland and Ita'y as "kaska," a sort of porridge from crushed *emmer*; grist; "pot barley;" bread; also used for feed.

4.4   Triticum sat. spelta.

1.5   Name: English, spelt; German, *spelz* or *dinkel*; French *epeautre*.

2.5   Characteristics: Grows fully as tall as wheat; heads loose, narrow, rather long, bearded or bald; very brittle rachis; spikelets two to five-grained; far apart in head; hardy; non-shattering; constancy in fertility; retains chaff in threshing.

3.5   Distribution: The oldest grain cultivated in ancient Egypt, Greece and the Roman Empire. With *emmer* is the principal bread grain of southwest German Empire; raised widely in Russia, Switzerland, Belgium, France, Italy, Spain. In Canada and the United States it is known only to experimenters.

4.5   Varieties: Winter and spring varieties white-bearded; black-bearded; red; smooth; white.

5.5   Use: Flour is placed in same rank as common wheat flour; fed to stock.

5.4   Triticum sat. compactum.

1.5   Name: Club or square head wheats; also "hedgehog wheat," "dwarf wheat."

2.5   Characteristics: Little more than two feet high, being a dwarf; heads very short, often squarely formed; commonly white, at times red; bearded or bald; spikelts very close, three or four-grained; grain short and small, red or white; great yielding power; stiff straw; non-shattering; eary maturity; drought resistant.

3.5   Distribution: Pacific coast and Rocky Mountain states of the United States; Chile; Turkestan; Abyssinia; to slight extent in Switzerland, Russia, and a few other districts of Europe.

4.5   Varieties: Generally known as "club" or "square head"; dwarf; hedgehog.

5.5   Use: Yield the flour desired in certain localities; crackers; breakfast foods.

6.4   Triticum sat. turgidum.

1.5   Name: Poulard or pollard wheats; English (a misnomer), rivet; German, *bauchiger Weizen*; French, *ble petanielle*; also known as English wheat; Egyptian wheat.

2.5   Characteristics: Rather tall; broad velvety leaves; stems thick and stiff; heads long, often square; bearded; spikelets compact, two to four-grained; grains hard and light color; resistant to rust and drought.

3.5   Distribution: France, Egypt, Italy, Turkey, Greece, Southern Russia, other Mediterranean and Black Sea districts, and experimentally in the United States.

4.5   Varieties: Poulard; composite wheats (T. compositum), known as Miracle, Egyptian or Mummy wheats, having branched or compound heads whose grains develop unequally.

5.5   Use: Macaroni and other pastes; bread; mixed with bread wheats to produce flour desired by certain French markets.

7.4   Triticum sat. durum.

1.5   Name: Durum, macaroni, or flint wheats.

2.5   Characteristics: Hardest grain and longest beard known among wheats; plants tall; leaves smooth with hard cuticle; heads slender, compact, at times very short; always bearded; grains glassy, sometimes rather transparent, yellowish, long; very sensitive to changes of environment; high gluten content; drought and rust resistant; spikelets two to four-grained.

3 5   Distribution: Practically the only wheat of Algeria, Spain, Greece, Mexico, and Central America; extensively raised in south and east Russia, Asia Minor, Turkestan, Egypt, Tunis, Sicily, Italy, India, Chile, Argentina, United States, and Canada.

4.5   Varieties:

  1.6   Gharnovka, Velvet Don, and Arnautka (Azov Sea region, Russia) United States.

  2.6   Kubanka (east of Volga river, Russia), United States.

  3.6   Saragolla (southeast Italy).

  4.6   Goose wheat (Canada, Dakota).

  5.6   Trigo candeal and Anchuelo (Argentina).

  6.6   Nicaragua (Central America, Texas).

  7.6   There are perhaps several dozen other varieties.

5.5   Uses: Macaroni; semolina; noodles; all kinds of pastries; bread; it is coming to be used for all purposes, in some regions, as ordinary wheat flour.

8.4   Triticum vulgare.

  1.5   Name: This is the common bread wheat.

  2.5   Characteristics: Well known; heads rather loosely formed; bearded or bald; chaff usually smooth but may be velvety; spikelets generally three-grained, but may be two, and rarely four; stem usually hollow; all the characteristics vary widely (see varieties).

  3.5   Distribution: Practically over the whole globe, within the limits already given (see varieties).

  4.5   Varieties: (Carleton's division, based not on botanical but on environmental characteristics).

    1.6   Soft winter wheats: Grain amber to white; produced by moist mild climate of even temperature; found in eastern United States, western and northern Europe, Japan, and in portions of China India, Australia, and Argentina.

    2.6   Hard winter wheats: Usually red-grained; usually bearded; relatively high gluten content; grown on black soils in climate characterized by extremes of temperature and moisture. Found chiefly in Kansas, Nebraska, Iowa, Missouri, and Oklahoma in the United States (the wheat of Crimean origin known as "Turkey red"), in Argentina (the Italian wheat, Barletta), in Hungary and Roumania, in southern and southwestern Russia, and to some extent in Canada, northern India, Asiatic Turkey, and Persia.

    3.6   Hard spring wheats: What has been said of the hard winter wheats also applies to this group, the difference being that the growing season is shorter, and the winter too severe for winter varieties. They are found in central and western Canada, the north central states of the United States (these are the fife and blue-stem wheats), east Russia and western and southern Siberia.

    4.6   White wheats: Soft and very starchy; grains harder and much drier than those of the soft winter wheats; fall or spring sown, even in same locality; grown chiefly in the Pacific coast and Rocky Mountain states of the United States, in Australia, in Chile, in Turkestan, and the Caucasus.

    5.6   Early wheats: Grain soft or semi-hard, amber to red; main characteristic is that they ripen early. Found in Australia and India, have a slight representation in California, and include some of the dwarf wheats of Japan.

  5.5   Districts in the United States (Carleton's division).

    1.6   Soft wheat.

      1.7   Present average yield per acre, about 14⅔ bushels.

      2.7   Chief varieties grown.

1.8 Fultz.		5.8 Jones' Winter Fife.
2.8 Fulcaster.		6.8 Red Wonder.
3.8 Early Red Clawson.		7.8 Gold Coin.
4.8 Longberry.		8.8 Blue Stem.

      3.7   Needs of the grower.

        1.8   Harder-grained, more glutinous varieties.

        2.8   Hardier winter varieties for the most northern portions.

        3.8   Early maturity.

        4.8   Rust resistance.

2.6 Semi-hard winter wheat.
  1.7 Present average yield per acre, about 14 bushels.
  2.7 Chief varieties grown.
    1.8 Fultz.
    2.8 Poole.
    3.8 Rudy.
    4.8 Mediterranean.
    5.8 Valley.
    6.8 Nigger.
    7.8 Dawson's Golden Chaff
    8.8 Early Red Clawson.
  3.7 Needs of the grower.
    1.8 Hardness of grain.
    2.8 Rust resistance.
    3.8 Hardy winter varieties.
3.6 Southern wheat.
  1.7 Present average yield per acre, about 9¾ bushels.
  2.7 Chief varieties now grown.
    1.8 Fultz.
    2.8 Fulcaster.
    3.8 Red May.
    4.8 Rice.
    5.8 Everett's High Grade.
    6.8 Boughton.
    7.8 Currel's Prolific
    8.8 Purple Straw.
  3.7 Needs of the grower.
    1.8 Rust resistance.
    2.8 Early maturity.
    3.8 Resistance to late spring frosts.
    4.8 Stiffness of straw.
4.6 Hard spring wheat.
  1.7 Present average yield per acre, about 13 bushels.
  2.7 Chief varieties.
    1.8 Saskatchewan Fife.
    2.8 Scotch Fife.
    3.8 Power's Fife.
    4.8 Wellman's Fife.
    5.8 Hayne's Blue Stem.
    6.8 Bolton's Blue Stem.
    7.8 Minnesota 163.
  3.7 Needs of the grower.
    1.8 Early maturity.
    2.8 Rust resistance.
    3.8 Drouth resistance.
    4.8 Hardy winter varieties.
5.6 Hard winter wheat.
  1.7 Present average yield per acre, about 12¾ bushels.
  2.7 Chief varieties grown.
    1.8 Turkey.
    2.8 Kharkov.
    3.8 Big Frame.
  3.7 Needs of the grower.
    1.8 Drouth resistance.
    2.8 Hardy winter varieties.
    3.8 Early maturity.
6.6 Durum wheat.
  1.7 Present average yield per acre, 11½ bushels.
  2.7 Chief varieties.
    1.8 Nicaragua.
    2.8 Turkey.
    3.8 Arnautka.
    4.8 Kubanka.
  3.7 Needs of the grower
    1.8 Durum varieties.
    2.8 Drouth resistance.
    3.8 Rust resistance.
    4.8 Early maturity.
7.6 Irrigated wheat.
  1.7 Present average yield per acre, about 21 bushels.
  2.7 Chief varieties.
    8 Sonora.

2.8 Taos.
3.8 Little Club
4.8 Defiance.
5.8 Turkey.
3.7 Needs of the grower.
    1.8 Increase of gluten content.
    2.8 Early maturity.
8.6 White wheat.
    1.7 Present average yield per acre, about $14\frac{7}{8}$ bushels.
    2.7 Chief varieties.
        1.8 Australian.
        2.8 California Club.
        3.8 Sonora.
        4.8 Oregon Red Chaff.
        5.8 Foise.
        6.8 Palouse Blue Stem.
        7.8 Palouse Red Chaff.
        8.8 White Winter.
        9.8 Little Club.
    3.7 Needs of the grower.
        1.8 Early maturity.
        2.8 Non-shattering varieties.
        3.8 Hardy winter varieties in the colder portions.

The distribution of these wheats in the United States in 1900 is shown in Map on page 9.

BIBLIOGRAPHY

This bibliography contains practically all of the works to which reference has been made in this volume. In addition it contains many other works that have been found of value. While it is not put forth as a complete list of all publications on wheat, it should, nevertheless, serve as a good foundation in all research work on this subject, for it is a fairly exhaustive list of American publications, and also contains many foreign works. An alphabetical list of all authors is first given, including periodicals containing articles of which the author is not stated, as well as miscellaneous official and unofficial publications. This list gives opportunity for looking up the works of any given author. For the purpose of aid in research, certain classifications of works will be found after the alphabetical list. All articles from encyclopedias and dictionaries are grouped together. Under each bureau or division of the United States Department of Agriculture are grouped the publications of that bureau or division. The next three groups are those of the United States census, the Department of Commerce and Labor, and consular reports. Then follows an alphabetical list of the state experiment stations of the United States, with station publications listed chronologically under each state. The publications of the Canadian Department of Agriculture are also grouped together. Finally, there is given a topical index of authors. In general, this index contains only those works which permit of definite classification, and it is arranged on the basis of individual works. Each work is placed under only one topic, the topic which it covers most definitely. The name of an author, however, appears as many times under different topics as he has written works on different phases of wheat. This topical index, and, to a certain extent, the classification under the United States Department of Agriculture, will facilitate a topical study of wheat, while the classification of experiment station works will aid in a geographical study.

Works of special merit are designated with*. Authors the whole of whose publications are of unusual value are designated with†. There are a few works that are inaccessible to the author, but should be contained in a bibliography of this nature. They are designated with §. New York is abbreviated N. Y., and London L.

OUTLINE OF BIBLIOGRAPHY.

AUTHORS.

*Adams, Cyrus C. A commercial geography. N. Y., 1902.
Adams, Edward F. The modern farmer in his business relations. San Francisco, 1899.
Albini, Giuseppi. Considerazioni sul valore nutritivo del pane integrale, in Rendiconto dell' accademia delle scienze fisiche e matematiche, serie 3a.-Vol. iv.-(Anno xxxvii), Naples, 1898.
Aldrich, W. Future wheat farming. Social Economist, 6:224, 1894.
Allen, E. W. The feeding of farm animals. U. S. Dept. Agr., Farmers' Bul. 22, 1895.
—— Subject list and abstracts of recent work in agricultural science. U. S. Dept. Agr., Exp. Sta. Record, Vol. 14, No. 11, 1903.
—— Some ways in which the Department of Agriculture and the Experiment Station supplement each other. Yearbook U. S. Dept. Agr., 1905, p. 167.
Allen, Grant. The pedigree of wheat. Pop. Sci. Mo., 22:662, 1883.
Allgemeines Statistisches Archiv, Tuebingen, 2:153-206, 517-614, 1891-2; 3:217-273, 1893, Russlands Bedeutung fur den Weltgetreidemarkt.
All the Year Round, L. 1:66, 1859, Farming by steam.
Andrew, A. P. Influence of the crops upon business in America. Quarterly Jour. of Econ., 20:323-53, 1906.
Andrews, C. C. Conditions and needs of spring wheat culture in the Northwest. U. S. Dept. Agr., Special Rept. 40, 1882.
Andrews, Frank. Crop export movement and port facilities on the Atlantic and Gulf coasts. U. S. Dept. Agr., Bu. of Sta., Bul. 38, 1905.
Annals of Botany, 18:321, 1904. On the fertilization, alternation, and general cytology of the uredineæ.
*Ardrey, R. L. American agricultural implements. Chicago, 1894.
Arnold, A. Mowing and reaping machines. Amer. Cycl. 12:16, 1875.
Atkinson, Edw. The wheat growing capacity of the United States. Pop. Sci. Mo., 54:145, 1898.
Atlantic Monthly, Boston, 45:33, 1880. The bonanza farms of the West.
Atwater, Helen W. Bread and the principles of breadmaking. U. S. Dept Agr., Farmers' Bul. 112, 1900.
Atwater, W. O. Organization of agricultural experiment stations in the United States. U. S. Dept. Agr., Off. Exp. Sta., Bul. 1, 1889.
Austin, O. P. Commercial Russia in 1904. Mo. Sum. Com. & Fin., Feb., 1904, p. 2755.
Baker, E. L. Transportation of wheat in the Argentine Republic. U. S. Cons. Rept., 49:460, 1895.
Baker, R. S. The movement of wheat. McClure's Mag., 14:124, 1899.
Baker, Willis E. Transportation of wheat in the Argentine Republic. U. S. Cons. Rept., 49:460, 1895.
Balz, Sylvester. Forage plants and cereals. S. D. Agr. Col. Exp. Sta., Bul. 96, 1906.
Barrow, D. N. Report for 1901, of North Louisiana Experiment Station.
Beal, F. E. L. Birds that injure grain. Yearbook U. S. Dept. Agr., p. 345, 1897.
—— Food of the bobolink, blackbirds and grackles. U. S. Dept. Agr., Div. Biolog. Sur., Bul. 13, 1900.
Beal, W. H. Farmyard manure. U. S. Dept. Agr., Farmers' Bul. 21, 1894.

———— Some practical results of experiment station work. Yearbook U. S. Dept.
    Agr., p. 589, 1902.
Beals, Edward A. Rainfall and irrigation. Yearbook U. S. Dept. Agr., p. 627,
    1902.
*Becker, Max. Der argentinische Weizen im Weltmarkte. Jena, 1903. (Bibli-
    ography on Argentina.)
Bedford, S. A. Reports Experimental Farm for Manitoba, 1899-1902, in Repts.
    Exp. Farms, Canada, 1899-1902.
Bennett, Alfred W. Wheat rust and barberry rust. Nature, 2:318, 1870.
Bennett, R. L. Wheat experiments. Ark. Agr. Exp. Sta., Bul. 29, 1894.
Bessey, Charles E. Wheat and some of its products. Neb. Agr. Exp. Sta., Bul.
    32, 1894.
Bessey, Ernst A. Progress of plant breeding in the United States. Yearbook
    U. S. Dept. Agr., p. 465, 1899.
Bicknell, Frank W. Wheat production and farm life in Argentina. U. S. Dept.
    Agr., Bu. of Statis., Bul. 27, 1904.
———— Agriculture in Argentina. Yearbook U. S. Dept. Agr., 1904, p. 271.
Bigelow, F. H. Work of the meteorologist for the benefit of agricultural commerce
    and navigation. Yearbook U. S. Dept. Agr., p. 71, 1899.
Birkbeck, Morris. Notes on a journey through France, 1814. L., 1815.
Blount, A. E. Cereals, etc. N. Mex. Agr. Col. Exp. Sta., Bul. 6, 1892.
Böhm, Otto. Die Kornhauser. Münchener Volkswirtschaftliche Studien, 26,
    1898.
Bolley, H. L. Wheat rust. Ind. Agr. Exp. Sta., Bul. 26, 1889.
———— Effect of seed exchange upon the culture of wheat. N. D. Agr. Col. Exp.
    Sta., Bul. 17, 1895.
———— Treatment of smut in wheat. N. D. Agr. Col. Exp. Sta., Bul. 19, 1895.
*———— New studies upon the smut of wheat, etc. N. D. Agr. Col. Exp. Sta., Bul.
    27, 1897. (Bibliography of station publications on cereal smuts.)
———— The stinking smut of wheat—the commercial side of the question. Pro-
    ceedings Tri-State Grain Growers' Ass'n., p. 85, 1900.
———— Department of Botany. Rept. N. D. Agr. Col. Exp. Sta., 1901, 1903.
———— Wheat rust. Science, n.s. 22:50-1, 1905.
*———— Rust problems. N. D. Agr. Col. Exp. Sta., Bul. 68, 1906.
Bollman, Lewis. The wheat plant. Rept. U. S. Dept. Agr., p. 65, 1862.
*Bond, Fred. Irrigation laws of the Northwest Territories of Canada and Wyom-
    ing. U. S. Dept. Agr., Off. Exp. Sta., Bul. 96, 1901.
Bonsteel, J. A. The use of soil surveys. Yearbook U. S. Dept. Agr., 1906, p. 181.
Boss, Andrew. Wheat varieties, breeding and cultivation. Univ. of Minn. Agr.
    Exp. Sta., Bul. 62, 1899.
Botanical Gazette, 25:30, 1898. A general review of the principal results of
    Swedish research into grain rust.
———— 39:241, 1905. Several reproductions of the rusts.
Bovey, C. C. The fallacy of exporting wheat. Rev. of Revs., 25:588, 1902.
Bradstreets, N. Y. 26:142, 1898, The Indian wheat outlook.
                  166,   Spanish wheat duties reduced.
                  166,   The Australian wheat yield.
                  230,   Siberian wheat.
                  262,   Co-operative wheat exporting.
                  349,   Wheat culture in North Canada.
                  610,   Is a dearth of wheat in sight?
                  654,   The world's wheat crop.
              27:406, 1899, Productive California wheat.
                  494,   Calendar of the world's wheat harvest.
              28:686, 1900, Wheat exports.
                  753,
                  830,   The wheat tides.
                  830,   The proposed German duty on wheat.
              30:149, 1902, Manitoba wheat versus American millers.
                  469,   Wheat yields.
                  469,   An English view of American wheat.
                  495,   The wheat crop of the world.
Brandenburg, Broughton. Moving this year's record grain crop. Harper's
    Weekly, 49:1632, 1905.
Breazeale, J. F. Effect of concentration of nutrient solution upon wheat cultures
    Science, n.s. 22:146-9, 1905.
†Brewer, Wm. H. First century of the Republic. N. Y., 1876.
———— Cereal production. 10th U. S. Census, Vol. 3, 1880.

†Briggs, Lyman J.   Electrical instruments for determining the moisture, temper-
    ature, and soluble salt content of soils.   U. S. Dept. Agr.. Div. of Soils.
    Bul. 15, 1899.
———   An electrical method for determining the moisture content of arable
    soils.   U. S. Dept. Agr., Div. of Soils, Bul. 6, 1897.
———   An electrical method for determining the temperature of soils.   U. S.
    Dept. Agr., Div. of Soils, Bul. 7, 1897.
———   The mechanics of soil moisture.   U. S. Dept. Agr., Div. of Soils. Bul.
    10 1897.
———   The movement and retention of water in soils.   Yearbook U. S. Dept.
    Agr., p. 399. 1898.
———   Field operations of the Division of Soils.   U. S. Dept. Agr., Rept. 64,
    1899.
———   Objects and methods of investigating certain physical properties of
    soils.   Yearbook U. S. Dept. Agr., p. 397, 1900.
British Almanac Companion. L., p. 62, 1839.   Enumeration and notice of acts of
    Parliament for the regulation of the trade of wheat.
British Manufacturing Industries. L 2d ed. Vol. 10 1877. Agricultural machinery.
British South African Gazette, July, Aug., Sept., 1903.   South African agriculture.
    in U. S. daily Cons. Repts., No. 1768, 1903.
Broomhall. J. S.   Corn trade year book, 1895.
Brown. A. Crum.   Justus Liebig.   Ency. Brit.. 9th ed. 14:565, 1882.
§Bruner, Lawrence.   The more destructive locusts of America north of Mexico.
    U. S. Dept. Agr., Div. of Entomol., Bul. 28, o.s., 1893.
Buchholz, E.   Homerischen Realien, Vol. I Leipsic 1871.
Buffalo. Chamber of Commerce, report 1905.
Buffum. B. C.   Results of three years' experiments in cost and profit of growing
    wheat.   Univ. Wy. Exp. Sta., Bul. 25, 1895.
———   The stooling of grains.   Univ. Wy. Exp. Sta., Bul. 37, 1898.
———   Some experiments with subsoiling.   Univ. Wy. Exp. Sta., Bul. 41 1899.
———   The use of water in irrigation in Wyoming.   U. S. Dept. Agr.. Off.
    Exp. Sta., Bul. 81, 1900.
———   Wheat growing on the Laramie Plains.   Univ. of Wy. Exp. Sta., Bul.
    60, 1903.
Bunker, Wm. H.   Report to the Chamber of Commerce of San Francisco of the
    Honorary Commission of foreign commerce, 1900.
———   Report of the Trans-Mississippi Commercial Congress at Cripple Creek.
    Colorado, July, 1900 (to San Francisco Chamber of Commerce).
———   Reports to the Chamber of Commerce of San Francisco, of July 10th,
    1902, reviewing legislative acts of the 57th Congress, 1st session.
———   Address to Chamber of Commerce, San Francisco, July 21, 1903.
Bürgel. Martin.   Russisches Getreide.   Jahrbuch f. Gesetzgebung, Verwaltung und
    Volkswirthschaft im deutschen Reich, Bd. 24, Heft II. s. 205, 1900.
Burrows, Alvin T.   Hot waves.   Yearbook U. S. Dept. Agr., p. 325, 1900.
Burtis, F. C.   Okla. Agr. Exp. Sta.. Bul. 65, 1905.
Caird, James.   High farming under liberal covenants the best substitute for
    protection.   Edinburg, 1849.
———   High farming vindicated.   Edinburg, 1850.
———   India, the land and the people.   L., 1884.
———   Agriculture in the reign of Queen Victoria, Vol. II.   London and Phila-
    delphia, 1887.
California State Agricultural Society, report 1905.
Cameron, Frank K.   Field operations of the Division of Soils.   U. S. Dept. Agr.,
    Rept. 64, 1899.
———   The chemistry of the soil as related to crop production.   U. S. Dept.
    Agr., Bu. Soils, Bul. 22. 1903.
Candolle, Alphonse de.   Origin of cultivated plants.   N. Y. 1885.
†Carleton, Mark Alfred.   Preliminary report on rusts of grain.   Kan. State Agr.
    Col. Exp. Sta., Bul. 38. 1893.
———   Improvements in wheat culture.   Yearbook U. S. Dept. Agr., p.
    489, 1896.   Also in Rept. Kan. State Bd. Agr., Vol. 21, No. 81, 1902, p. 50
*———   Cereal rusts of the United States.   U. S. Dept. Agr., Div. Veg.
    Phys. & Path., Bul. 16, 1899.   (Bibliography on rust, 74 works.)
*———Russian cereals adapted for cultivation in the United States.   U. S.
    Dept. Agr., Div. Botany, Bul. 23, 1900.
*———The basis for the improvement of American wheats.   U. S. Dept. Agr
    Div. Veg. Phys. & Path., Bul. 24, 1900.

———————— Successful wheat growing in semi-arid districts. Yearbook U. S. Dept. Agr.. p. 529. 1900.

*———————— Macaroni wheats. U. S. Dept. Agr., Bu. Plant Indus., Bul. 3, 1901.

———————— Emmer; a grain for the semi-arid regions. U. S. Dept. Agr., Farmers' Bul. 139, 1901.

———————— Wheat improvements in Kansas. Rept. Kan. State Bd. Agr., Vol. 21, No. 81, p. 56, 1902.

———————— Investigations of rusts. U. S. Dept. Agr., Bu. of Plant Indus., Bul. 63, 1904.

*———————— The commercial status of durum wheat. U. S. Dept. Agr., Bu. of Plant Indus., Bul. 70, 1904.

*———————— Lessons from the grain-rust epidemic of 1904. U. S. Dept. Agr., Farmers' Bul. 219, 1905.

Casson. Herbert N. The Romance of the Reaper. N. Y., 1908.

*Chamberlain, Joseph S. The commercial status of durum wheat. U. S. Dept. Agr., Bu. Plant Indus., Bul. 70, 1904.

Chambers' Journal, Edinburg, 21:4, 1854. Steam among farmers.

†Chase. Leon Wilson. Farm Machinery and Farm Motors N. Y., 1908.

Cheyney, Edw. P. Social changes in England. Boston, 1895. Penn. Univ. Publications, Philology, Literature and Archæology, series, Vol. 4, No. 2.

Chief Grain Inspectors' National Association, Proceedings 1st, 2d and 3d annual conventions, 1902-3.

Chilcott, E. C. Macaroni wheat in South Dakota. S. D. Agr. Col. Exp. Sta.; Bul. 77, 1902.

———————— Crop rotation for South Dakota. S. D. Agr. Col. Exp. Sta., Bul. 79, 1903.

———————— Some soil problems for practical farmers. Yearbook U. S. Dept. Agr., pp. 441-52, 1903.

*Chittenden, F. H. Some insects injurious to stored grain. Yearbook U. S. Dept. Agr., p. 277, 1894. Also in Farmers' Bul. 45, 1897.

Church, A(rthur) H(erbert). Food grains of India. L., 1886.

Clark, V. A. Seed selection according to specific gravity. N. Y. Agr. Exp. Sta. Bul. 256, 1904.

Clarke, S. A. The Inland Empire's harvests. Outlook, 66:286, 1900.

Coburn, F. D. Kansas and her resources.

———————— Kansas wheat growing. Rept. Kan. State Bd. Agr., Vol. 21, No. 81, 1902.

———————— Kansas statistics. Rept. Kan. State Bd. Agr., Vol. 21, No. 84, 1902.

Code, W. H. Use of water in irrigation. U. S. Dept. Agr., Off. Exp. Sta. Bul. 86, 1899.

Coffin, Fred F. B. Final report of the mid-plains division of the artesian and underflow investigation between the 97th meridian of longitude and the foothills of the Rocky Mountains. Sen. Ex. Doc., 1st sess., 52 Cong., 1891-'92. Vol. 4 Irrigation part 4, serial No. 2899.

Cole, John S. Crop rotation. S. D. Agr. Col. Exp. Sta., Bul. 98, 1906.

Colman, Norman J. Introduction to "Organization of Agricultural Experiment Station in the United States." U. S. Dept. Agr., Off. Exp. Sta., Bul. 1, p. 5. 1899.

Colquhoun. Peter. Treatise on the wealth, power and resources of the British Empire. L. 1814.

Comer, C. M. Wheat. S. C. Agr. Exp. Sta., Bul. 27, 1898.

Commercial and Financial Chronicle. N. Y. 75:292, 1902, The Deering Harvester Company.

Commercial Review, Portland, Oregon, July 1, 1901 & 1903.

Conant, Charles A. The uses of speculation. Forum, 31:698-712, 1901.

Conner, C. M. Field experiments with wheat. Mo. Agr. Col. Exp. Sta., Bul. 21, 1893.

Conner, John B. Eighth Biennial Report (14th Volume) of the Indiana Department of Statistics. 1899-1900.

Conrad, J. Getreidepreise. Handwörterbuch der Staatswissenschaften, 3:888, 1892.

Cooley, R. A. Entomological Department. Rept. Mont. Agr. Col. Exp. Sta., 1902.

Corn Trade Year Book, Liverpool & L., 1895, J. S. Broomhall.

Cottrell H. M. Wheat as a food for farm animals. Rept. Kan. State Bd. Agr., Vol. 21. No. 81, p. 117, 1902.

Coville, F. V. Division of Botany. Yearbook U. S. Dept. Agr. p. 90, 1897.

———————— Report of Botanist of Bu. of Plant Indus., 1901.

Crawford, R. F. An inquiry into wheat prices and wheat supply. Jour. Roy. Sta. Soc., 58:75, 1895.

———— Notes on the food supply of the United Kingdom, Belgium, France and Germany. Jour. Roy. Sta. Soc., 62:597, 1899.

Crookes, Wm. Address. Rept. Brit. Ass'n for Adv. of Sci., 1898.

———— The wheat problem. L., 1900.

*Crosby, D. J. Agricultural Experiment Stations in foreign countries. U. S. Dept. Agr., Off. Exp. Sta., Bul. 112, 1902.

———— Work and expenditures of the Agricultural Experiment Stations. Rept. Off. Exp. Sta., p. 23, 1902.

———— The usual illustrative material in teaching agriculture in rural schools. Yearbook U. S. Dept. Agr., 1905, p. 257.

Culbertson, Harvey. Irrigation investigations in western Texas. U. S. Dept. Agr., Off. of Exp. Sta., Bul. 158, 1904.

Cunningham, Bryson. Grain storage and manipulation at the quayside. Cassier, 28:40-54, 1905.

Current Literature, 36:548, 1904. Different breads and their values.

Dabney, Charles W. Agriculture, United States. Ency. Brit., 10th ed., 1:209, 1902.

Dabney, John C. The superior value of large, heavy seed. Yearbook U. S. Dept. Agr., p. 305, 1896.

Dalrymple, Wm. Treatise on the culture of wheat. L., 1801.

Dalton, J. C. Aliment or food. Amer. Cy., 1:314, 1873.

*Darwin, Charles. The variation of animals and plants under domestication. N. Y., 1868.

Daubeny, Charles Giles Bridle. Lectures on Roman husbandry. Oxford, 1857.

†Davidson, J. Brownlee. Farm Machinery and Farm Motors, N. Y., 1908.

Davis, C. W. The wheat supply of Europe and America, 1891. Arena, 3:641, 1891.

———— Some new views of "options," "futures," and "hedging," 1892.

———— Wheat: Crookes vs. Atkinson, Dodge, et al. Forum, 27:101, 1899.

———— Our present and prospective food supply. In "The Wheat Problem" by Crookes, p. 155, 1900.

Davis, Horace. Wheat in California. Overland Mo., 1:442, 1868; n.s., 32:60, 1898.

*Dehérain, P. P. Science in wheat growing. Pop. Sci. Mo., 50:101, 1896.

Dennis, J. S. Irrigation laws of the Northwest Territories of Canada and of Wyoming. U. S. Dept. Agr., Off. Exp. Sta., Bul. 96, 1901.

*DeVries, Hugo. Plant breeding. Comments on the experiments of Nilsson and Burbank, Chicago, 1907.

†Dewey, Lyster Hoxie. The Russian thistle. U. S. Dept. Agr., Farmers' Bul., 10, 1893.

*———— The Russian Thistle. U. S. Dept. Agr., Div. Botany, Bul. 15, 1894.

*———— Legislation against weeds. U. S. Dept. Agr., Div. Botany, Bul. 17, 1896.

———— Migration of weeds. Yearbook, U. S. Dept. Agr., p. 263, 1896.

Dike, G. P. The Argentine wheat farm. Outlook, 64:119, 1900.

Dorsey, Clarence W. Field operations of the Division of Soils. U. S. Dept. Agr., Div. of Soils, Rept. 64, 1899.

Douglas, E. S. A model bakery in London. The World Today, 10:213-14, 1906.

Drill, Robert. Soll Deutschland seinen ganzen Getreidebedarf selbst produzieren? Stuttgart, 1895.

Duvel, J. W. T. The vitality of buried seeds. U. S. Dept. Agr., Bu. of Plant Indus., Bul. 83, 1905.

Eastman, Philip. This year's big wheat harvest in Kansas. Rev. of Revs., 28:193, 1903.

Edgar, William C. The story of a grain of wheat. N. Y., 1903.

———— England's fight for free bread. The Northwestern Miller, Vol. 55, No. 1, p. 21, 1903.

Edwards, S. Fred. Some essential soil changes produced by micro-organisms. Mich. Agr. Col. Exp. Sta., Bul. 218, 1904.

Eighty years progress of the United States, Hartford, 1877.

Elliott, C. G. Some engineering features of drainage. Yearbook U. S. Dept. Agr. p. 231, 1902.

Ellis, William. The modern husbandman. L., 1744.

*†Emery, Henry Crosby. Legislation against futures. Pop. Sci. Quar., 10:62, 1895.

*———— Speculation on the stock and produce exchanges of the United States. N. Y., 1896.

*———— The results of the German exchange act of 1896. Pol. Sci. Quar., 13:286, 1898.

*———— Futures in the grain market. Econ. Jour., 9:44, 1899.

England, Chas. All grain inspection departments should be conducted upon civil
     service principles. Amer. Elevator & Grain Trade, 22:200, 1903.
§†Eriksson, J. Zeitschrift fur Pflanzenkrankenheiten, Vol. 4, Stuttgart, 1894.
———— Die Getreiderost, Stockholm, 1896.
Ewell, Ervin E. Every farm an experiment station. Yearbook U. S. Dept. Agr.,
     p. 291, 1897.
Fairchild, David. Cultivation of wheat in permanent alfalfa fields. U. S. Dept.
     Agr., Bu. of Plant Indus., Bul. 72, 1905.
Fairchild, David G. Saragolla wheat. U. S. Dept. Agr. Bu. Plant Indus., Bul.
     25, 1903.
Fairfield, W. H. Some experiments with subsoiling. Univ. Wy. Exp. Sta., Bul.
     41, 1899.
———————— Experiments in wheat culture. Univ. Wy. Exp. Sta., Bul. 48, 1901.
Farm Implement News, Chicago, July 17, 1902.
Farm Machinery Daily, St. Louis, Oct. 23, 1903, Proceedings of the tenth annual
     convention of the National Association of Agricultural Implement and
     Vehicle Manufacturers.
Fields, John. Reports of wheat raisers. Okla. Agr. Exp. Sta., Bul. 47, 1900.
———————— Report Oklahoma Agricultural Experiment Station, 1902
Fitch, Chas. H. Manufacturers of interchangeable mechanism: vii., Agricultural
     Implements. 10th U. S. Census, Vol. 2, 1880.
*Fitzherbert, Anthony (?). Book of husbandry. L., 1882. Reprinted from edition
     of 1534 by W. W. Skeat.
Fleming, O. The supply of wheat. Economic Review, 15:212-14; 332-5; 474-6,
     1905.
Fletcher, James. Report of the Entomologist and Botanist, 1898-1902. In Repts.
     of Can. Exp. Farms, 1898-1902.
———————— Injurious insects of the year in Canada. U. S. Dept. Agr., Div. of
     Entomology, Bul. 40, n.s., 1903.
Flint, Charles L. Progress in Agriculture. In Eighty Years Progress of the
     U. S., 1867.
Foering, John O. Proceedings of the first annual convention of the Chief Grain
     Inspectors' National Association, Feb. 1902.
———————— Same, second annual convention. Oct. 1902.
———————— Same, third annual convention, Oct. 1903. In Amer. Elevator & Grain
     Trade, Vol. 22, No. 4, Oct. 15, 1903.
Forbes, R. H. Sixteenth Annual Report, Univ. of Ariz. Agr. Exp Sta., 1905.
Ford, W. C. The question of wheat. I, England. Pop. Sci. Mo., 52:760, 1898.
———— The question of wheat. II, France. Pop. Sci. Mo., 53:1, 1898.
———— The question of wheat. III, Russia. Pop. Sci. Mo., 53:351, 1898.
*———— Wheat in commerce. Mo. Summary of Com. & Fin., March, 1898, p. 1386.
Fortier, S. Report Montana Agricultural College Experiment Station, 1902.
Fortune, Robert. Wanderings in China. L., 1847.
Foster, Luther. Experiments in wheat culture. Univ. Wy. Exp. Sta., Bul. 48,
     1901.
Fowler, Eldridge M. Agricultural machinery and implements. In C. M. Depew's
     100 Yrs. Amer. Commerce, 2:352, 1895.
Fream, Wm. The complete grazier. 13th ed., L., 1893, p. 706.
———————— Agriculture: United Kingdom. Ency. Brit. 10th ed., 1:209, 1902.
*Freeman, E. M. Minnesota plant diseases, 1905
———————— The use of the seed plat in the prevention of diseases in wheat. Proc.
     Amer. Breeder's Ass'n, 2:49, 1906.
Fuchs, Carl Johannes. Der Englische Getreidehandel und seine Organisation.
     Jahrbücher für Nationalökonomie, 54:1, 1890.
———————— Der Waren-Terminhandel. Leipzig, 1891.
Galloway, B. T. Water as a factor in the growth of plants. Yearbook U. S. Dept.
     Agr., p. 165, 1894.
———————— Division of Vegetable Physiology and Pathology. Yearbook U. S.
     Dept. Agr., p. 99, 1897.
———————— Progress in the treatment of plant diseases in the United States.
     Yearbook U. S. Dept. Agr., p. 191, 1899.
———————— Industrial progress in plant work. Yearbook U. S. Dept. Agr., p.
     219, 1902.
Gardner, Frank D. An electrical method of determining the moisture content
     of arable soils. U. S. Dept. Agr., Div. of Soils, Bul. 6, 1897.
———————— Field operations of the Division of Soils. U. S. Dept. Agr., Rept. 64,
     pp. 36, 77.

———— The electrical method of moisture determination in soils. U. S. Dept. Agr., Div. of Soils, Bul. 12, 1898.
Garman, H. Red rust of wheat. Ky. State Agr. Col. Exp. Sta., Bul. 77, 1898.
———— A new wheat fly. Ky. State Agr. Col. Exp. Sta., Bul. 30, 1890.
Garnier, Russel M. History of the English landed interest. L., 1892.
———— Annals of the British peasantry. L., & N. Y., 1895.
Gauss, Robert. Breeding drought-resistant crops. Proc. Amer. Breeders' Ass'n, 2:106, 1906.
Georgeson, C. C. Experiments with wheat. Kan. State Agr. Col. Exp. Sta., Buls. 20, 1891; 33, 1892; 40, 1893; 59, 1896.
———— Report of Alaska Agr. Exp. Sta., 1904.
Gibbins, H. de B. Industry in England. L., 1896.
Gibson, J. Agriculture in Wales. L., 1879.
Giffin, R. The coming scarcity of wheat. Nature 61:169, 1899.
Gilbert, J. H. On agricultural chemistry, especially in relation to the numerical theory of Baron Liebig, L., 1851.
———— On some points in the composition of wheat grains, its products in the mill and bread, L., 1857.
———— On the home produce, imports, consumption and price of wheat, L., 1852-53. Jour. Roy. Sta. Soc., 43:313, 1880.
Girard, A. Le froment et sa mouture. Paris, 1903.
Googe, Barnaby. The whole art and trade of husbandry. L., 1614.
Grandeau, L. Le movement agricole. Journal des Economistes, 5e s., 34:192; 35:188; 36:182, 1898.
Greathouse, Charles H. Historical sketch of the United States Department of Agriculture. U. S. Dept. Agr., Div. of Pub., Bul. 3, rev. ed., 1898.
———— Development of agricultural libraries. Yearbook U. S. Dept. Agr., p. 491, 1898.
———— State publications on agriculture. Yearbook U. S. Dept. Agr., 1904, p. 521.
*Gregory, J. W. Final report of the mid plains division of the artesian and underflow investigation between the 97th meridian of longitude, and the foothills of the Rocky Mountains. Sen. Ex. Doc., 1st sess., 52d Cong., 1891-92, Vol. 4, Irriga. Part. 4, serial No. 2899.
Grey, John. A view of the past and present state of agriculture in Northumberland. Berwick, 1841.
Grimes, H. S. Report of the annual convention of the Grain Dealers' National Association. Amer. Elev. & Grain Trade, 22:177, 1903.
Grisdale, J. H. Report of the Agriculturalist. Rept. Can. Exp. Farms, p. 263, 1901; p. 129, 1902.
Grosvenor, W. M. The world's wheat situation. Banker's Mag., (N. Y.) 50:26, 1894.
Hackney, Herbert. Kansas wheat and its products. Rept. Kan. State Bd. Agr., Vol. 21, No. 81, p. 83, 1902.
Hadley, A. T. The world's wheat harvest. Nation, 47:306, 1888.
Hall, A. D. Artificial fertilizers. Sci. Am. S., Vols. 63 and 64, 1907.
Halsted, Byron D. Conditions of growth of the wheat rust. Science, 3:457, 1884.
Hamilton, John. The farmers' institutes. Yearbook U. S. Dept. Agr., pp. 109-158, 1903.
Handwörtenbuch der Staatswissenschaften, Jena.
    3:861, 1892, Die altere Getreidehandelspolitik und Allgemeines, W. Lexis.
    869,   Der Getreidehandel in den Vereinigten Staaten von Amerika, M. Sering.
    872,   Der Getreidehandel in Russland, Jollos.
    878.   Statistic des Getreidehandel, v. Juraschek.
    888,   Getreidepreise, J. Conrad.
    893,   Getreideproduktion, A. Wirminghaus.
    899,   Getreidezölle, H. Paasche.
    4:249, 1892, Hagelschadenversicherung, Emminghaus. Supplemenstband.
    1:345, 1895, Getreidehandel.
Hansbrough, H. C. Address. Proc. Tri-State Grain Growers' Ass'n, p. 122, 1900.
Harper, J. N. Protein-content of the wheat kernel. Ky. Agr. Exp. Sta., Bul. 113, 1904.
Hart, E. B. The nature of the principal phosphorus compound in wheat bran. N. Y. Agr. Exp. Sta., Bul. 250, 1904.
Harter, L. L. The variability of wheat varieties in resistance of toxic salts. U. S. Dept. Agr., Bu. of Plant Indus., Bul. 79, 1905. (Contains bibliography on subject.)

*Hartlib, Samuel.   Legacy of husbandry.   L., 1655.
Harwood, W. S.   Breeding new wheats.   World's Work, 2:745, 1901.
Hassall, Arthur Hill.   Food: Its adulterations and the methods for their detection.
    L., 1876.
Haworth, E.   Gypsum as a soil fertilizer.   Rept. Kans. State Bd. Agr., Vol. 21,
    No. 81, p. 206, 1903.
*Hay, Robert.   Final geological reports of the artesian and underflow investiga-
    tion between the 97th meridian of longitude and the foothills of the Rocky
    Mountains.   Sen. Ex. Doc. 1st sess. 52d Cong., 1891-92, Vol. 4, Irriga.,
    Part 3, serial No. 2899.
†Hays, Willet M.   Grain and forage crops.   N. D. Agr. Exp. Sta., Bul. 10, 1893;
    Bul. 40, 1894; Bul. 46, 1895; Bul. 50, 1896.
———— The Russian thistle, or Russian tumble weed.   Univ. of Minn. Agr. Exp.
    Sta., Bul. 33, 1894.
*———— Progress in plant and animal breeding.   Yearbook U. S. Dept. Agr., pp.
    217-232, 1901.
———— Wheat varieties, breeding and cultivation.   Univ. Minn. Exp. Sta., Class
    Bul. 62, 1899.
———— Minnesota No. 163 wheat.   Univ. Minn. Exp. Sta., Class Bul. 8, 1900.
*———— Plant breeding.   U. S. Dept. Agr., Div.Veg. Phys. & Pathol., Bul. 29, 1901.
———— Winter wheat in Minnesota.   Univ. Minn. Exp. Sta., press Bul. 17, 1903.
Hayward, A. I.   Wheat.   Md. Agr. Exp. Sta., Bul. 14, 1891.
Haywood, J. K.   Analysis of waters and interpretation of results.   Yearbook U. S.
    Dept. Agr., p. 283, 1902.
§Heer, O.   Die Pflanzen der Pfahlbauten.   Zurich, 1866.
§Henley, Walter.   H.s Husbandry w. an anon. husbandry, seneschaucie & R.
    Grosseteste's rules.   Transcripts, tr. & glossary by E. Lamond, intr. by W.
    Cunningham (R. Hist. Soc. London).
Hervey de Saint Denys, M. J. L.   Recherches sur l'agriculture des Chinois.   Paris,
    1850.
Hess, Enos H.   Pa. Agr. Exp. Sta., Bul. 46, 1899.
———— Variety tests of wheat.   Pa. Agr. Exp. Sta., Bul. 55, 1901.
Hickman, J. Fremont.   Field experiments with wheat.   O. Agr. Exp. Sta., Bul.
    82, 1897.
———— Field experiments with wheat.   O. Agr. Exp. Sta., Bul. 129, 1901.
Hicks, Gilbert H.   Standards of the purity and the vitality of agricultural seeds.
    U. S. Dept. Agr. Div. Botany, Cir. 6, 1896.
———— The superior value of large, heavy seed.   Yearbook U. S. Dept. Agr.,
    p. 305, 1896.
———— The germination of seeds as affected by certain chemical fertilzers.
    U. S. Dept. Agr., Div. Botany, Bul. 24, 1900.
Hill, J. J.   Address.   Proc. Tri-State Grain Growers' Ass'n, p. 157, 1900
Hill, John, Jr.   Gold bricks of speculation.   N. Y. Com., Sept. 21-27, 1903.
*Hinton, Richard J.   Report on irrigation, including article, "Facts and conditions
    relating to irrigation in various countries."   Sen. Ex. Doc., 1st sess., 52d
    Cong., 1891-92, Vol. 4, serial No. 2899, p. 337.
Hitchcock, A. S.   Preliminary report on rusts of grain.   Kan. State Agr. Col. Exp.
    Sta., Bul. 38, 1893.
———— Prevention of grain smuts.   Kan. State Agr. Col. Exp. Sta., Bul. 99,
    1900.
———— Botanical notes on wheat and spelt.   Kan. State Agr. Col. Exp. Sta.,
    Bul. 99, 1900.
Hoffman, C. B.   Milling in the Kansas wheat belt.   Rept. Kan. State Bd. Agr.,
    Vol. 21, No. 81, p. 89, 1902.
*†Holmes, Edwin S., Jr.   Wheat growing and general agricultural condition in
    the Pacific coast regions of the United States.   U. S. Dept. Agr., Div. Sta.,
    Misc. series, Bul. 20, 1901.
———— Wheat ports on the Pacific coast.   Yearbook U. S. Dept. Agr., p.
    567, 1901.
Holmes, George K.   Progress of agriculture in the United States.   Yearbook U. S.
    Dept. Agr., p. 307, 1899.
*———— The course of prices of farm implements and machinery.   U. S. Dept.
    Agr., Div. Sta., Misc. ser., Bul. 18, 1901.
———— Practices in crop rotation.   Yearbook U. S. Dept. Agr., p. 519, 1902.
Home, Francis.   The principles of agriculture and vegetation.   Edinburgh, 1757.
Hopkins, Cyril G.   Soil treatment for wheat in rotation.   Univ. Ill. Exp. Sta., Bul.
    88, 1903.

———— Soil treatment for wheat on the poorer lands of the Illinois wheat belt. Univ. Ill. Agr. Exp. Sta., Cir. 97, 1905.

Hoskyn, Chandos Wren. Short inquiry into history of agriculture. L., 1849.

Hourwich, I. A. Wheat growing in Russia. Jour. Pol. Econ., 12:257, 1904.

†Howard, L. O. The chinch bug. U. S. Dept. Agr., Div. Entomol., Bul. 17, 1888. (Bibliography on chinch bug, 64 works.)

———— Danger of importing insect pests. Yearbook U. S. Dept. Agr., p. 529. 1897.

———— Recent laws against injurious insects in North America. U. S. Dept. Agr., Div. Entomol., Bul. 13, n. s., 1898.

———— The joint worm. U. S. Dept. Agr., Bu. of Entomol., Cir. 66, 1905.

———— Progress in economic entomology in the United States. Yearbook, U. S. Dept. Agr., p. 135, 1899.

———— Experimental work with fungous diseases of grasshoppers. Yearbook U. S. Dept. Agr. p. 459, 1901.

Humboldt, F. W. H. A. Essai politique sur le royaume de la Nouvelle Espagne. Paris, 1811.

Hummel, J. A. Soil investigations. Univ. Minn. Agr. Exp. Sta., Bul. 89, 1905.

*Hunt, Thomas F. Cereals in America. N. Y. 1904.

Hunt's Merchants' Magazine, N. Y., 16:293, 1847. The cost of raising wheat.

Huston, H. A. Forms of nitrogen for wheat. Ind. Agr. Exp. Sta., Bul. 41, 1892.

Hutchinson, B. P. Speculation in wheat. N. Amer. Rev., 153:414, 1891.

Hutchinson, W. L. Soils of Mississippi. Miss. Agr. Exp. Sta., Bul. 66, 1901.

———— Analyses of commercial fertilizers. Miss. Agr. Exp. Sta., Bul. 77, 1902.

†Hyde, John. Statistics of agriculture. 11th U. S. Cen., 1890.

———— Division of Statistics. Yearbook U. S. Dept. Agr., p. 258, 1897.

———— The fertilizer industry. U. S. Dept. Agr., Div. Sta., Misc. Ser., Bul. 13, 1898.

———— America and the wheat problem. N. Amer. Rev., 168:191, 1899. Also in "The Wheat Problem" by William Crookes, p. 189, 1900.

Industrial Commission, Washington, reports:
    Vol. 4, 1900, Transportation.
        6, 1901, Distribution of farm products.
        9, 1901, Transportation.
        10, 1901, Agriculture.
        11, 1901, Agriculture.

Ingersoll, C. L. Wheat and some of its products. Neb. Agr. Exp. Sta., Bul. 32 1894.

Inman, A. H. Domesday and feudal statistics. L., 1900.

Insurance Times, N. Y. 1:391, 1868.
        4:473, 1871.
        11:261, 1878.

Interstate Commerce Commission, Washington, reports, 1900, 1902.

Irving, H. Use of water in irrigation. U. S. Dept. Agr., Off. Exp. Sta., Bul. 86, p. 131, 1899.

Jackson, John R. On the botanical origin of wheat. Intellectual Observer, 11:262, 1867.

Jeffery, J. A. Soil moisture, its importance and management. Mich. Agr. Col. Exp. Sta., Bul. 219, 1904.

Jensen, G. H. Toxic limits and stimulation effects of some salts and poisons on wheat. Bot. Gazette, 43:11, 1907

Jesse, E. Pedigree wheat. Once a Week, 9:332, 1863.

Johnson, C. T. The use of water in irrigation. U. S. Dept. Agr., Off. Exp. Sta., Bul. 86, p. 47, 1899.

———— Practical irrigation Yearbook U. S. Dept. Agr., p. 491, 1900.

†Johnson, S. W. How crops grow. N. Y., 1868.

———— How crops feed. N. Y., 1870.

Johnson, Willis G. The Hessian fly in Maryland. Md. Agr., Exp. Sta. Bul. 58, 1898.

Jollos,—Der Getreidehandel in Russland. Handwörterbuch der Staatswissenschaften, 3:872, 1892.

Journal of Political Economy, Chicago:
    1:68, 1892, The price of wheat since 1867.
    1:365, 1893, The food supply and the price of wheat.

Journal of the Royal Agricultural Society of England. L., 1st s., 12:587, 1851. Report to H. R. H., the President of the commission for the exhibition of the works of industry of all nations on agricultural implements.

Judd, Sylvester D. Birds as weed destroyers. Yearbook U. S. Dept. Agr., p. 221, 1898.

Juraschek v. ——. Statistic des Getreidehandel. Handwörterbuch der Staatswissenschaften, 3:878, 1892.
Kansas State Board of Agriculture, report, 1902, Vol. 21, Nos. 81, 84.
Kapp, Friedrich Die Amerikanische Weizenproduktion. Volkswirthschaftliche Zeitfragen, Jahrgang 2, Heft 6, 1880.
Kaufman, E. E. Farmers' Institute Annual, N. D., 1902.
Kearney, Thomas H. Crops used in the reclamation of alkali lands in Egypt. Yearbook U. S. Dept. Agr., p. 573, 1902.
Kedzie, R. C. The ripening of wheat. Rept. Mich. Bd. Agr., 1881-82.
—— Composition of wheat and straw. Mich. Agr. Exp. Sta., Bul. 101, 1893.
Keyser, Alvin. Winter wheat. Univ. of Neb. Agr., Exp. Sta., Bul. 89, 1905.
—— Variation in wheat hybrids, Proc. Amer. Breeders' Ass'n, 2:84, 1906.
—— Methods in wheat breeding. Proc. Amer. Breeders' Ass'n, 2:186, 1906.
King, F. H. Irrigation in humid climes. U. S. Dept. Agr., Farmers' Bul. 46, 1896.
—— Some results of investigations in soil management. Yearbook U. S. Dept. Agr., pp. 159-174, 1903.
King-Parks, Henry. On the supposed germinating powers of mummy wheat. Jour. Sci., 22:604, 1885.
Klippart, John H. An essay on the origin, growth, diseases, varieties, etc., of the wheat plant. Annual O. Agr. Rept., 1857.
Knappen, T. M. Reciprocity with Canada. Amer. Elev. & Grain Trade, 22:195, 1903.
Knight, Edward. Reaping machine. Knight's New Mechan. Dict., p. 743, 1884.
—— Reaper. Amer. Mechan. Dict., 3:1888, 1876.
Kuczynski, R. R. Freight rates on Argentine and North American wheat. Jour. Pol. Econ., 10:333, 1902.
Ladd, E. F. North Dakota soils. N. D. Agr. Col. Exp. Sta., Bul. 24, 1896.
—— Maintaining our soil fertility. Proc. Tri-State Grain Growers' Ass'n, p 142, 1900.
—— Humus and soil nitrogen and studies with wheat. N. D. Agr. Col. Exp. Sta., Bul. 47, 1901.
—— Chemical Department. Annual Repts. N. D. Agr. Col. Exp. Sta., 1901-1903
—— Alkali lands. N. D. Farmers' Inst. Annual, p. 129, 1902.
—— Analysis of formaldehyde sold in North Dakota. N. D. Agr. Col. Exp. Sta., Bul. 60, 1904.
Ladd, Story B. Patent growth of the industrial arts. 12th U. S. Cen., Vol. 10, Part 4, 1900.
Lamphere, George N. The history of wheat raising in the Red river valley, Proc. Tri-State Grain Growers' Ass'n, p. 179, 1900.
Lamprecht, Karl. Beitrage zur Geschichte des französischen Wirthschaftslebens im elften Jahrhundert. Leipzig, 1878.
Latham, Wilfred. The states of the River Plate. L., 1868.
Latta, W. C. Field experiments with wheat. Ind. Agr. Exp. Sta., Bul. 41, 1892.
Laut, Agnes C. The American invasion of Canada's wheat belt. Century, 65:481, 1903.
Lawes, J. B. On the growth of wheat upon the same land for four successive years. L., 1845.
—— On agricultural chemistry, especially in relation to the universal theory of Baron Liebig. L., 1851.
—— On some points in the composition of wheat grain, its products in the mill, and bread. L., 1857.
—— On the home produce, imports, consumption and price of wheat, 1852-53. Jour. Roy Sta. Soc., 43:313, 1880.
Leclerc, J. A. The effect of climatic conditions on the composition of durum wheat. Yearbook, U. S. Dept. Agr., 1906, p. 199.
Leiter, Joseph. Wheat and its distribution. Cosmopolitan, 26:114, 1898.
Lévy, Raphael Georges. Les marchés à terme. Annales des Sciences Politiques, 16: 1-17, 1901.
Lexis, W. Die altere Getreidehandelspolitik und Allgemeines. Handwörterbuch der Staatswissenschaften, 3:861, 1892.
Lindet, L. Le froment et sa mouture. Paris, 1903.
Lippert, Gustav. Getreidepreise und Getreidezölle. Zeitschrift für Volkswirthschaft, Socialpolitik und Verwaltung, 8:276, 1899.
Lippert, Julius. Kulturgeschichte, Stuttgart, 1:584, 1886.
Livingston, B. E. Relation between growth of roots and of tops in wheat. Botanical Gazette, 41:139-43, 1906.
Lorenz, Ch. Deutschlands Getreideproduktion, Brodbedarf und Brodbeschaffung. Volkswirthschaftliche Zeitfragen, Jahrgang, 3, Heft 6, 1881.

Loughridge, R. H.   Tolerance of alkali by various cultures.   Univ. of Calif. Agr. Exp. Sta., Bul. 133, 1901.
Lowe, Joseph.   Present state of England in regard to agriculture, trade and finance. L., 1824.
Lyon, T. L.   The adaptation and improvement of winter wheat.   Neb. Agr. Exp. Sta., Bul. 72 1902.
——— Winter wheat.   Univ. of Neb. Agr. Exp. Sta., Bul. 89, 1905.
*——— Improving the quality of wheat.   U. S. Dept. Agr., Bu. Plant Indus., Bul. 78, 1905.   (Bibliography in foot notes.)
——— Some correlated characters in wheat and their transmission.   Proc. Amer. Breeders' Ass'n, 2 29 1906.
*——— Examining and grading grains.   N. Y., 1907.
McClure, W. F.   New wheat center of the world.   Independent, 61:191-8, 1906.
McCormick, Robert.   Memorial of.   Chicago 1885.
McCulloch, J. R.   Statistical account of the British Empire.   L., 1839.
——— The probable future price of wheat.   Bankers' Mag., (L.) 10:213, 1850.
MacDonald, James.   Wheat.   Chambers' Ency., n.ed., 10:625, 1892.
McDougall, John.   Indian wheats.   Jour. of Soc. of Arts, 37:637, 1889.
Mackay, Angus.   Reports for Experiment Farm for the Northwest Territories, 1900-02.   Rept. Can. Exp. Farms, 1900-02.
Macmillan,   n. s.1 45-52, 1905.   Bread.
Marcosson, I. F.   Harvesting the wheat.   World's Work, 9:5459-77, 1904.
Marks, Charles E.   The New York curb market.   Sat. Eve. Post, 176:36, 1903.
*Marlatt, C. L.   The principal insect enemies of growing wheat.   U. S. Dept. Agr., Farmers' Bul. 132, 1901.
——— Report of acting entomologist.   Annual Rept. U. S. Dept. Agr., p. 189, 1902.
——— The annual loss occasioned by destructive insects in the United States. Yearbook U. S. Dept. Agr., 1904, p. 461.
*Marquis, J. Clyde.   The Economic Significance of the Cereal Rust Fungi, 1904. (A thesis deposited at Purdue University, La Fayette, Ind.   Contains bibliography.)
Marston, R. B.   Our urgent need of a reserve of wheat.   19th Cen., 43:879, 1898.
Mason, Frank H.   Working of the German law against speculation in grain.   U. S. 'Consular reports, 64:438-444, 1900.
Masters, Maxwell T.   Wheat.   Ency. Brit., 24:531, 1888.
Matson, C. H.   The grain buyers' trust.   Review of Reviews, 25:201-5, 1902.
Maxwell, George H.   Address to National Association of agricultural implement and vehicle manufacturers, 1901.
†Mead, Elwood.   The use of water in irrigation.   U. S. Dept. Agr., Off. Exp. Sta., Bul. 86, 1899.
——— Rise and future of irrigation in the United States.   Yearbook U. S. Dept. Agr., p. 591, 1899.
——— The scope and purpose of the irrigation investigations of the office of experiment stations.   Rept. Off. Exp. Sta., 1901.
——— Some typical reservoirs in the Rocky Mountain states.   Yearbook U. S. Dept. Agr., p. 415, 1901.
——— Review of irrigation investigations for 1902.   Rept. Off. Exp. Sta., p. 359, 1902.
*——— Irrigation institutions.   N. Y., 1903.
——— The relation of irrigation to dry farming.   Yearbook U. S. Dept. Agr., 1905, p. 423.
†Means, Thomas H.   An electrical method for determining the soluble salt content of soils.   U. S. Dept. Agr., Div. of Soils, Bul. 8, 1897.
——— The soluble mineral matter of soils.   Yearbook U. S. Dept. Agr., p 495, 1898.
——— Field operations of the Division of Soils.   U. S. Dept. Agr., Div. of Soils, Rept. 64, 1899.
——— Crops used in the reclamation of alkali lands in Egypt.   Yearbook U. S. Dept. Agr., p. 573, 1902.
Merrill, L. A.   Arid farming in Utah.   Utah Agr. Col. Exp. Sta., Bul. 91, 1905.
Merrill, L. H.   The digestibility and nutritive value of bread.   U. S. Dept. Agr Off. Exp. Sta., Bul. 85, 1900.
——— Entire wheat flour.   Me. Agr. Exp Sta., Bul. 103, 1904.
Mertens, Oscar.   Russlands Bedeutung fur den Weltgetreidemarkt.   Allgemeines Statistisches Archiv., Vol. II, pp. 153-206, 517-614, 1891-92; Vol. III, pp. 217-273, 1893.
Metcalf, H.   Organisms on the surface of grain.   Science, n.s., 22:439-41, 1905.

Miller, E. . . .   Kansas wheat the pest in the world.   Rept. Kan. State Bd. Agr.,
          Vol. 21, No. 81, p. 20 1902.
*Miller, Merritt Finley.   The evolution of reaping machines.   U. S. Dept. Agr.,
          Off. Exp. Sta., Bul. 103, 1902.   (Bibliography on subject, 45 works.)
Miller, N. G.   Variety tests of wheat.   Pa. State Col. Agr., Exp. Sta., Bul., 76, 1906.
Minnesota, Bureau of Labor Statistics Rep. 1891-2, Part 2, Inventions in flour
          making machinery, and the prices of wheat, flour, etc.
Minnesota, Insurance commissioners, annual report, 1903, Part 1.
Minnesota, Railroad and warehouse commission, annual reports, 1902, 1906.
Montgomery, E. G.   Examining and grading grains.   N. Y., 1907.
Monthly Summary of Commerce and Finance of the U. S., Jan., 1900.   The grain
          trade of the United States and the world's wheat supply and trade.
Moore, George T.   Bacteria and the nitrogen problem.   Yearbook U. S. Dept.
          Agr., p. 333, 1902.
———— Soil inoculations for legumes.   U. S. Dept. Agr., Bu. of Plant Indus., Bul.
          71, 1905.
Moorhouse, L. A.   Okla. Agr. Exp. Sta., Bul. 65, 1905.
Moran, M. P.   Address.   Proc. Tri-State Grain Growers' Ass'n, p. 118, 1900.
Mortimer, John.   The whole art of husbandry.   (2d ed.)   L., 1708.
§Morton, John Chalmers.   Cyclopedia of agriculture.   Glasgow, 1851-1855.
———— The Farmers' Calendar.   (6th ed.)   L., 1884.
Morton, J. Sterling.   Report of the Secretary of Agriculture of the United States.
          Yearbook U. S. Dept. Agr., 1894-96.
Müller, Louis.   Our tariff in its relation to the grain trade.   Annals Amer. Acad.
          of Pol. & Soc. Sci., 29:528, 1907.
Nation, 84:463 1907.   Dollar Wheat.
          84:469, 1907.   Price of wheat.
National Association of Agricultural Implement Manufacturers, Proceedings 10th
          annual convention, in Farm Machinery Daily, Oct. 23, 1903.
Nature, 34 629, 1886.   Hybrid wheat.
Neftel, Knight.   Flour milling.   10th U. S. Cen., Vol. 3, 1880.
Nelson, Milton O.   Progress of Northwest Canada.   Northwestern Miller, 55:35,
          1903.
Nelson, S. A.   The A, B, C of Wall Street.   N. Y., 1900.
Nesbit, Christopher.   Milling in Hungary.   Northwestern Miller, 55:40, 1903.
*Nettleton, Edwin S.   Artesian and underflow investigation.   Sen. Ex. Doc.,
          1st sess., 52d Cong., 1891-92, Vol. 4, Irriga.   Part 2, serial No. 2899.
Newell, Frederick H.   Agriculture by irrigation.   11th U. S. Cen., 1890.
———— Irrigation on the Great Plains.   Yearbook, U. S. Dept. Agr., p. 167, 1896.
Newman, C. L.   Wheat experiments.   Ark. Agr. Exp. Sta., Bul. 62, 1900.
Newman, J. S.   Wheat.   S. C. Agr. Exp. Sta., Bul. 37, 1898.
———— Experiments with wheat.   S. C. Agr. Exp. Sta., Bul. 56, 1900.
New York Produce Exchange, reports, 1902-1906.
North Dakota, Commissioner of Agriculture and Labor, report, 1903.
North Dakota Farmers' Institute Annual, Fargo, 1902.
Noyes, A. D.   The price of wheat.   Nation, 63:378, 1896.
———— The scarcity of wheat.   Nation, 66:356 1898.
———— The farmer and the wheat crop.   Nation, 66:417, 1898.
———— The predicted wheat famine.   Nation, 67:237, 1898.
Options and futures, U. S. Senate, committee on the judiciary, hearings authorized
          by resolution of the Senate of March 3, 1892.
Osborn, Herbert.   The Hessian fly in the United States.   U. S. Dept. Agr., Div.
          Entomol., Bul. 16, n.s., 1898.   (Bibliography, 147 works.)
Osborne, Thomas B.   The proteids of the wheat kernel.   Amer. Chem. Jour.,
          15:392-471, 1893.
Paasche, H.   Getreidezölle.   Handwörterbuch der Staatswissenschaften,   3:899,
          1892.
———— Getreidezölle.   Handwörterbuch der Staatswissenschaften, Supplement
          Band, 1:361, 1895.
Paley, F. A.   Waste in wheat crops.   Science, 7:174, 1886.
Passy, Frederic.   La Speculation.   Journal des Economistes, Ser. 6, 4:227-32, 1904.
Paton, James.   Baking.   Ency. Brit., 9th ed., 3:250, 1875.
Patten, A. J.   The nature of the principal phosphorus compound in wheat bran.
          N. Y. Agr. Exp. Sta., Bul. 250, 1904.
Payne, J. E.   Unirrigated lands of eastern Colorado.   Col. Agr. Col. Exp. Sta.,
          Bul. 77, 1903.
†Payne, Will.   A Day in Broking.   Century, 36:340, 1888.
———— The Chicago Board of Trade.   Cen., 65:745, 1903,

Penny Magazine. L., 6:386, 1837. On the culture of wheat within the tropics.
Perels, Emil. Ueber die Bedeutung des Maschinenwessens fur die Landwirth-schaft. Berlin, 1867.
Peter, A. M. Protein content of the wheat kernel. Ky. Agr. Exp. Sta., Bul. 113, 1904.
Peters, Edward T. Influence of rye on the price of wheat. Yearbook U. S. Dept. Agr., p. 167, 1900.
Peterson, C. W. The wheat-growing capacities of the Northwest Territories. Canadian Mag., 14.137, 1899.
Peterson, Leo. The Commercial Review, Portland, Ore., July 1, 1901; July 1, 1903.
Pettit, Rufus H. Some insects of the year 1897. Mich. State Agr. Col. Exp. Sta., Bul. 160, 1898.
Philpot, J. H. The sacred tree. L., 1897.
Pickett, J. F. Experiments with wheat. S. C. Agr. Exp. Sta., Bul. 56, 1900.
Pieters, A. J. Seed selling, seed growing and seed testing. Yearbook U. S. Dept. Agr., p. 549, 1899.
———— Agricultural seeds—where grown and how handled. Yearbook U. S. Dept. Agr., p. 223, 1901.
———— The business of seed and plant introduction. Yearbook U. S. Dept. Agr., 1905, p. 291.
Plat, Hugh. The jewel house of art and nature. L., 1653.
Plumb, C. S. Univ. of Tenn. Agr. Exp. Sta., Vol. III, Bul. 2, 1890.
———— The geographical distribution of cereals in North America. U. S. Dept. Agr., Div. of Biolog. Sur., Bul. 11, 1898.
Poggi, T. Sul costo di Produzione del frumento in Italia. Atti del R. Istituto Veneto di Scienze, Lettere ed Arti, Tomo lvi, 7th s., T. ix., 1898.
Popenoe, E. A. Some insect enemies of wheat in Kansas. Rept. Kan. State Bd Agr., Vol. 21, No. 81, p. 110, 1902.
Powers, Le Grand. Crops and irrigation. 12th U. S. Cen., Vol. 6, pt. 2, 1900.
Poynting, J. H. Fluctuations in wheat. Jour. Roy. Sta. Soc., 47:35, 1884.
Pusey, Ph. Report on agricultural implements. Jour. Roy. Agr. Soc., Vol. 12, 1st s., p. 587, 1851.
Quarterly Review, L., 164:445, 1887. Competition in wheat growing
Railroad Gazette, N. Y.
    35:715, 1903, Train load and train mile cost.
    722,   The diversion of grain from Atlantic ports.
    760,   Summary of the case against the Erie Canal enlargement.
Rein, Johannes. Geographische und naturwissenschaftliche Abhandlungen. Leipzig, 1892.
Reports from Her Majesty's representatives on legislative measures respecting gambling in "option" and "future" contracts, Commercial No. 5, 1898, L.
*Retrospective exhibit of harvesting machinery by the Deering Harvester Company in the Palace of Agriculture; 56th Cong., 2d Sess., Sen. Doc. No. 232, Paris Exposition, 1900, Rept., Vol. 3, Sen. Docs., Vol. 29, serial No. 4057, p. 456.
Richardson, Clifford. An investigation of the composition of American wheat and corn. Reviewed by J. Wrightson in Nature, 29:173, 1883.
———— What science can teach about wheat. U. S. Dept. Agr., Misc. Special Rept. 2, 1883.
———— An investigation of the composition of American wheat and corn. U. S. Dept. Agr., Bu. of Chem., Bul. 4, 1884.
———— Report of chemist. Rept. of Com. of Agr., U. S., 1884.
*Riley, C. V. Destructive locusts. U. S. Dept. Agr., Div. Entomol., Bul. 25, 1891.
Risser, A. K. Variety tests of wheat. Pa. State Col. Agr. Exp. Sta., Bul. 67, 1904.
Roberts, I. P. Reaping and mowing machines. Johnson's Univ. Cy., 7:17, 1895.
Rogers, J. E. T. History of agriculture and prices in England, 1259-1582, Oxford, 1882.
Roper, S. C. D. Production of wheat in Canada. Can. Mag., 3:468, 1894.
———— The wheat lands of Canada. Pop. Sci. Mo., 55:766, 1899.
Rose, Joshua. Agricultural machinery. Appleton's Cy. of applied mechanics, 1:2, 1880.
Ross, D. W. Use of water in irrigation. U. S. Dept. Agr., Off. Exp. Sta., Bul. 86, p. 219, 1899.
Rubinow, I. M. Russia's wheat surplus. U. S. Dept. Agr., Bu. of Statis., Bul 42, 1906.
Saturday Review, London:
    57:411, 1884, The Indian wheat trade.
    478,   Wheat rings.

58:142, 1884, Cheap wheat.
61:265, 1886, The extreme cheapness of wheat.
64:559, 1887, The wheat crop.
    662,   American and Indian wheat.
67: 13, 1889, The wheat market.
87:103, 1899, The wheat question.
88:286, 1899, The wheat crop.
Saunders, Charles E. Report of the cerealist. Can. Dept. Agr., 1904, 1905.
———— The milling and chemical value of the grades of wheat in the Manitoba inspection division crop of 1904. Can. Dept. Agr., Cen. Exp. Farm, Bul. 50, 1905.
———— Evidence before the select standing committee on agriculture and colonization. Canada, 1904, 1905.
———— Results obtained in 1905 from trial plats of grain, etc. Can. Dept. Agr., Cen. Exp. Farm, Bul. 53, 1905.
———— Some observations on heredity in wheat. Proc. Amer. Breeders' Ass'n, 1:77, 1905.
———— A natural hybrid in wheat. Proc. Amer. Breeders' Ass'n, 1:137, 1905.
Saunders, D. A. Some destructive insects. S. D. Agr. Col. Exp. Sta., Bul. 81, p. 63, 1903.
†Saunders, William. Trials with grain. Can. Dept. Agr., Cen. Exp. Farm, Buls. 26, 1897; 29, 1898; 34, 1899; 36, 1900; 39, 1901; 41, 1902; 44, 1903.
———— Evidence before select standing committee on agriculture and colonization. Canada, 1900, 1902, 1903, 1904, 1905.
———— Report of director and acting agriculturist of the Canada Experimental Farms, 1898 to 1902.
———— Some results of cross-fertilization and decrease in vitality of grain by age. Ottawa, 1903.
———— Wheat growing in Canada. Toronto, 1904. Reprint Canadian Mag., 22:561-8, 1904.
———— Results obtained in 1905 from trial plats of grain, etc. Can. Dept. Agr., Cen. Exp. Farm, Bul. 53, 1905.
Sayons, A. E. Le marche a terme en grains a Londres. Jour. des Econ. 5e s, T. xxxviii., p. 78. 1899.
Scaramelli, Francois. Manufacture of semolina and macaroni. U. S. Dept. Agr., Bu. Plant Indus., Bul. 20, 1902.
Schrader, O. Weizen und spelz. Reallexikon der indogermanischen Altertumskunde, 2:947. Strassburg, 1901.
Schulte, J. I. Illustrations of the influence of experiment station work on culture of field crops. Yearbook U. S. Dept. Agr., 1905, p. 407.
*Schumacher, H. Die Getreidebörsen in den vereinigten Staaten von Amerika. Jahrbücher fur Nationalökonomie Bd. 66, 1896.
Schweitzer, Paul. Soils and fertilizers. Mo. Agr. Col. Exp. Sta., Bul. 19, 1892.
Science, N. Y., 10:253· 134, 1887. Indian wheat.
Scientific American. N. Y.
    Dec. 16, 23 1854, History of reaping machines.
    July 25,   1896, Agricultural machinery.
    Feb. 3,   1900, An early reaping machine.
    Feb.24, 83:53, 1900, Schweitzer system of bread making in Paris.
    Dec.23, 93:508, 1905, Effect of colcred light on grain.
    June 1, 96:450, 1907, Our 735,000,000-bushel wheat crop.
    Supplement:
    49:20,198, 1900, Schweitzer system of bread making in Paris.
    57:23,735, 1904, Experiments on wheat.
    61:25,263, 1906, Bleaching flour.
    62:25,641, 1906, Vitrolizing wheat.
    62:25,897, 1906, The agricultural division of the world.
    63:26,096, 1907. Castelin automobile plow.
*†Scofield, Carl S. Algerian durum wheats. U. S. Dept. Agr., Bu. Plant Indus., Bul. 7, 1902.
*———— The commercial grading of corn. U. S. Dept. Agr., Bu. Plant Indus., Bul. 41, 1903.
*———— The description of wheat varieties. U. S. Dept. Agr., Bu. Plant Indus., Bul. 47, 1903.
———— Accurate methods of grain grading. Amer. Elev. & Grain Trade, 22:192, 1903.
Selby, A. D. Some diseases of wheat and oats. O. Agr. Exp. Sta., Bul. 97, 1898.

†Sering, Max. Die Landwirthschaftliche Konkurrenz Nordamerikas in Gegenwart und Zukunft. Leipzig, 1887.
——— Der Getreidehandel in den vereinigten Staaten von Amerika. Handwörterbuch der Staatswissenschaften, 3:869, 1892.
Shamel, A. D. The effect of inbreeding in plants. Yearbook U. S. Dept. Agr., 1905 p. 337.
Shaw, G. W. Univ. of Calif. Agr. Exp. Sta., Cir. 16, 1905.
Shaw, R. S. Agricultural Department. Rept. Mont. Agr. Col. Exp. Sta., p. 21, 1902.
Sheldon, John L. Tubercles on legumes with and without cultures. Univ. of W. Va. Agr. Exp. Sta., Bul. 105, 1906.
Shepard, James H. The artesian well waters of South Dakota. S. D. Agr. Col. Exp. Sta., bul. 81, p. 43, 1903.
——— Macaroni wheat. Yearbook U. S. Dept. Agr., pp. 329-336, 1903.
——— Macaroni wheat and bread. S. D. Agr. Col. Exp. Sta., Bul. 92, 1905.
——— Macaroni or durum wheats. S. D. Agr. Col. Exp. Sta., Bul. 99, 1906.
Shepperd, J. H. Variety tests and changing seed wheat. N. D. Agr. Col. Exp. Sta., Bul. 39, 1898.
——— Wheat farming experiments and soil moisture studies. N. D. Agr. Col. Exp. Sta., Bul. 48, 1901.
——— Agricultural Department. Rept. N. D. Agr. Col. Exp. Sta., 1901-1903.
——— Root systems of field crops. N. D. Agr. Col. Exp. Sta., Bul. 64, 1905.
——— Rept. N. D. Edgerley sub. Exp. Sta., 1905.
Shipley, Arthur Everitt. Insects injurious to wheat. Ency. Brit., 24:534, 1888.
Shutt, Frank T. Report of chemist, 1898 to 1902. Rept. Can. Exp. Farm, 1898-1902.
——— The milling and chemical value of the grades of wheat in the Manitoba inspection division crop of 1904. Can. Dept. Agr., Cen. Exp. Farm, Bul. 50, 1905.
Skinner, Frank C. Agricultural machinery. Ency. Brit., 10th ed., 1:166, 1902.
Skinner, Robert P. Manufacture of semolina and macaroni. U. S. Dept. Agr., Bu. Plant Indus., Bul. 20, 1902.
Smith, Adam. Wealth of nations. L., 1811.
Smith, B. H. Formaldehyde: its composition and uses. Yearbook U. S. Dept Agr., 1905, p. 477.
Smith, C. B. Agricultural education in France. Yearbook U. S. Dept. Agr., p. 115, 1900.
——— Relation of the United States Department of Agriculture to the Farmer. Miss. Agr. Exp. Sta., Bul. 80, p. 25, 1903.
Smith, C. D. Shrinkage of farm products. Mich. State Agr. Col. Exp. Sta., Bul. 191, 1901.
Smith, Charles W. Commercial gambling. L., 1893.
Smith, Jared G. Commercial plant introduction. Yearbook U. S. Dept. Agr., p. 131, 1900.
Smith, Kingsland. The milling capacity of Great Britain. Northwestern Miller, v. 55, No. 1, p. 49, 1903.
——— History of flour milling. Northwestern Miller, Aug. 8, 1906-March 20, 1907.
Smith, Rollin E. The mighty river of wheat. Munsey's Mag., 25:17, 1901.
Smith, T. T. V. New wheat fields in the Northwest. 19th Cen., 6:10, 1879.
Smith, William G. Recent investigations on rust of wheat. Nature, 62:352, 1900.
Snow, B. W. Agricultural progress and the wheat problem. Forum, 28:94, 1899.
†Snyder, Harry. Wheat. Univ. of Minn. Agr. Exp. Sta., Bul. 29, 1893.
——— Humus in its relation to soil fertility. Yearbook U. S. Dept. Agr., p. 131, 1895.
——— Human food investigations. Univ. of Minn. Agr. Exp. Sta., Bul. 54, 1897.
——— The proteids of wheat flour. Univ. of Minn. Agr. Exp. Sta., Bul .63, 1899.
——— Studies on bread and bread making. U. S. Dept. Agr., Off. Exp. Sta., Bul. 67, 1899.
——— Studies on bread and bread making. U. S. Dept. Agr., Off. Exp. Sta., Bul. 101, 1901.
——— Wheat flour and bread. Yearbook U. S. Dept. Agr., pp. 347-362, 1903.
——— Wheat and flour investigations. Univ. of Minn. Agr. Exp. Sta., Bul. 85, 1904.
——— Soil investigations. Univ of Minn. Agr. Exp. Sta., Bul. 89, 1905; Bul. 94, 1906.
——— Cereal breakfast foods. U. S. Dept. Agr., Farmers' Bul. 249, 1906.
——— What science does for farm crops. Harper's Mo., 115:729, 1907.
Social Economist, N. Y. 8:201, 1895, British wheat production under free trade.

Soule, Andrew M.   The influence of climate and soil on the composition and milling qualities of winter wheat.   Univ. Tenn. Agr. Exp. Sta.. Bul. V. 16 ,No. 4, Oct., 1903.

Spectator, L.   59:1164, 1884, The unprecedented price of wheat.
          79:267, 1897, The rise in wheat.

Spillman, W. J.   Systems of farm management in the United States.   Yearbook U. S. Dept. Agr., p. 343, 1902.

Stabler, Edward.   Overlooked pages of reaper history.   Chicago, 1897.

Stannard, J. D.   Practical irrigation.   Yearbook U. S. Dept. Agr., p. 491, 1900.

Stedman, J. M.   The Hessian fly in Missouri.   Mo. Agr. Col. Exp. Sta., Bul. 62 1903

Stevens, Albert Clark.   Futures in wheat.   Quar. Jour. Econ., 2:37, 1887.

————   The utility of speculation.   Pol. Sci. Quar., 7:419, 1892.

————   The world's wheat crops and the cause for low prices.   Banker's Mag. (N. Y), 50:782, 1895.

*Storer, F. H.   Agriculture in some of its relations with chemistry.   N. Y., 1887.

Street, J. P.   Protecting the farmer against fraud.   Rev. of Revs., 35:213-6, 1907.

Swift, R. B.   Who invented the reaper?   (No publisher) 1897.

†Swingle, Walter Tennyson.   Treatment of smuts of oats and wheat.   U. S. Dept. Agr., Farmers' Bul. 5, 1892.   (Bibliography.)

————   The grain smuts.   Yearbook U. S. Dept. Agr., p. 409, 1894.   Also in U. S. Dept. Agr., Farmers' Bul. 75, 1898.

————   Hybrids and their utilization in plant breeding.   Yearbook U. S. Dept Agr., p. 383, 1897.

————   The prevention of stinking smut of wheat and loose smut of oats.   U. S. Dept. Agr., Farmers' Bul. 250, 1906.

Tait, C. E.   Pumping plants in Texas.   U. S. Dept. Agr., Off. Exp. Sta., Bul. 158, 1904.

Taylor, C.   Insects destructive to wheat.   Harper's Mag., 20:38, 1859.

Teele, R. P.   The organization of irrigation companies.   Jour. Pol. Econ., 12:61, 1904.

————   Preparing land for irrigation.   Yearbook U. S. Dept. Agr., 1903, pp. 239-250.

Teller, G. L.   Wheat and its mill products.   Ark. Agr. Exp. Sta., Bul. 42, 1896.

————   Progress of investigation in chemistry of wheat.   Ark. Agr. Exp. Sta., Bul. 53, 1898.

Ten Eyck, A. M.   Variety tests and changing seed wheat.   N. D. Agr. Col. Exp. Sta., Bul. 39, 1898.

————   A study of the root systems of wheat.   N. D. Agr. Col. Exp. Sta., Bul. 36, 1899.

————   Wheat farming experiments and soil moisture studies.   N. D. Agr. Col. Exp. Sta., Bul. 48, 1901.

————   Agricultural Department.   Rept. N. D. Agr. Col. Exp. Sta., 1901-1903.

Thomas, John Jacobs.   Farm implements and the principles of their construction and use.   L. & N. Y., 1860.

Thompson, C. W.   Movement of wheat growing.   Quar. Jour. of Econ., 18:570-84, 1904.

Thompson, O. A.   Rept. N. D. Edgerley sub-Exp. Sta., 1905.

Thompson, S. P.   When wheat fails.   Harper's Weekly, 51:874, 1907.

Thorne, Charles E.   Forty years of wheat culture in Ohio.   O. Agr. Exp. Sta., Bul. 2d s., V. 4, No. 8, art. 12, 1891.

————   The maintenance of fertility.   O. Agr. Exp. Sta., Bul. 110, 1899.

————   The Hessian fly in Ohio.   O. Agr. Exp. Sta., Bul. 36, 1902.

Thornton, Wm. Thos.   Agriculture.   Ency. Brit., 9th ed., 1:291, 1875.

Thorpe, T. B.   Wheat and its associations.   Harper's Mag., 15.301, 1857.

Thun, Alphons.   Landwirthschaft und Gewerbe in Mittelrussland seit Aufhebbung der Leibeigenschaft.   Leipzig. 1880.

Tisserand, E.   An ideal department of agriculture and industries.   Yearbook U. S. Dept. Agr., p. 543, 1896.

Townsend, Charles O.   Some important wheat diseases.   Md. Agr. Exp. Sta., Bul. 58, 1898.

Transactions of the California State Agricultural Society.   Sacramento, 1901.

Transactions of the New York Agricultural Society.   Albany, 1852, 1855, 1857, 1860, 1866.

Traphagen, F. W.   Department of Chemistry.   Rept. Mont. Agr. Col. Exp. Sta., p. 57, 1902.

Tri-State Grain Growers' Association, Proceedings, 1900, Fargo.

*†True, A. C.   A brief account of the experiment station movement in the United States.   U. S. Dept. Agr., Off. Exp. Sta., Bul. 1, p. 73, 1889.

——— Education and research in agriculture in the United States. Yearbook U. S. Dept. Agr., p. 81, 1894.

——— Agricultural education and research in Belgium. Yearbook U. S. Dept. Agr., p. 361, 1896.

——— Popular education for the farmer in the United States. Yearbook U. S. Dept. Agr., p. 279, 1897.

——— Agricultural education in the United States. Yearbook U. S. Dept. Agr., p. 157, 1899.

——— Agricultural experiment stations in the United States. Yearbook U. S. Dept. Agr., p. 513, 1899.

——— Report of the office of experiment stations, 1902. Rept. U. S. Dept. Agr., p. 241, 1902.

——— Work and expenditures of the agricultural experiment stations. Rept. Off. Exp. Sta., p. 23, 1902.

——— Progress in secondary education in agriculture. Yearbook U. S. Dept. Agr., p. 481, 1902.

*——— Agricultural experiment stations in foreign countries. U. S. Dept. Agr., Off. Exp. Sta., Bul. 112, 1902.

Tubeuf, Karl. Freiherr von Pflanzenkrankheiten durch kryptogame Parasiten verursacht. Berlin, 1895.

*Tull, Jethro. The horse-hoeing husbandry. L., 1822.

Turner, R. J. Seventh biennial report of the commissioner of agriculture and labor of North Dakota, June, 1903.

Tusser, Thomas. Five hundred points of good husbandry. (1st ed., 1557). L., 1812.

*Ulrich, J. C. Irrigation in the Rocky Mountain states. U. S. Dept. Agr., Off. Exp. Sta., Bul. 73, 1899.

Upham, Warren. The glacial lake Agassiz. Monographs U. S. Geol. Sur., V. 25, 1895.

Vanatter, Phares O. Influence of climate and soil on the composition and milling qualities of winter wheat. Univ. Tenn. Agr. Exp. Sta., Bul. V. 16, No. 4, Oct., 1903.

Veblen, T. B. The price of wheat since 1867. Jour. Pol. Econ., 1:68, 1892.

——— The food supply and the price of wheat. Jour. Pol. Econ., 1:365, 1893.

Vernon, J. J. Irrigation investigations at New Mexico Experiment Station. U. S. Dept. Agr., Off. of Exp. Sta., Bul. 158, 1904.

Voorhees, Clark C. The proteids of the wheat kernel. Amer. Chem. Jour., 15:392-471, 1893.

Voorhees, E. B. Commercial fertilizers. U. S. Dept. Agr., Farmers' Bul. 44, 1896.

——— Effect of commercial fertilizers. Sci. Am. S., 63:25921, 1907.

Voorhees, L. A. Studies on bread and bread making. U. S. Dept. Agr., Off. Exp. Sta., Bul. 67, 1899.

Waldron, L. R. Weed studies. N. D. Agr. Col. Exp. Sta., Bul. 62, 1904.

Walford, Cornelius. Agricultural insurance. Ins. Cy., 1:43, 1871; 5:272, 461, 536, 586, 1878.

Walker, Francis A. Remarks on the statistics of agriculture. 10th U. S. Cen. Vol. 3, 1880.

Wallace, John R. Mill operatives in New Zealand. Northwestern Miller, Vol. 55, No. 1, p. 39, 1903.

Walsh, G. E. Transportation of the world's wheat. Arena 24:516, 1900.

Waltershausen, W. Sartorius v. Ueber den Sicilianischen Ackerbau. Göttingen, 1863.

Warner, G. T. Landmarks in English industrial history. L., (Glasgow), 1899.

Washburn, F. L. Distribution of the chinch bug in Minnesota. U. S. Dept. Agr. Div. Entomol., Bul. 40, n.s., 1903.

——— Injurious insects of 1903. Univ. of Minn. Agr. Exp. Sta., Bul. 84, 1903.

Washington, Bureau of Statistics, Agriculture, and Immigration, Rept. 1903.

Waters, H. J. Wheat, varieties and change of seed. Mo. Agr. Col. Exp. Sta., Bul. 15, 1891.

Watson, G. C. Wheat. Johnson's Univ. Cy., 8:731, 1895.

——— Variety tests of wheat. Pa. State Col. Agr. Exp. Sta., Bul. 46, 1899; Bul. 55, 1901; Bul. 67, 1904; Bul. 76, 1906.

†Webber, Herbert J. Hybrids and their utilization in plant breeding. Yearbook U. S. Dept. Agr., p. 383, 1897.

——— Improvement of plants by selection. Yearbook U. S. Dept. Agr., p. 355, 1898.

——— Progress of plant breeding in the United States. Yearbook U. S. Dept. Agr., 1899, p. 465.

———— The improvement of cereals. Proc. Tri-State Grain Growers' Ass'n, p. 56, 1900.

*Webster, F. M.   The chinch bug.   U. S. Dept. Agr., Div. Entomol., Bul. 15, n.s., 1898.

———— Farm practice in the control of field-crop insects.   Yearbook U. S. Dept. Agr., 1905, p. 465.

———— The joint worm.   U. S. Dept. Agr., Bu. of Entomol., Cir. 66, 1905.

———— The Hessian fly.   U. S. Dept. Agr., Bu. of Entomol., Cir. 70, 1906.

———— The spring grain-aphis.   U. S. Dept. Agr., Bu of Entomol., Cir. 85, 1907.

Weldon, W. F. R.   Professor De Vries on the origin of species, Biometrika, 1:365, 1902.

Whalen, W. H.   Wheat growing and dairying for North Dakota.   N. D. Agr. Col. Exp. Sta., Bul. 8, 1892.

Wheeler, W. A.   Preliminary experiments with vapor treatments for the prevention of the stinking smut of wheat.   S. D. Agr. Col. Exp. Sta., Bul. 89, 1904.

———— Forage plants and cereals.   S. D. Agr. Col. Exp. Sta., Bul. 96, 1906.

Whelpley, J. D.   An international wheat corner   Fortnightly Review, 74:208, 1900.

White, H.   The wheat crisis.   Nation, 39:259, 1884.

———— Wheat and cotton.   Nation, 41:544, 1885.

———— India and America as wheat growers.   Nation, 45:430, 1887.

White, W. A.   The business of a wheat farm.   Scribner's Mag., 22:531, 1897.

†Whitney, Milton.   Soils in their relation to crop production.   Yearbook U. S. Dept. Agr., p. 129, 1894.

———— Reasons for cultivating the soil.   Yearbook U. S. Dept. Agr., p. 123, 1895.

———— An electrical method for determining the moisture content of arable soils.   U. S. Dept. Agr., Div. of Soils, Bul. 6, 1897.

———— An electrical method of determining the temperature of soils.   U. S. Dept. Agr., Div. of Soils, Bul. 7, 1897.

———— An electrical method of determining the soluble salt content of soils.   U. S. Dept. Agr., Div. of Soils, Bul. 8, 1897.

———— Division of Soils.   Yearbook U. S. Dept. Agr., p. 122, 1897.

———— Field operations of the Division of Soils.   U. S. Dept. Agr., Rept.64, 1899.

———— Soil investigations in the United States.   Yearbook U. S. Dept Agr., p. 335, 1899.

———— The chemistry of the soil as related to crop production.   U. S. Dept. Agr., Bu. of Soils, Bul. 22, 1903.

———— Report of the Bureau of Soils.   Rept. U. S. Dept. Agr., p. 155, 1902.

Widtsoe, John A.   Arid farming in Utah.   Utah Agr. Col. Exp. Sta., Bul. 91, 1905.

*Wiedenfeld, Kurt.   Getreidehandel.   Handwörterbuch der Staatswissenschaften, Supplement Band, 1:345-357, 1895.

*———— Getreideproduktion.   Handwörterbuch der Staatswissenschaften, Supplement Band, 1:357-361, 1895.

*———— Die Organisation des deutschen Getreidehandels und die Getreidepreisbildung im 19 Jahrhundert.   Jahrbuch f. Gesetzgebung, Verwaltung und Volkswirthschaft im Deutschen Reich, Vol. 24, Heft II, s. 165, 1900.

†Wiley, Harvey W.   The economical aspects of agricultural chemistry.   Proc. Amer. Ass'n for Adv. Sci. 35th meeting, p. 125, 1886.

———— Mineral phosphates as fertilizers.   Yearbook U. S. Dep. Agr., p. 177, 1894.

———— Soil ferments important in agriculture.   Yearbook U. S. Dept. Agr., p. 69, 1895.

———— Potash and its function in agriculture.   Yearbook U. S. Dept. Agr., p, 107, 1896.

———— Division of chemistry.   Yearbook U. S. Dept. Agr., p. 76, 1897.

———— The relation of chemistry to the progress of agriculture.   Yearbook U. S. Dept. Agr., p. 201, 1899.

———— Report of the chemist of the United States Department of Agriculture, 1899-1900, 1901-1902.

———— Flouring and grist mill products.   12th U. S. Cen., Vol. 9, pt. 3, 1900.

———— The influence of environment on the chemical composition of plants. Yearbook U. S. Dept. Agr., p. 299, 1901.

Wilkinson, J. Gardner.   A popular account of the ancient Egyptians.   Vol. II, N. Y., 1854.

Williams, C. G.   Experiments with winter wheat.   O. Agr. Exp. Sta., Bul. 165, 1905.

Williams, J. C.   Wheat situation in the United States.   Science, U. S., 21:458-9 1905.

Williams, J. R. Production of wheat in the United States. Hunt's Merchants' Mag., 12:307, 1845.
Willis, H. Parker. The adjustment of crop statistics. Jour. Pol. Econ., 11:1, 363, 540, 1902-3.
Wilson, James. Report of the Secretary of Agriculture. Rept. U. S. Dept. Agr., 1897-1906.
*Wilson, J. M. Irrigation laws of the Northwest Territories of Canada and of Wyoming. U. S. Dept. Agr., Off. Exp. Sta., Bul. 96, 1901.
Wirminghaus, A. Getreideproduktion. Handwörterbuch d. Staatswissenschaften, 3:893, 1892.
Woods, Albert F. Water as a factor in the growth of plants. Yearbook U. S. Dept. Agr., p. 165, 1894.
———— The present status of the nitrogen problem. Yearbook U. S. Dept. Agr., p. 125, 1906.
———— Work in vegetable physiology and pathology. Yearbook U. S. Dept. Agr., p. 261, 1898.
———— The relation of plant physiology to the development of agriculture. Yearbook U. S. Dept. Agr., 1904, p. 119.
Woods, Chas. D. The digestibility and nutritive value of bread. U. S. Dept. Agr. Off. Exp. Sta., Bul. 85, 1900.
———— Wheat flour and bread. Yearbook U. S. Dept. Agr., pp. 347-362, 1903.
Worst, J. H. Report of director. Rept. N. D. Agr. Col. Exp. Sta., p. 9, 1903.
Wrightson, John. Review of "An investigation of the composition of American wheat and corn," by C. Richardson. Nature, 29:173, 1883.
———— Wheat production in India. Nature, 32:79, 1885.
———— Agricultural machinery. Brit. Mfg. Indus., Vol. 10, Stanford, 1867.
———— Entire wheat flour. Me. Agr. Exp. Sta., Bul. 103, 1904.
———— Cereal breakfast foods. U. S. Dept. Agr., Farm. Bul. 249, 1906.
Youatt, Wm. The complete grazier. 12th ed., p. 706, L., 1893.
Young, Arthur. Annals of agriculture and other useful arts. Bury St. Edmunds 1787.
Zanitz, C. A. Breeding cereals. Proc. Amer. Breeders' Ass'n, 2:118, 1906.
Zeitschrift fur Volkswirthschaft, Socialpolitik und Verwaltung, Wien and Leipzig. 8:276, 1899, Getreidepreise und Getreidezölle, Gustav Lippert.

ENCYCLOPEDIAS AND DICTIONARIES

American Cyclopedia, 1:314, 1873, Aliment or food.
    10:767, 1875, Macaroni.
    12: 16, 1875, Mowing and reaping machines. A. Arnold.
    15:413, 1875, Straw.
    16:585, 1875, Wheat.
American Mechanical Dictionary, 3:1888, 1876, Reaper. E. H. Knight.
Appleton's Cyclopedia of Applied Mechanics, 1:2, 1880, Agricultural Machinery. J. Rose.
Chamber's Encyclopedia, 6:762, 1892, Macaroni.
    8:595, 1892, Reaping.
    9:765, 1892, Straw.
    10:625, 1892, Wheat. J. MacDonald.
Encyclopedia Britannica, * 1:291, 1875, Agriculture. W. T. Thornton.
  Ninth Edition.
    3:250, 1875, Baking. J. Paton.
    9:343, 1879, Manufacture of Flour. J. Paton.
    14:565, 1882, Justus Leibig. A. Crum Brown.
    24:531, 1888, Wheat. M. T. Masters.
    24:534, 1888, Insects injurious to wheat. A. E. Shipley.
  Tenth Edition, * 1:166, 1902, Agricultural machinery. F. C. Skinner.
    * 1:178, 1902, Agriculture: United Kingdom. W. Fream
    * 1:209, 1902, Agriculture: United States. C. W. Dabney
Insurance Cyclopedia, 1:43, 1871; 5:272, 461, 536, 586, 1878, Agricultural Insurance. C. Walford.
International Encyclopedia, 15:465, 1893, Wheat.
Johnson's Universal Cyclopedia, 1:784, 1893, Bread
    3:433, 1894, Flour.
    5:416, 1894, Macaroni.
    7: 17, 1895, Reaping and Mowing Machines. I. P. Roberts.
    8:731, 1895, Wheat. G. C. Watson.

Knight's New Mechanical Dictionary, p. 743, 1884, Reaping Machine. E. H
    Knight.
New International Encyclopedia, 3:377, 1902, Bread.
                              7:516, 1903, Flour.
                              9:138, 1903, Harvest and Harvesting.
                              11:626, 1903, Macaroni.
Reallexikon der Indogermanischen Altertumskunde, 2:947, 1901. Weizen und Spelz.
    O. Schrader.

OFFICIAL PUBLICATIONS

United States Department of Agriculture

Annual Report, p. 332, 1847, On wheat, tobacco, and spelt, as cultivated on the
                              Rhine.
              65, 1862, The wheat plant.   Lewis Bollman.
             369, 1873, Wheat culture in Japan.
               1, 1902, Report of the   Secretary   of   Agriculture.   James
                        Wilson.
             155,   Report of the Bureau of Soils.   M. Whitney.
             189,   Report of acting entomologist, C. L. Marlatt.
             241,   Report of the Office of the Experiment Stations.
                        A. C. True.

Bureau of Chemistry:
    Report of the United States Commissioner of Agriculture, 1884.
    Report of Chemistry.   C. Richardson.
    Bul.   4, 1884, An investigation of the composition of American wheat and
              corn.   Clifford Richardson.
    Report of Chemist, 1899-1902.   H. W. Wiley.
    Bul. 99, 1905, Proceedings of the 22d annual convention of the association
              of official agricultural chemists.

Bureau of Plant Industry:
    Bul.   3, 1901, Macaroni wheats.   M. A. Carleton.
    Report of Chief, 1901-1902,   B. T. Galloway.
    Report of Botanist, 1901.   F. V. Coville.
    Bul.   7, 1902, Algerian durum wheats.   C. S. Scofield.
          20, 1902, Manufacture of semolina and macaroni.   R. P. Skinner.
          25, 1903, Saragolla wheat.   D. G. Fairchild.
          40, 1903, Injurious insects of the year in Canada.   J. Fletcher.
          41, 1903, The commercial grading of corn.   C. S. Scofield.
          47, 1903, The description of wheat varieties.   C. S. Scofield.
          63, 1904, Investigations of rusts.   M. A. Carleton.
          70, 1904, The commercial status of durum wheat.   M. A. Carleton,
                    J. S. Chamberlain.
          71, 1905, Soil inocculations for legumes.   G. T. Moore.
          72, 1905, Cultivation of wheat in permanent alfalfa fields.   David
                    Fairchild.
          78, 1905, Improving the quality of wheat.   T. L. Lyon. (Bibliography
                    in foot notes.)
          79, 1905, The variability of wheat varieties in resistance of toxic salts
                    L. L. Harter (Bibliography).
          83, 1905, The vitality of buried seeds.   J. W. T. Duvel.

Division of Biological Survey:
    Bul. 11, 1898, The geographical distribution of cereals in North America.
                    C. S. Plumb.
          13, 1900, Food of the bobolink, blackbirds and grackles.   F. E. L. Beal.

Division of Botany:
    Bul. 15, 1894, The Russian thistle.   L. H. Dewey.
    Cir.   6, 1896, Standards of the purity and vitality of agricultural seeds.
                    G. H. Hicks.
    Bul. 17, 1896, Legislation against weeds.   L. H. Dewey.
          23, 1900, Russian cereals adapted for cultivation in the United States.
                    M. A. Carleton.
          24, 1900, The germination of seeds as affected by certain chemical fer-
                    tilizers.   G. H. Hicks.

Division of Entomology:
    Bul. 17, 1888, The chinch bug.   L. O. Howard.
          25, 1891, Destructive locusts.   C. V. Riley.
        § 28, 1893, The more destructive locusts of America north of Mexico.
                    L. Bruner.

13, 1898, Recent laws against injurious insects in North America. L. O. Howard.

15, 1898, The chinch bug. F. M. Webster.

16, 1898, The Hessian fly in the United States. H. Osborn.

Report of Acting Entomologist, 1902. C. L. Marlatt.

Bul. 40, 1903, The distribution of the chinch bug in Minnesota. F. L. Washburn.

Injurious insects of the year in Canada. J. Fletcher.

46, 1904, Proceedings of the 16th annual meeting of the Association of Economic Entomologists.

52, 1905, Proceedings of the 17th annual meeting of the Association of Economic Entomologists.

60, 1906, Proceedings of the 18th annual meeting of the Association of Economic Entomologists.

Cir. 66, 1905, The joint worm. F. M. Webster.

70, 1906, The Hessian fly. F. M. Webster.

85, 1907, The spring grain-aphis. F. M. Webster.

Division of Publications. Rev. ed., Bul. 3, 1898.

Historical sketch of the United States Department of Agriculture. C. H Greathouse.

Division of Soils:

Bul. 4, 1896, Methods of the mechanical analysis of soils and the determination of the amount of moisture in soils in the field.

5, 1896, Texture of some important soil formations.

6, 1897, An electrical method of determining the moisture content of arable soils. M. Whitney, F. D. Gardner, etc.

7, 1897, An electrical method of determining the temperature of soils. M. Whitney, L. J. Briggs.

8, 1897, An electrical method of determining the soluble soil content of soils. M. Whitney, T. H. Means.

10, 1897, The mechanics of soil moisture. L. J. Briggs.

12, 1898, The electrical method of moisture determination in soils. F. D. Gardner.

Report No. 64, 1899, Field operations of the Division of Soils. M. Whitney, T. H. Means, etc.

Bul. 15, 1899, Electrical instruments for determining the moisture, temperature and soluble salt content of soils. L. J. Briggs.

Report of Chief of Bureau, 1902. M. Whitney.

Bul. 22, 1903, The chemistry of the soil as related to crop production. M. Whitney, F. K. Cameron.

Division of Statistics, Misc. series:

Bul. 12, 1896, Freight charges for ocean transportation of the products of agriculture.

13, 1898, The fertilizer industry. J. Hyde.

18, 1901, The course of prices of farm implements and machinery G. K. Holmes.

20, 1901, Wheat growing and general agricultural conditions in the Pacific coast region of the United States. E. S. Holmes, Jr.

Division of Statistics:

Crop Cir. 1898-1899.

The Crop Reporter, Vols. 1-10, 1899-1908.

Bul. 27, 1904, Wheat production and farm life in Argentina. F. W. Bicknell.

38, 1905, Crop export movement and port facilities on the Atlantic and Gulf coasts. Frank Andrews.

42, 1906, Russia's wheat surplus. I. M. Rubinow.

Division of Vegetable Physiology and Pathology:

Bul. 16, 1899, Cereal rust of the United States. M. A. Carleton.

24, 1900, The basis of the improvement of American wheats. M. A. Carleton.

29, 1901, Plant breeding. W. M. Hays.

Report of Chief, 1900. B. T. Galloway.

Farmers' Bulletins:

No. 5, 1892, Treatment of smuts of oats and wheat. W. T. Swingle.

10, 1893, The Russian thistle. L. H. Dewey.

21, 1894, Barnyard manure. W. H. Beal.

22, 1895, The feeding of farm animals. E. W. Allen.

44, 1896, Commercial fertilizers. E. B. Voorhees.

45, 1897, Some insects injurious to stored grain. F. H. Chittenden.

46, 1896, Irrigation in humid climates.   F. H. King.
56, 1897, Experiment station work, I.
75, 1898, The grain smuts.   W. T. Swingle.
79, 1898, Experiment station work, VI.
105, 1899, Experiment station work, XII.
112, 1900, Bread and the principles of bread-making.   Helen W. Atwater.
132, 1901, The principal insect enemies of growing wheat.   C. L. Marlatt.
186, 1904, Experiment station work, XXIII.
219, 1905, Losses from the grain rust epidemic of 1904.   M. A. Carleton.
233, 1905, Experiment station work, XXXI.
237, 1905, Experiment station work, XXXII.
249, 1906, Cereal breakfast foods.   C. D. Woods, H. Snyder.
250, 1906, The prevention of stinking smut of wheat and loose smut of
    oats.   W. T. Swingle.
251, 1906, Experiment station work, XXXIV.

Office of Experment Stations:
 Bul. 1, 1889, Organization of agricultural experiment stations in the
     United States.

   11, 1892.
   67, 1899, Studies on bread and bread making.   H. Snyder and L. A.
     Voorhees.
   73, 1899, Irrigation in the Rocky Mountain states.   J. C. Ulrich.
   86, 1899, The use of water in irrigation.   E. Mead, C. T. Johnson, etc.
   81, 1900, The use of water in irrigation in Wyoming.   B. C. Buffum.
   85, 1900, The digestibility and nutritive value of bread.   C. D. Woods,
     L. H. Merrill.
   96, 1901, Irrigation laws of the Northwest Territories of Canada and
     of Wyoming.   J. S. Dennis, Fred Bond, etc.
   101, 1900, Studies on bread and bread making.   H. Snyder.
 Report, 1901, including "The scope and purpose of irrigation investigations
    of the Office of Experiment Stations," by E. Mead.
 Bul. 103, 1902, The evolution of reaping machines.   F. M. Miller.
   112, 1902, Agricultural Experiment Stations in foreign countries.   A. C.
     True, D. J. Crosby.
 Report of the Director, 1902.   A. C. True.
 Annual Report, 1902.   A. C. True, D. J. Crosby, E. Mead.
 Experiment Station Record, Vol. II., No. 12; Vol. XIV., No. 11, 1903.
 Bul. 158, 1905, Irrigation, and drainage investigations, 1904.   J. J. Vernon,
    Harvey Culbertson, C. E. Tait.
 Bul. 164, 1906, Proceedings of the 19th annual convention of the Association
    of American Agricultural Colleges and Experiment Stations.
*Senate Executive Documents, 1st Session, 52d Congress, 1891-92:
 Vol. 4, Irrigation, Parts 1-4, serial No. 2899.
 Part I, Report on Irrigation. J. R. Hinton.
    Facts and conditions relating to irrigation in various countries
    p. 377.   J R. Hinton.
   II, Artesian and underflow investigation.   E. S. Nettleton.
   III, Final geological reports of the artesian and underflow investigation
    between the 97th meridian of longitude and the foothills of the
    Rocky Mountains.   R. Hay.
    Report of the mid-plains division.   J. W. Gregory, F. F. B. Coffin.
Special Reports:
 No. 40, 1882, Conditions and needs of spring wheat culture in the North-
    west.   C. C. Andrews.
 Miscellaneous, No. 2, 1883, What science can teach about wheat.   C. Rich-
    ardson.

Yearbook United States Department of Agriculture:
 1894, Report of the Secretary of Agriculture.   J. S. Morton, p. 9.
    Education and research in argiculture in the United States.   A. C.
    True, p. 81.
    Soils in their relation to crop production.   M. Whitney, p. 129.
    Water as a factor in the growth of plants.   B. T. Galloway, A. F.
    Woods, p. 165.
    Mineral phosphates as fertilizers.   H. W. Wiley, p. 177.
    The most important insects injurious to stored grain.   F. H. Chitten-
    den, p. 277.
    The grain smuts: Their causes and prevention.   W. T. Swingle, p. 409.

1895, Report of the Secretary of Agriculture.   J. S. Morton, p. 9.
       Soil ferments important in agriculture.   H. W. Wiley, p. 69.
       Reasons for cultivating the soil.   M. Whitney, p. 123.
       Humus in its relation to soil fertility.   H. Snyder, p. 131.
1896, Report of the Secretary of Agriculture.   J. S. Morton, p. 9.
       Potash and its function in agriculture.   H. W. Wiley, p. 107.
       Irrigation on the Great Plains.   F. H. Newell, p. 167.
       Migration of weeds.   L. H. Dewey, p. 263.
       The superior value of large, heavy seed.   G. H. Hicks, J. C. Dabney,
       p. 305.
       Agricultural education and research in Belgium.   A. C. True, p. 361.
       Improvement in wheat culture.   M. A. Carleton, p. 489.
       An ideal department of agriculture and industries.   E. Tisserand,
       p. 543.
1897, Report of the Secretary of Agriculture.   J. Wilson, p. 9.
       Division of Chemistry.   H. W. Wiley, p. 76.
       Division of Botany.   F. G. Coville, p. 90.
       Division of Vegetable Physiology and Pathology.   B. T. Galloway,
       p. 99.
       Division of Soils.   M. Whitney, p, 122.
       Division of Statistics.   J. Hyde, p. 258.
       Popular education for the farmer in the United States.   A. C. True,
       p. 279.
       Every farm an experiment station.   E. E. Ewell, p. 291.
       Birds that injure grain.   F. E. L. Beal, p. 345.
       Hybrids and their utilization in plant breeding.   W. T. Swingle,
       H. J. Webber, p. 383.
       Danger of importing insect pests.   L. O. Howard, p. 529.
1898, Report of the Secretary of Agriculture.   J. Wilson, p. 9.
       Birds as weed destroyers.   S. D. Judd, p. 221.
       Work in Vegetable Physiology and Pathology.   A. F. Woods, p. 261.
       Improvement of plants by selection.   H. J. Webber, p. 355.
       The movement and retention of water in the soils.   L. J. Briggs, p. 399.
       The soluble mineral matter of soils.   T. H. Means, p. 495.
1899, Report of the Secretary of Agriculture.   J. Wilson, p. 9.
       Work of the meteorologist for the benefit of agricultural commerce
       and navigation.   F. H. Bigelow, p. 71
       Progress in economic entomology in the United States.   L. O. How-
       ard, p. 135.
       Agricultural education in the United States.   A. C. True, p. 157.
       Progress in the treatment of plant diseases in the United States.   B. T.
       Galloway, p. 191.
       The relation of chemistry to the progress of agriculture.   H. W. Wiley,
       p. 201.
       Progress of agriculture in the United States.   G. K. Holmes, p. 307.
       Soil investigations in the United States.   M. Whitney, p. 335.
       Progress of plant breeding in the United States.   H. J. Webber, E. A
       Bessey, p. 465.
       Development of agricultural libraries.   C. H. Greathouse, p. 491.
       Agricultural experiment stations in the United States.   A. C. True,
       p. 513.
       Seed selling, seed growing and seed testing.   A. J. Pieters, p. 549.
       Rise and future of irrigation in the United States.   E. Mead, p. 591.
1900, Report of the Secretary of Agriculture.   J. Wilson, p. 9.
       Agricultural education in France.   C. B. Smith, p. 115.
       Commercial plant introduction.   J. G. Smith, p. 131.
       Influence of rye on the price of wheat.   E. T. Peters, p. 167.
       Hot waves: Conditions which produce them and their effect on agri-
       culture.   A. T. Burrows, p. 325.
       Objects and methods of investigation of certain physical properties of
       soils.   L. J. Briggs, p. 397.
       Practical irrigation.   C. T. Johnston, etc., p. 491.
       Successful wheat growing in semi-arid districts.   M. A. Carleton, p.529.
1901, Report of the Secretary of Agriculture.   J. Wilson, p. 9.
       Progress in plant and animal breeding.   W. M. Hays, p. 217.
       Agricultural seeds—where grown and how handled.   A. J. Pieters,
       p. 233.
       Influence of environment on the chemical composition of plants.

H. W. Wiley, p. 299.

Some typical reservoirs in the Rocky Mountain states. E. Mead p. 415.

Experimental work with fungous diseases of grasshoppers. L. O Howard, p. 459.

Wheat ports on the Pacific coast. E. S. Holmes, Jr., p. 567.

1902, Report of the Secretary of Agriculture. J. Wilson, p. 9.

Industrial progress in plant work. B. T. Galloway, p. 219.

Some engineering features of drainage. C. G. Elliott, p. 231.

Analysis of waters and interpretation of results. J. K. Haywood, p. 283.

Bacteria and the nitrogen problem. G. T. Moore, p. 333.

Systems of farm management in the United States. W. J. Spillman, p. 343.

Progress in secondary education in agriculture. A. C. True, p. 481.

Practices in crop rotation. G. K. Holmes, p. 519.

Crops used in the reclamation of alkali lands in Egypt. T. H. Kearney, T. H. Means, p. 573.

Some practical results of experiment station work. W. H. Beal, p.589.

Rainfall and irrigation. E. A. Beals, p. 627.

1903, Report of the Secretary of Agriculture. J. Wlson, p. 9.

The farmers' institutes. John Hamilton, p. 109.

Some results of investigations in soil management. F. H. King, p. 159.

Preparing land for irrigation. R. P. Teele, p. 239.

Macaroni wheat J. H. Shepard, p. 329.

Wheat for bre. Harry Snyder, C. D. Wood, p. 347.

Some soil problems for practical farmers. E. C. Chilcott, p. 441.

1904, Report of the Secretary of Agriculture, p. 9.

The relation of plant physiology to the development of agriculture. A. F. Woods, p. 119.

Agricultural development in Argentina. F. W. Bicknell, p. 271.

The annual loss occasioned by destructive insects in the United States. C. L. Marlatt, p. 461.

State publications on agriculture. C. H. Greathouse, p. 521.

1905, Report of the Secretary of Agriculture, p. 9.

Some ways in which the Department of Agriculture and the Experiment Station supplement each other. E. W. Allen, p. 167.

The use of illustrative material in teaching agriculture in rural schools. D. J. Crosby, p. 257.

The business of seed and plant introduction. A. J. Pieters, p. 291

The effect of inbreeding in plants. A. D. Shamel, p. 377.

Illustrations of the influence of experiment station work on culture of field crops. J. I. Schulte, p. 407.

The relation of irrigation to dry farming. Elwood Mead, p. 423.

Farm practice in the control of field-crop insects. F. M. Webster, p. 465.

Formaldehyde: Its composition and uses. B. H. Smith, p. 477.

1906, Report of the Secretary of Agriculture, p. 9.

The present status of the nitrogen problem, p. 125.

The use of soil surveys, p. 181.

The effect of climatic conditions on the composition of durum wheat, p. 199.

UNITED STATES CENSUS

8th, 1860, Agriculture.

9th, 1870, Vol. 3.

10th, 1880, 2, Statistics of manufacturers.

Manufactures in interchangeable mechanism; VII, Agricultural implements. C. H. Fitch.

3, Remarks on the statistics of agriculture. F. A. Walker.

General statistics—tabular statements.

Cereal production. W. H. Brewer.

Flour milling. K. Neftel.

11th, 1890, Part 1, Fire insurance. T. A. Jenny.

Manufacturing industries in the United States.

Statistics of agriculture. J. Hyde.

Agriculture by irrigation. F. H. Newell.

12th 1900, Vol. 6, Part 2, Crops and irrigation. Le Grand Powers.
　　　　　　9,　　3, Flouring and gristmill products. H. W. Wiley.
　　　　　　10,　　4, Agricultural implements J. D. Lewis.
　　　　　　　　　Patent growth of the industrial art. S. B. Ladd.

UNITED STATES DEPARTMENT OF COMMERCE AND LABOR
Bureau of Manufactures, Tariff Series No. 2, 1907.

UNITED STATES CONSULAR REPORTS
49:460, 1895, Transportation of wheat in the Argentine Republic. W. E. Baker.
Daily Reports:
　　　1903, Agricultural implements in foreign countries. Nos. 1743, 1745, 1747,
　　　　　1750,1752, 1753, 1754, 1757, 1763, 1764, 1768, 1782.
　　　　　Macaroni wheat in foreign countries. Nos. 1796, 1808, 1820, 1838.
64:438-444, 1900, Working of the German law against speculation in grain. F. H.
　　　　　Mason.

EXPERIMENT STATION PUBLICATIONS
Experiment Stations:
Alaska: Annual Report, 1904. C. C. Georgeson.
Arizona: Timely hints for farmers, No. 20, 1900.
　　　　　Annual report, 1901, 1905.
　　　　　Irrigation at the station farms, 1898, 1901, 1902.
Arkansas: Bul. 29, 1894, Wheat experiments. R. L. Bennett.
　　　　　42, 1896, Concerning wheat and its mill products. G. L. Teller.
　　　　　53, 1898, Progress of investigation in chemistry of wheat. G. L.
　　　　　　　Teller.
　　　　　62, 1900, Wheat experiments. C. L. Newman.
California: Bul. 133, 1901, Tolerance of alkali by various cultures. R. H. Lough-
　　　　　ridge.
　　　　　Circular 16, 1905.
Colorado: Bul. 77, 1903 Unirrigated lands of eastern Colorado. J. E. Payne.
Illinois: Bul. 88, 1903. Soil treatment for wheat in rotations, etc. C. G. Hopkins.
　　　　　Circular 97, 1905, Soil treatment for wheat on the poorer lands of the
　　　　　Illinois wheat belt. C. G. Hopkins.
Indiana: Bul. 26, 1889, Wheat rust. H. L. Bolley.
　　　　　41, 1892, Field experiments with wheat. W. C. Latta.
　　　　　　Forms of nitrogen for wheat. H. A. Huston.
　　　　　114, 1906, Winter wheat.
Kansas: Bul. 20, 1891, Experiments with wheat. C. C. Georgeson.
　　　　　33, 1892, Experiments with wheat. C. C. Georgeson.
　　　　　40, 1893, Experiments with wheat. C. C. Georgeson.
　　　　　59, 1896, Experiments with wheat. C. C. Georgeson.
　　　　　38, 1893, Preliminary Report on rusts of grain. A. S. Hitchcock.
　　　　　　M. A. Carleton.
　　　　　99, 1900, Prevention of grain smuts. A. S. Hitchcock.
　　　　　　Botanical notes on wheat and spelt. A. S. Hitchcock.
Kentucky: Bul. 30, 1890, A new wheat fly. H. Garman.
　　　　　57, 1895, Wheat experiments.
　　　　　77, 1898, Red rust of wheat. H. Garman.
　　　　　83, 1899, Wheat.
　　　　　89, 1900, Wheat.
　　　　　113, 1904, Protein content of the wheat kernel. J. N. Harper.
　　　　　　A. M. Peter.
Louisiana: North Report, 1901. D. N. Barrow.
Maine: Bul. 103, 1904, Entire wheat flour. C. D. Woods, L. H. Merrill.
Maryland: Bul. 56, 1898, Wheat and lime experiments.
　　　　　58, 1898, The Hessian fly in Maryland. W. G. Johnson.
　　　　　　Some important wheat diseases. C. O. Townsend.
　　　　　14, 1891, Wheat. A. I. Hayward.
Michigan: Bul. 101, 1893, Composition of wheat and straw. R. C. Kedzie.
　　　　　160, 1898, Some insects of the year 1897. R. H. Pettit.
　　　　　191, 1901, Shrinkage of farm products. C. D. Smith.
　　　　　218, 1904, Some essential soil changes produced by micro-organ-
　　　　　　isms. S. F. Edwards.
　　　　　219, 1904, Soil moisture, its importance and management,
　　　　　　J. A. Jeffery.

Minnesota· Bul. 29, 1893, Wheat. Harry Snyder.
                33, 1894, The Russian Thistle or Russian tumble weed. W. M.
                        Hays.
                40, 1894, Grain and forage crops. W. M. Hays.
                46, 1895, Grain and forage crops. W. M. Hays.
                50, 1896, Grain and forage crops. W. M. Hays and others.
                54, 1897, Human food investigations. Harry Snyder.
                62, 1899 Wheat, varieties, breeding and cultivation. W. M.
                        Hays, A. Boss.
                63, 1899, The proteids of wheat flour. Harry Snyder.
       Class Bul. 8, 1900, Minnesota No. 163 wheat. W. M. Hays, A. Boss.
       Press Bul. 17, 1903, Winter wheat in Minnesota. W. M. Hays.
                24,   Seed grain.
            Bul. 84, 1903, Injurious insects of 1903. F. L. Washburn.
                85, 1904, Wheat and flour investigations. Harry Snyder.
                89, 1905, Soil investigations. Harry Snyder, J. A. Hummel.
                94, 1906, Soil investigations. Harry Snyder.
Mississippi: Bul. 66, 1901, Soils of Mississippi. W. L. Hutchinson.
                77, 1902, Analyses of commercial fertilizers. W. L. Hutchinson.
            Annual Report, 1902.
            Bul. 80, 1903, Farmers' Institute Bulletin, 1902.
Missouri: Bul. 15, 1891, Wheat varieties and change of seed. H. J. Waters.
                19, 1892, Soils and fertilizers. P. Schweitzer.
                21, 1893, Field experiments with wheat. C. M. Conner.
                62, 1903, The Hessian fly in Missouri. J. M. Stedman.
Montana: Annual Report, 1902.
Nebraska: Bul. 32, Vol. 6, Art. 6, 1894, Wheat and some of its products. C. L.
                        Ingersoll, C. E. Bessey.
                72,   14,   2, 1902, The adaptation and improvement of
                        winter wheat. T. L. Lyon.
                89,   17,   5, 1905, Winter wheat—co-operative experi-
                        ments with the U. S. Dept of Agr. T. L.
                        Lyon, A. Keyser.
New Mexico: Bul. 6, 1892, Cereals. A. E. Blount.
New York: Bul. 194, 1901, The Hessian fly and its ravages in New York.
                216, 1904.
                250, 1904, The nature of the principal phosphorus compounds in
                        wheat bran. A. J. Patten, E. B. Hart.
                256, 1904, Seed selection according to specific gravity. V. A.
                        Clark.
North Dakota: Bul. 8, 1892, Wheat growing and dairying for North Dakota.
                        E. F. Ladd, W. H. Whalen.
                10, 1893, Grain and forage crops. W. M. Hays.
                17, 1895, Effect of seed exchange upon the culture of wheat.
                        H. L. Bolley.
                19, 1895 Treatment of smut and wheat. H. L. Bolley.
                24, 1896, North Dakota soils. E. F. Ladd.
                27, 1897, New studies upon the smut of wheat, oats and
                        barley, etc. H. L. Bolley.
                39, 1898, Variety tests and changing seed wheat. J. H.
                        Shepperd, A. M. Ten Eyck.
                36, 1899, A study of the root systems of wheat, etc. A. M.
                        Ten Eyck.
                47, 1901, Humus and soil nitrogen and climatic studies with
                        wheat. E. F. Ladd.
                48, 1901, Wheat farming experiments and soil moisture
                        studies. J. H. Shepperd, A. M. Ten Eyck.
            Annual Reports, 1901-1905.
            Bul. 60, 1904, Analysis of formaldehyde sold in North Dakota.
                        E. F. Ladd.
                62, 1904, Weed studies. L. R. Waldron.
                64, 1905, Root systems of field crops. J. H. Shepperd.
                68, 1906, Rust problems. H. L. Bolley, F. J. Pritchard.
Ohio: Bul. 2d Ser., Vol. 4, No. 8, Art. 12, 1891, Forty years of wheat culture in
                        Ohio. C. E. Thorne.
                82, 1897, Field experiments with wheat. J. F. Hickman.
                97, 1898, Some diseases of wheat and oats. A. D. Selby.
                110, 1899, The maintenance of fertility. C. E. Thorne.

129, 1901, Field experiments with wheat. J. F. Hickman.
136, 1902, The Hessian fly in Ohio. C. E. Thorne.
165, 1905, Experiments with winter wheat. C. G. Williams.

Oklahoma: Bul. 47, 1900, Reports of wheat raisers. J. Fields.
Annual Report, 1902-3. J. Fields.
Press Buls. 99, 100, 1903, Wheat experiments.
Bul. 65, 1905. F. C. Burtis, L. A. Moorhouse.

Pennsylvania: Bul. 39, 1897, Variety tests of wheat.
46, 1899, Variety tests of wheat. G. C. Watson, E. H. Hess.
55, 1901, Variety tests of wheat. G. C. Watson, E. H. Hess.
67, 1904, Variety tests of wheat. G. C. Watson, A. K. Risser.
76, 1906, Variety tests of wheat. G. C. Watson, N. G. Miller.

Rhode Island: Report, 1894.

South Carolina: Buls. 37, 1898; 56, 1900, Wheat. J. S. Newman, etc.

South Dakota: Bul. 77, 1902, Macaroni wheat in South Dakota. E. C. Chilcott.
79, 1903, Crop rotation for South Dakota. E. C. Chilcott.
81, 1903, The artesian well waters of South Dakota. J. H. Shepard.
Some destructive insects. D. A. Saunders.
89, 1904, Preliminary experiments with vapor treatments for the prevention of the stinking smut of wheat. W. A. Wheeler.
92, 1905, Macaroni wheat and bread. J. H. Shepard.
96, 1906, Forage plants and cereals. W. A. Wheeler, S. Balz.
98, 1906, Crop rotation. J. S. Cole.
99, 1906, Macaroni or durum wheats. J. H. Shepard.

Tennessee: Bul. 2, Vol. 3, 1890. C. S. Plumb.
4, 16, 1903, Influence of climate and soil on the composition and milling qualities of winter wheat. A. M. Soule, P. O. Vanatter.

Utah: Bul. 91, 1905, Arid farming in Utah. J. A. Widtsoe, L. A. Merrill.

West Virginia: Bul. 105, 1906, Tubercles on legumes with and without cultures. J. L. Sheldon.

Wyoming: Bul. 22, 1895.
25, 1895, Results of three years' experiments in cost and profit of growing wheat. B. C. Buffum.
37, 1898, The stooling of grains. B. C. Buffum.
41, 1899, Some experiments with subsoiling. B. C. Buffum, W. H. Fairfield.
48, 1901, Experiments in wheat culture. L. Foster, W. H. Fairfield.
60, 1903, Wheat growing on the Laramie Plains. B. C. Buffum.

CANADA DEPARTMENT OF AGRICULTURE

Annual Reports of the Experimental Farms of Canada, 1898-1905.

Central Experimental Farm, Ottawa:
Buls. 26, 1897; 29, 1898; 34, 1899; 36, 1900; 39, 1901; 41, 1902; 44, 1903, Results from trial plots of grain, etc. W. Saunders.

Bul. 50, 1905, The milling and chemical value of the grades of wheat in the Manitoba Inspection Division crop of 1904. C. E. Saunders, F. T. Shutt.
53, 1905 Results obtained in 1905 from trial plats of grain, etc. W. Saunders, C. E. Saunders.

Cerealist, reports, 1904, 1905. C. E. Saunders.
Evidence of Dr. Charles E. Saunders before the Select Standing Committee on Agriculture and Colonization, 1904, 1905.

Evidence of Dr. William Saunders, Director Canada Experimental Farms, before Select Standing Committee on Agriculture and Colonization, 1902,1900, 1903, 1904, 1905.

Government of the Province of Saskatchewan.
Bul. 4, Condition of the crops at harvest time, Sept. 20, 1906.

TOPICAL INDEX OF AUTHORS

Saunders, W..............
Seed,
  Change of:
    Bolley, H. L..........
    Carleton, M. A........
    Shepperd, J. H........
    Smith, J. G...........
    Ten Eyck, A. M.......
    Waters, H. J..........
  General:
    Duvel, J. W. T........
    Hicks, G. H..........
    Pieters, A. J..........
  Selection:
    Clark, V. A.
    Dabney, J. C..............
    Hicks, G. H..............
    Lyon, T. L...............
    Webber, H. J.............
  Spring wheat:
    Andrews, C. C.............
  Winter wheat:
    Hays, W. M...............
    Keyser, A................
    Lyon, T. L...............
    Williams, C. G............
Diseases.
  General:
    Eriksson, J...............
    Freeman, E. M............
    Galloway, B. T...........
    Selby, A. D..............
    Townsend, C. O...........
    Tubeuf, K................
    Woods, A. F..............
  Rust:
    Annals of Botany.........
    Bennett, A. W............
    Bolley, H. L.............
    Botanical Gazette.........
    Carleton, M. A...........
    Eriksson, J...............
    Garman, H................
    Halsted, B. D............
    Hitchcock, A. S..........
    Marquis, J. C............
    Smith, W. G..............
  Smut:
    Bolley, H. L.............
    Hitchcock, A. S..........
    Ladd, E. F...............
    Scientific American.......
    Smith, B. H..............
    Swingle, W. T............
    Wheeler, W. A............
Dry farming.
  See semi-arid regions.........
Durum wheat.
  Carleton, M. A. ....
  Chamberlain, J. S............
  Chilcott, E. C...............
  Le Clerc, J. A...............
  Scofield, C. S...............
  Shepard, J. H...............
Duties.
  Bradstreet's.................
  British Almanac Companion...
  Caird, J.....................

Knappen, T. M..............
Lippert, G.................
Paasche, H.................
Elevators.
  Bohm, O. ....
  Cunningham, B..............
  Minn. R.R. & Warehouse Com.
Emmer.
  Carleton, M. A.............
Evolution.
  Allen, G...................
  Candolle, A. de............
  Darwin, C..................
  Jackson, J. R..............
  Jesse, E...................
  Weldon, W. F. R............
Exports.
  Andrews, F.................
  Bovey, C. C................
  Bradstreet's...............
Farmer, business relations of.
  Adams, E. F................
Fertilizers.
  Beal, W. H.................
  Hall, A. D.................
  Haworth, E.................
  Hutchinson, W. L...........
  Hyde, J....................
  Schweitzer, P..............
  Street, J. P...............
  Thorne, C. E...............
  Voorhees, E. B.............
  Wiley, H. W................
Food, animal.
  Allen, E. W................
  Balz, S....................
  Cottrell, H. M.............
  Wheeler, W. A..............
Food, human.
  Bread:
    Atwater, H. W............
    Current Literature.........
    Douglas, E. S............
    Johnson's Univer. Cy......
    Macmillan................
    Merrill, L. H.............
    New Internat. Ency........
    Paton, J..................
    Scientific American........
    Shepard, J. H.............
    Snyder, H.................
    Voorhees, L. A............
    Woods, C. D...............
  Breakfast foods:
    Albini, G.................
  General:
    American Cy...............
    Dalton, J C...............
    Hassall, A. H.............
    Snyder, H.................
  Macaroni:
    American Cy...............
    Chamber's Ency...........
    Johnson's Univer. Cy......
    New Internat. Ency........
    Scaramelli, F.............
    Skinner, R. P.............

INDEX

Page